Civil Engineering: Development, Design and Analysis

Civil Engineering: Development, Design and Analysis

Edited by
Armando Ruiz

Willford Press
www.willfordpress.com

Published by Willford Press,
118-35 Queens Blvd., Suite 400,
Forest Hills, NY 11375, USA

ISBN: 978-1-68285-350-4

Cataloging-in-Publication Data

Civil engineering : development, design and analysis / edited by Armando Ruiz.
p. cm.
Includes bibliographical references and index.
ISBN 978-1-68285-350-4
1. Civil engineering. 2. Engineering design. 3. Construction projects. 4. Concrete construction. I. Ruiz, Armando.
TA145 .C58 2017
624--dc23

For information on all Willford Press publications
visit our website at www.willfordpress.com

Willford Press

Printed in China.

Contents

Preface

This book has been a concerted effort by a group of academicians, researchers and scientists, who have contributed their research works for the realization of the book. This book has materialized in the wake of emerging advancements and innovations in this field. Therefore, the need of the hour was to compile all the required researches and disseminate the knowledge to a broad spectrum of people comprising of students, researchers and specialists of the field.

This book on civil engineering deals with the processes which are necessary for the construction and maintenance of built-environment. The field of civil engineering has seen innumerable advancements in the sectors of design, production, and research. The topics in this book serve as an invaluable source of knowledge related to safe and sustainable urban design, the effects of hazards on structures, etc. The various sub-fields of civil engineering along with technological progress that have future implications are glanced in this book. With state-of-the-art inputs by acclaimed experts of this field, this book targets students and professionals. A number of latest researches have been included to keep the readers up-to-date with the global concepts in this area of study.

At the end of the preface, I would like to thank the authors for their brilliant chapters and the publisher for guiding us all-through the making of the book till its final stage. Also, I would like to thank my family for providing the support and encouragement throughout my academic career and research projects.

Editor

1

Application of Jute Fiber for the Improvement of Subgrade Characteristics

Md. Akhtar Hossain*, Md. Shakhawat Hossain, Md. Kamrul Hasan

Department of Civil Engineering, Rajshahi University of Engineering & Technology, Rajshahi, Bangladesh

Email address:
akhtar412002@yahoo.com (Md. A. Hossain), shakhawat09ruet@gmail.com (Md. S. Hossain), hasankamrul.ce@gmail.com (Md. K. Hasan)

Abstract: Improvement in subgrade has always been an area of concern to highway and geotechnical engineers. In case of a highway, a weak subgrade results in greater thickness of pavement layer which increases the cost of pavement construction. To strengthen the subgrade soil, the use of jute fiber is advantageous because they are cheap, locally available, biodegradable, and eco-friendly. Jute fiber mixed randomly with subgrade soil has significant impact on the improvement of subgrade characteristics over the last decade. Keeping this in view an experimental study was conducted on locally available soil reinforced with jute fiber. In this study the soil samples were prepared at its maximum dry density corresponding to optimum moisture content in the CBR mold with and without reinforcement. The percentage of jute fiber by dry weight of soil was taken as 0.3%, 0.6%, 0.9%, and 1.2%. In the present investigation the length of fiber was taken as 15 mm and 30 mm, and two different diameters (4 mm and 8 mm) were considered for each fiber length. From the laboratory tests the effect of jute fiber content on density of soil (optimum moisture content and maximum dry density) was observed for each length and diameter of jute fiber. It was observed that the optimum moisture content increases and the maximum dry density decreases with the increase in jute fiber content for each length and diameter of jute fiber. The laboratory CBR values of soil and soil reinforced with fiber were determined. The effects of aspect ratio (length/diameter ratio) of jute fiber on CBR value of soil were also investigated. Test results indicate that CBR value of soil increases with the increase in length of jute fiber. It was also observed that increasing the diameter of jute fiber further increases the CBR value of reinforced soil, and this increase is substantial at fiber content of 1.2% for aspect ratio of 3.75 (length = 30 mm, diameter = 8 mm).

Keywords: Subgrade, Jute Fiber, Length, Diameter, CBR Value

1. Introduction

Transportation plays a vital rule in the development of economy, industry, society and culture of any country. To achieve development in those fields, a satisfactory mode of transport and communication is indispensable. This can be achieved through a transportation system, which is economically viable. Economy in road network can be achieved through economical pavement design. Practically it is not possible to have good subgrade always. Poor subgrade necessitates a greater pavement thickness resulting in increased construction cost. With the aim of reducing pavement thickness on poor subgrade new techniques of construction and soil stabilization have been continuously explored. Poor natural soils make them practically unsuitable for many civil engineering construction activities including road pavements. In such cases natural soils are being treated with different kinds of materials to improve their engineering properties. Aggarwal and Sharma (2010) studied the application of jute fiber in the improvement of subgrade characteristics. From this study it was concluded that jute fiber reinforcement reduces the maximum dry density and increases the optimum moisture content of the subgrade soil. The CBR value of the subgrade soil increases up to 250% with the inclusion of bitumen coated jute fiber [1]. Fiber reinforced soil has been used in many countries in the recent past and further research is in progress for many hidden aspects of it. Fiber reinforced soil is effective in all types of soil (i.e. sand, silt, clay). Use of natural material such as jute, coir, sisal and bamboo, as reinforcing materials in soil is prevalent for a long time and they are abundantly used in many countries like India, Philippines, and Bangladesh etc. The main advantage of these materials is they are locally available and are very cheap. They are biodegradable hence do not create disposal problem in environment. Processing of these materials into a usable form is an employment generation activity in rural areas of these countries. If these materials are used effectively, the rural economy can get uplift and also the cost of construction can be reduced, if the material use leads to beneficial effects in engineering construction. Singh and Bagra (2013) studied the application of jute fiber for the improvement of subgrade soil. They concluded that the CBR value of subgrade soil increases with the increase in jute fiber content. They also conclude that the CBR value also increases with increase in length and diameter of jute thread [2]. Tera *et al.* (2011) proposed that of all the natural fiber jute has highest tensile strength and withstand rotting and heat [3]. Gray and Ohashi (1983) conducted a series of direct shear tests on dry sand reinforced with different synthetic, natural and metallic fiber to evaluate the effects of parameters such as fiber orientation, fiber content, fiber area ratios, and fiber stiffness on contribution to shear strength [4]. We have several examples of reinforcing the soil like Great Wall of China (earliest example of reinforced earth using branches of trees as tensile materials), Ziggurats of Babylon (woven mats of read were used) etc. In USA & Europe, the erosion & stability of slopes for highway and railway embankment is controlled using reinforced soil [5]. Singh (2011) studied the strength characteristics of soil by using Geosynthetic [6]. Sivakumar and Vasudevan (2008) studied the strength and stiffness response of soil reinforced with coir-fiber [7]. Singh (2012) used the coir fiber to improve the laboratory CBR value of soil [8]. Sharma *et al.* (2011) studied the strength characteristics of soil reinforced with coir fiber [9]. Lawton *et al.* (1993) used multioriented geosynthetic to improve the laboratory CBR value of soil [10]. Consoli *et al.* (2002) used plastic waste to improve the engineering behavior of soil [11]. Ranjan *et al.* (1996) studied the probabilistic analysis of soil mixed randomly with fiber [12]. Rao *et al.* (2006) studied the behavior of load deformation of fiber reinforced gravel beds overlying soft clay [13]. Michalowski *et al.* (2002) studied the strength anisotropy of sand reinforced with fiber [14]. Gray *et al.* (1986) studied the comparative behavior of sand with fabric reinforcement and fiber reinforcement [15].

This paper presents the influence of jute fiber on the CBR value of the studied soil. A number of CBR value tests have been conducted on soil, and soil reinforced with varying amount of jute fiber. The effects of aspect ratio (length/diameter ratio) of jute fiber on CBR value of reinforced soil have also been investigated. The laboratory test results of reinforced and unreinforced soils were also compared.

2. Material Used

The soil used in this study is collected from Rajapur of Chapai Nawabgonj district, Bangladesh. The soil sample was collected from a depth of 60 cm after removing the top surface soil from natural ground surface. The jute fibers used in this study was purchased from the market in the diameter of 4 mm and 8 mm and in the length of 15 mm and 30 mm. The fiber was coated with bituminous layer in order to protect it from decay. The bitumen used for the coating of jute fiber was of 80-100 grades.

The materials used for the present study are soil, jute fiber, and bitumen.

2.1. Soil

Index properties of the soil were determined (Table 1) and classification of soil was done as per Textural Classification. The soil is classified as Clay Loam.

Table 1. *Index properties of soil.*

Natural Water Content (%)	9.35
Liquid Limit (%)	24.82
Plastic Limit (%)	17.87
Plasticity Index (%)	6.95
Maximum Dry Density (gm./cc)	1.96
Optimum Moisture Content (%)	11.11
Specific Gravity (Gs)	2.68
CBR (at 2.5 mm) (%)	5.22

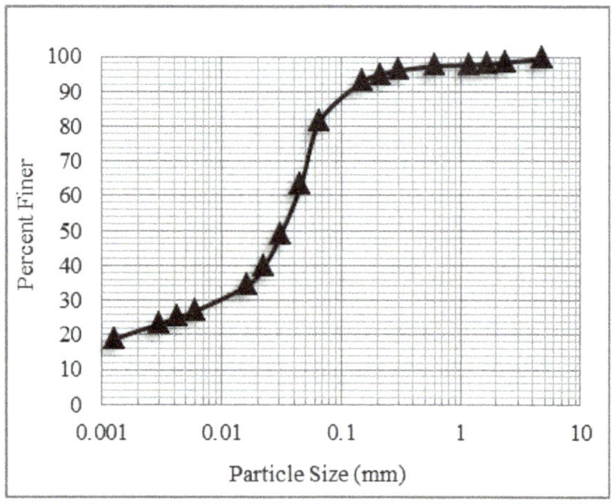

Figure 1. *Particle size distribution curve of soil.*

Textural Classification:
Clay Size (<0.005 mm) = 26%
Silt Size (0.005-0.05 mm) = 40%
Sand (0.05-2.0 mm) = 32%
Gravel (>2 mm) = 2%

2.2. Jute

Figure 2. *View of jute fiber (D = 4 mm, L = 30 mm).*

The jute fiber used was procured from the local market. The diameter of the thread was 4 mm and 8 mm and length was 15 mm and 30 mm. These fibers were generally available in the threaded form. These were mechanically woven fibers with very fine threads. The physical appearance of jute thread is shown in Figure 2. Jute industry is one of the oldest industries in Bangladesh, has traditionally been used for packaging. However its versatility is coming to light now after the world had started looking for natural options to save the environment. Jute is a coarse natural hast fiber. The major components of jute fiber are shown in Table 2.

Table 2. *Properties and constituents of jute (after Aggarwal and Sharma, 2011)*

Color	Yellowish brown
Specific Gravity (G_S)	1.12
Diameter Used (mm)	4 and 8
Length Used (mm)	15 and 30
Hoiocelluiose (%)	83-87
Lignin (%)	12-14
Wax (%)	0.4 to 0.81
Ash (%)	0.5 to 1.04
Nitrogen (%)	0.4

2.3. Bitumen

Bitumen was used for coating the jute fiber to protect them from microbial attack & degradation. Bitumen coating was done in the hot state at a temperature of 220 °C.

Grade and some of the other properties of bitumen are tabulated in Table 3.

Table 3. *Properties of bitumen.*

Grade	80-100
Softening Point (°C)	52
Flash Point (°C)	290
Fire Point (°C)	310
Specific Gravity (G_B)	1.02

3. Experimental Program

A series of proctor compaction tests and California Bearing Ratio tests have been carried out on soil mixed with jute fiber. The detailed procedure and results are as under.

3.1. Proctor Compaction Test

To assess maximum dry density (MDD) and optimum moisture content (OMC) Modified Proctor test is performed as per ASTMD-1557.

The jute fiber purchased from the market in diameter of 4 mm and 8 mm had been cut in the pieces of length nearly 1 m. These pieces of the jute were coated with bitumen in order to protect it from microbial attack & degradation. To coat the bitumen over jute fiber, bitumen was heated up to a temperature of nearly 220 °C. The thread of jute was dipped in it for about 2 to 3 second & extra bitumen was streamed off with the help of fork. The pieces in the way were left for cooling of 24 hour. After 24 hour the threads were further cut in the small pieces of length equal to 15 mm and 30 mm.

A series of modified proctor tests were carried out on the soil sample with 0.3% to 1.2% by weight of jute fiber coated with bitumen. While adding the jute fiber coated with bitumen in the soil mass, appropriate correction was applied to cater for the amount of bitumen.

From the result of modified proctor test it was observed that inclusion of jute fiber reduces the maximum dry density and increases the optimum moisture content for each length and diameter of jute fiber.

3.2. California Bearing Ratio Test

A series of California Bearing Ratio (CBR) tests were performed on the soil without jute fiber reinforcement and with different proportion of jute fiber based on the modified proctor test results. The test was performed for all the combination of the lengths and diameters of jute fiber and percentage.

4. Results and Discussions

Results of jute fiber reinforced soil are compared with that of unreinforced soil in the following sections.

Table 4. *CBR value of soil reinforced with jute fiber*

Fiber Length (mm)	Percentage of Fiber by Dry weight of Soil	Fiber Diameter 4 mm		Fiber Diameter 8 mm	
		CBR Value (%)	% increase in CBR Value	CBR Value (%)	% increase in CBR Value
	0%	5.22	-	5.22	-
	0.30%	5.76	10.34	6.30	20.69
15	0.60%	6.48	24.14	6.84	31.03
	0.90%	7.02	34.48	7.20	37.93
	1.20%	7.38	41.38	7.56	44.83
	0%	5.22	-	5.22	-
	0.30%	7.02	34.48	7.38	41.38
30	0.60%	7.92	51.72	8.82	68.97
	0.90%	8.46	62.07	9.96	90.80
	1.2%	9.36	79.31	10.56	102.30

4.1. Effect of Jute Content on the MDD and OMC

The Proctor tests results obtained from the tests conducted on the soil sample without jute fiber and with different percentage of jute fiber for various aspect ratios are discussed as below:

It is observed that inclusion of jute fiber reduces the MDD and increases the OMC.

With addition of 0.3% of jute fiber (aspect ratio = 3.75, length = 15 mm, diameter = 4 mm) the maximum dry density reduces from 1.96 gm./cc (0% jute content) to 1.92 gm./cc, whereas the optimum moisture content increases from 11.11% (0% jute content) to 11.19%.

MDD reduces from 1.92 gm. /cc (aspect ratio = 3.75, length = 15 mm, diameter = 4 mm) to 1.83 gm. /cc (aspect ratio = 3.75, length = 15 mm, diameter = 4 mm), whereas there is an increase in OMC from 11.19% to 14.28% with inclusion of 0.3% to 1.2% jute content in the subgrade soil.

Similar affects are observed for aspect ratio 1.875 (length = 15 mm, diameter = 8 mm), 7.50 (length = 30 mm, diameter = 4 mm) and 3.75 (length = 30 mm, diameter = 8 mm) (see Figure 3).

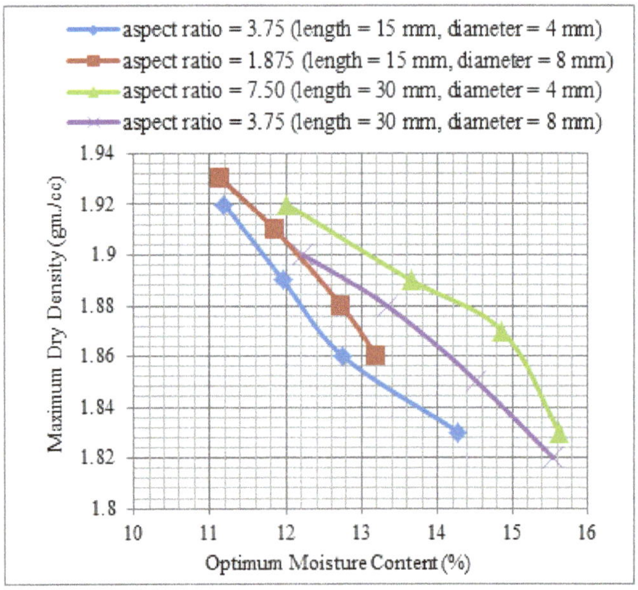

Figure 3. *Proctor test results on soil mixed with different percentage of jute fiber for various aspect ratios.*

4.2. Effect of Aspect Ratio on CBR Value

Table 4 presents the variation of CBR values with respect to aspect ratio (length/diameter ratio) of different percentage of jute fiber. The obtained results of Table 4 are plotted in Figure 4. It is observed from Figure 4 that when aspect ratio decreases the CBR value of soil increases for same percentage of jute content. For instance the CBR value of soil reinforced with jute fiber at 0.3% are 5.76% (aspect ratio = 3.75, length = 15 mm, diameter = 4 mm) and 6.30% (aspect ratio = 1.875, length = 15 mm, diameter = 8 mm). Similarly the CBR values of soil reinforced with fiber contents of 0.6%, 0.9 % and 1.2 %, increases from 7.92%, 8.46% and 9.36% to 8.82%, 9.96% and 10.56% respectively when aspect ratio decreases from 7.50 (length = 30 mm, diameter = 4 mm) to 3.75 (length = 30 mm, diameter = 8 mm).

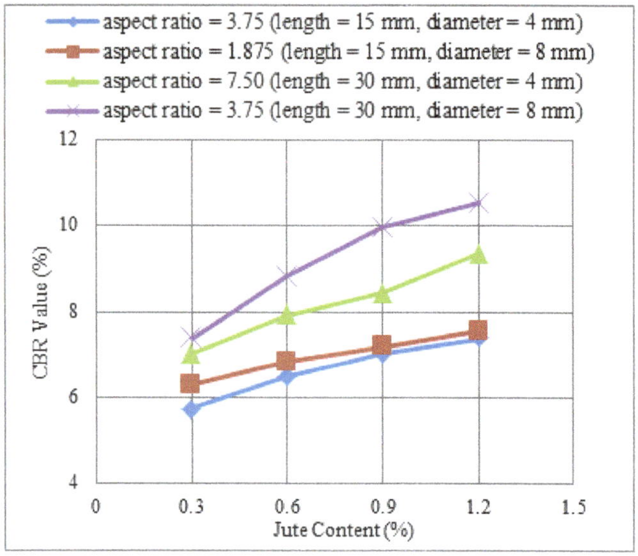

Figure 4. *Variation of CBR value with respect to jute content for different aspect ratios.*

5. Conclusions

A series of CBR tests and Modified Proctor tests were conducted in the present study. The laboratory test results are presented and discussed in the sections above. From the above discussion the following conclusions can be drawn:

(i) Jute fiber reinforcement reduces the maximum dry density and increases the optimum moisture content of the subgrade soil for each aspect ratio.

(ii) The CBR value of soil increases when the aspect ratio decreases for same percentage of jute fiber.

Acknowledgements

The authors are thankful to the Head of Civil Engineering Department for providing laboratory facilities for conducting the tests. The help and support extended by the Laboratory Assistant of geotechnical engineering lab are gratefully acknowledged.

References

[1] Aggarwal, P. and Sharma, B., "Application of Jute Fiber in the Improvement of Subgrade Characteristics". International Journal on Transportation and Urban Development, 2011; 1(1), 56-58.

[2] Singh, H. P. and Bagra, M., "Improvement in CBR Value of Soil Reinforced with Jute Fiber". International Journal of Innovative Research in Science, Engineering and Technology, 2013; Vol. 2, Issue 8, 3447-3452.

[3] Tera, S., and Reddy Jagannatha, H.P., "International Journal of Innovation, Management and Technology, 2011; 2 (3), 186-191.

[4] Gray, D.H., and Ohashi, H., "Mechanics of fiber reinforcing in sand". Journal of Geotechnical Engineering, 1983; 112 (8), 335-353.

[5] Rao, P.J. "Jute Geotextile for improving the performance of Highway Embankment on soft Marine Soil", Proc. Nat. Sem, on Jute based Geotextiles, New Delhi, 1996.

[6] Singh, H.P., "Strength Characteristics of Soil Reinforced With Geosynthetic". International Journal of Earth Sciences and Engineering, 2011; 4 (6), 969-971.

[7] Sivakumar B., G.L., and Vasudevan, A.K., "Strength and Stiffness Response of Coir -Reinforced Tropical Soil". Journal of Materials in Civil Engineering, 2008; 571-578.

[8] Singh, H.P., "Improvement in CBR Value of Soil Reinforced with Coir fiber". 3rd International Conference on Natural Polymer, (2012); held at Mahatma Gandhi University Kottayam, Kerala, India. 26-27 October Paper No118.

[9] Singh, H.P., Sharma, A., and Chanda, N., "Study of Strength Characteristics of Coir Reinforced Soil". International Conference on Advances in Material and Techniques for Infrastructure Development, 2011; held at NIT Calicut Kerala, India. Paper No.: G002 28-30 September.

[10] Lawton E.C., Khire, M.V. and Fox, N.S., "Reinforcement of soils by multioriented geosynthetic inclusion". Journal of Geotechnical Engineering, 1993; 119 (2), 257-275.

[11] Consoli, N.C., Montardo, J.P., Prietto, P.D.M. and Pasa, G.S., "Engineering behavior of sand reinforced with plastic waste". J. Geotech Geoenviron. Eng., 2002; 128 (6), 462-472.

[12] Ranjan, G., Vasan, R.M. and Charan, H.D., (1996), "Probabilistic analysis of randomly distributed fiber-reinforced soil". Journal of Geotechnical Engineering, 1996; 122 (6), 419-426.

[13] Rao, A.S. Rao, K.V.N., Sabitha, G. and Suresti K., "Load deformation behavior of fiber-reinforced gravel beds overlying soft clay". A National Conference on Corrective Engineering Practices in Troublesome Soils, 2006; 187 -190.

[14] Michalowski, R.L. and Cerma'k, J., "Strength anisotropy of fiber reinforced sand". Comput. Geotech., 2002; 29 (4), 279-299.

[15] Gray, D.H. and Al-Refeai, T., "Behaviour of fabric versus fibre-reinforcecd sand". Journal of Geotechnical Engineering, 1986; 112 (8), 804-820.

Study on Strategies of Energy Saving and Emission Reducing Building Design in Urban and Rural Areas

Hongyan Li, Jia Yu

Department of Architecture, Faculty of Human Settlement and Civil Engineering, Xi'an Jiao-Tong University, Xi'an, Shaanxi, China

Email address:
lihy04@126.com (Hongyan Li), 1152715229@qq.com (Jia Yu)

Abstract: In China, huge building energy consumption is the most prominent problem in the current situation of energy consumption. It has caused a series of energy and environmental problems. Therefore, it is extremely urgent and significant to study energy saving and emission reducing strategies in the aspects of building orientation, separation distance, solar radiation, wind direction, internal and external environment, etc. This paper discusses current situation and development trends of building design on energy saving and emission reduction in urban and rural areas of China. Through discovering existing problems in building energy consumption and architectural design in China, this paper proposes strategies of energy saving and emission reduction in architectural design and analyzes specific measures in architectural design through the typical case study.

Keywords: Energy Saving, Emission Reduction, Building Design Strategies

1. Introduction

In China, construction, industry and transportation form three major energy consumption fields. With the rapid development of urban construction, the traditional energy has been increasingly exhausted. As a big country in the field of architecture, a large amount of building energy consumption also become a huge burden on the national economy in China. In addition, the environmental pollution caused by energy crisis and climate change is increasingly serious. Under this background, the energy-saving and emission reduction buildings have become a popular trend in today's society. Thus, in order to have a harmonious future, it's critical to absorb the successful experiences from developed countries and promote the energy-saving and emission-reduction by means of architectural design. In this way, will China can realize the dream of energy conservation, environmental improvement and social sustainable development.

2. Current Situation and Development Potential of Energy Saving and Emission Reduction in Urban and Rural Buildings

2.1. Background of Building Energy Saving and Emission Reduction

In China, building energy consumption accounts for more than 27% of the total energy consumption and is increasing at an annual rate of 1 percentage points [1]. This phenomenon presents a trend of low level energy consumption, serious waste, and high potential energy consumption. Relevant statistics show that China's construction steel consumption accounted for about 25% of its total amount, cement usage accounted for 70% of the total amount of national cement, wood consumption accounted for 40% of the national total amount, the construction of water consumption accounted for 32% of the total. This has exacerbated the pressure to the environment [2].

Not only does the building have a high consumption of energy, but the energy utilization rate is also extremely low. Construction Department statistics show that China's urban and rural construction area of new housing construction is nearly 2 billion square meters per year, of which more than 80% are high energy consumption buildings. More than 95% of existing buildings are high energy consumption [3]. China's building energy efficiency is still relatively backward, so building energy efficiency technology has to be improved.

According to the experience from developed countries, energy consumption of building construction will exceed that of industry, transportation and other industries, and ultimately be in the first place of social energy consumption, so the study of energy-saving and emission reduction is imminent. Firstly, China has the world's largest construction market, so only by adhering to the energy saving development, clean development and security development, can we achieve reasonable and fast economic development. Moreover, building is one of the main sources of greenhouse gas emissions, which has a significant impact on climate change. The carbon dioxide gas emission from construction industry accounts for about 30% of the total emissions of greenhouse gases. Therefore, low-carbon housing is not only conducive to energy-saving and emission reduction, but also significant to deal with global climate change. Most importantly, reasonable building design considering energy-saving can benefit the construction of a resource-saving and environment-friendly society and a sustainable environment.

2.2. Problems of Urban and Rural Building Design Regarding Energy Saving and Emission Reduction

The problems and challenges can be generalized into five aspects.

2.2.1. Unreasonable Construction Layout

Some residential plans do not consider natural ventilation and residents often use air condition to adjust the indoor temperature and humidity, which increased the energy consumption.

2.2.2. Massive Use of Glass Curtain Wall Causes Large Energy Consumption

In many cities, the pursuit to beautiful and stylish building has lead to many office buildings with glass curtain wall. It causes light pollution in summer for strong ultraviolet radiation and no heat preservation in winter. People have to use high-power air conditioner to adjust the temperature throughout a year. The glass curtain wall brings about the so-called aesthetic feeling, but also causes the energy consumption doubled. For example, the Shanghai Pudong cadre's college main building is like a big glass columns without heat insulation (Figure 1). And its annually electricity fee is 2~3 times higher than other general buildings.

Figure 1. Shanghai Pudong cadre's college main building.

2.2.3. High Energy Consumption of Reinforced Concrete Structure

Relevant study shows that the energy consumption and carbon dioxide emissions from timber building structure are the least, followed by the light steel structure [4]. The energy consumption of concrete structure is the most in the three kinds of structures. The massive use of reinforced concrete structure has led to a greater increase in the construction energy consumption and caused more pollutant emissions.

2.2.4. Serious Thermal Phenomenon

In the architectural design, the building envelope plays an important role in the building insulation. It is the key to building energy efficiency. The kaleidoscopic designs easily cause the heat transfer coefficient of a part in construction far greater than in the adjacent parts. The thermal resistance becomes smaller and the thermal insulation performance of the building envelope is worse. Or the heat dissipation area of a part of the building is larger than the heating area, causing heat loss too much, which leads to the phenomenon of thermal bridge, resulting in a waste of energy.

2.2.5. Insufficient Utilization of Clean Energy

In the past, relying on combustion of traditional energy has caused a waste of energy and increased the environmental pollution. So it is imperative to use clean and convenient energy. For example, the use of solar energy is has a large room for improvement.

2.3. Development Potential of Energy Saving and Emission Reduction in China

There are three main reasons for high building energy consumption: poor heat insulation of building envelope, low heating system efficiency and low heat source efficiency. In the about 40 billion square meters existing buildings, there are widespread problems such as poor building envelope insulation and low heating air-conditioning system efficiency [5]. The energy saving potential is huge.

In addition, the energy saving potential of public building is also great. The energy consumption level of large public buildings is several times of ordinary buildings and the overall energy use efficiency is far lower than that of the developed countries. The energy saving potential of existing public building is large and energy saving can be approximately 30% (Figure 2). In addition, the proportion of new buildings using advanced technology to achieve energy saving exploration in 2012 of China is only about 26%, far lower than it in the United States and Japan [6].

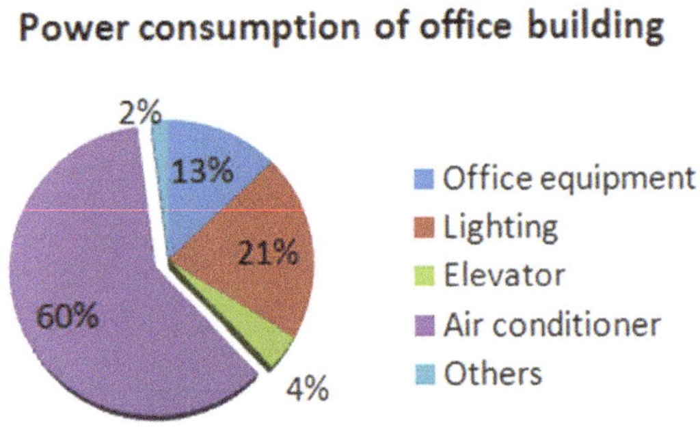

(Source: Air conditioning refrigeration in China)

Figure 2. *The component of electricity usage in an office building.*

In terms of carbon dioxide gas emissions, data shows that the existing building renewal can greatly reduce carbon dioxide emissions [7]. Market expansion space and energy saving prospects are huge in China.

3. Strategies of Energy Saving and Emission Reduction in Architectural Design

3.1. Planning and Design of Buildings

In architectural design, the layout, orientation, shape and material of the building itself can greatly affect the energy consumption in the building's construction and operation. Reasonable planning and design can not only reduce the energy consumption but also obtain pleasant indoor and outdoor environment, which is beneficial to people's health. First, construction planning (including shape, height, open space, etc.) and layout should benefit natural ventilation and lighting, besides, the design should consider local predominant wind direction and sunshine distance (Figure 3), and the location of doors and windows also should be set to make the air flow through the activity district as far as possible so that it can form indoor draught to increase the ventilation range.

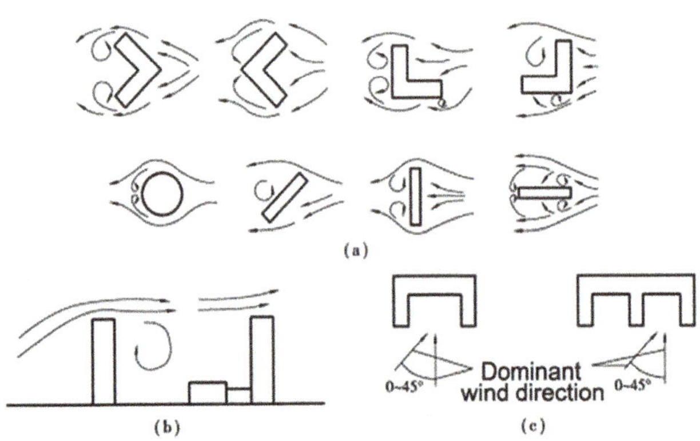

Figure 3. *Effect of architectural design on ventilation.*

In fact, Chinese ancient buildings have taken the natural ventilation design into account, such as Huizhou style architecture (Figure 4) or Lingnan architecture, which has a small courtyard surrounded by attic and forms a natural ventilation corridor. In this way, building itself becomes a ventilation chimney, which makes natural ventilating effect much more obvious.

Figure 4. *Courtyard styles in Huizhou architecture.*

Secondly, the orientation of building should depend on the location and relationship with the surrounding environment and roads. In most situations, Chinese construction is suitable to choose the south-north orientation or close to the south-north orientation because cooling load in this direction is smaller than the east-west orientation and the index of energy consumption can also be reduced by about 5%.

In addition, the shape coefficient is an important indicator of the thermal characteristics of building, which is related to the number of layers, size, shape and other factors. The larger the shape coefficient, the larger is the external envelope of building. When building volume is a certain value, the larger the outer protecting structure area, the bigger is the heat loss, and when the shape coefficient increases 0.01, the energy consumption indicator will increase about 2.5% [2]. However, in order to solve the sunshine requirements of many high-rise buildings, sometimes architects make concave or convex changes in the flat and elevation, which leads to buildings' shape coefficient not conducive to the building energy efficiency. Therefore, the shape coefficient should be considered in the construction design, and it is not encouraged to pursue a better shape at the cost of the building energy consumption.

3.2. Building Envelope Design

The energy consumption caused by building envelope dissipation, heating and cooling system accounts for the majority in the whole building energy consumption, so more attention should be paid to the building envelope insulation (Figure 5). To strengthen the roof and exterior wall heat preservation and heat insulation performance, it is necessary to design a reasonable external insulation system and to use energy-saving insulation materials and construction technology, which can help in an effective way to improve the insulation performance of the wall and also can solve the common phenomenon of cracking and water seepage meanwhile. In addition, light color surface material can be used to reflect sunlight and enhance the external wall and roof insulation in summer from a certain extent. Furthermore, outside window is also a critical part of the building envelope. In general, the area ratio of building walls and windows in the north is no more than 25%, south is no more than 35%, east and west is no more than 30% [8]. In this way, it can not only meet the requirements of indoor lighting and ventilation, but also reduce the energy consumption that is due to the poor heat preservation and heat insulation.

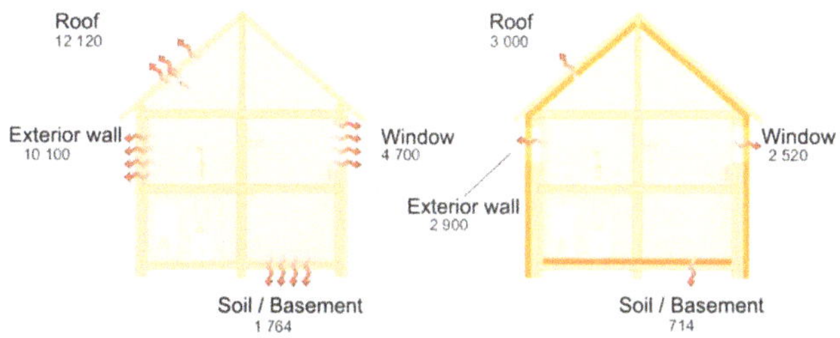

Figure 5. *Comparation of heat dissipation with building envelope insulation (Left) and without building envelope insulation (Right).*

3.3. Use New Type of Energy Saving and Environmental Friendly Building Materials

In China, many new energy-saving and environmental protection building materials have been developed from blankness, variety from single to diversity and quality from low to high, especially the new wall materials develop faster and more varieties, such as hollow clay brick (Figure 6), clay brick mixed with waste, composite panels, etc. For example, energy-saving brick used in rural houses has the advantages of saving energy and soil compared with clay brick, and it is also the superior new building materials. Moreover, comparing to cement, the use of ecological cement and coal ash are the powerful measure for energy saving and emission reduction (Figure 7). Ecological cement is a kind of ecological building materials made from various kinds of industrial wastes and coal ash is the waste of coal-fired power plant. Since coal ash has the characteristics of light, porous and potential water hardness, it can be a raw material of various building materials. The use of this kind of cement and coal ash can reduce the load of waste disposal, which not only solves the pollution caused by waste, but also makes living garbage and industrial waste become raw materials into useful construction resources, thereby reducing the cost of production.

Figure 6. *Hollow clay brick.*

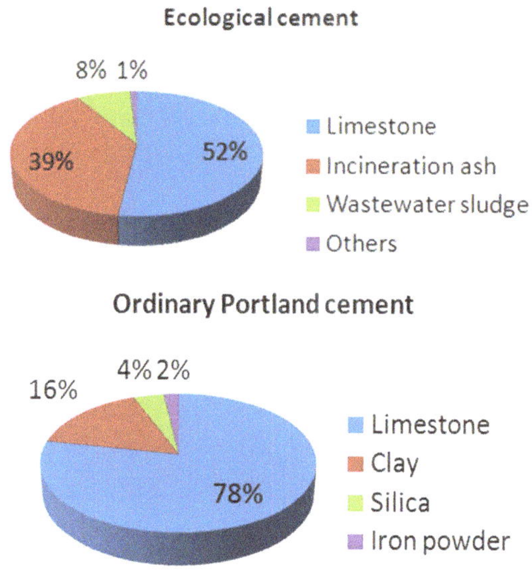

Figure 7. *Comparation of components of common cement and ecological cement.*

3.4. Architectural Shading Design

Too much sunshine into the interior can cause indoor too hot and increase the energy consumption of air-conditioning. So it is necessary to take certain shading measures in some rooms or outdoor to ensure the indoor comfortable environment. However, contribution of different shading technology to energy saving is not the same. The visible light and energy transmittance of ordinary transparent glass is almost 100%. About 75% of the light and heat will enter the room if internal shading is set, such as setting control blinds or curtains. But if the outer sunshade is installed, sunshine thermal effect will reduce more than 80% (Figure 8) [9]. In addition, it is necessary to promote a variety of measures such as roof shading, external wall shading, planting trees to reduce solar radiation.

Figure 8. *Examples of shading measures.*

3.5. Promotion of Three-Dimensional Afforesting

Building three-dimensional green space is a very efficient and cheap energy-saving method. Greening or planting crops in roof, facade and public building surface can save 30% energy of the air-conditioning in summer [10], which could not only create a comfortable indoor environment and significantly reduce urban heat island effect and dust emissions, but also beautify the environment (Figure 9).

Figure 9. *Three-dimension afforesting.*

4. The Case: Headquarters Building of Frankfurt Commercial Bank (Germany)

Commerzbank headquarters was built in 1997 completed by N Forster in 1994. This triangular tower is 53 stories and 298.74 meters high, which is not only the world's first high-rise ecological architecture, but also the world's highest ecological architecture (Figure 10). Except for a very small part of cold or hot weather, the whole building uses natural daylight, natural ventilation and temperature adjustment, reduces energy consumption to a minimum degree.

Figure 10. *Overall effect diagram of Commerzbank headquarters.*

4.1. Layout Analysis

The building plane is an equilateral triangle shape with 60 meters in each side (Figure 11). Forster reconfigured the internal and external environment of the traditional office towers. He made indoor and outdoor mutual fusion by introduced outdoor environment inside. He dispersed public transportation and other core (elevator, step ladder, toilet, etc.) in the center of the tower to the three corners of the triangle layout of the construction, and treated the structure's triangular vertex of the three independent frame tube as giant column and the eight story steel frame as the giant beam, which connected and formed the giant tube system. Thereby, it liberates the larger space in center to redesign. Trapezoidal sections between each of the two parts of the traffic are the main office building. The three trapezoidal formed a hollow triangular atrium. Forster arranged sky gardens up to 4 layers (about 14 meters) at these trapezoidal areas between every 8 layers. The gardens are scattered set and enable people in each layer office can enjoy the garden landscape.

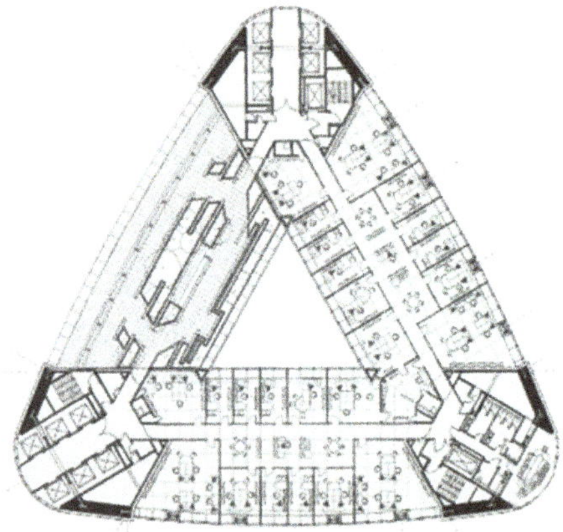

Figure 11. *Plan layout diagram of Commerzbank headquarters.*

4.2. Facade Analysis

In addition to the design of atrium and interior garden, the double deck design approach on construction also adds to the high-rise building some green space (Figure 12). The outer layer is fixed by a single layer of glass, and the inner layer is adjustable double Low-E insulating glass. The two layers form a hollow part of 165mm thick between them, which can allow the outdoor fresh air and the indoor stale air to exchange when the inner window is open. In the hollow part, window-shades which can be used to adjusted indoor environment are also set up. It can block the direct sunlight in hot summer and can reflect more sunlight into the interior in cold winter.

Figure 12. *Facade of Commerzbank headquarters.*

4.3. Daylight and Air Flow Analysis

The triangular atrium formed in the building is like a large chimney, which is organized by the natural ventilation of the office space in order to exert its chimney effect. After the wind tunnel test, a number of air gardens were set up in the three office space. And they are distributed in three directions of different elevation, as the air inlet and outlet of the "chimney", which organized the natural ventilation in office space effectively.

Triangular layout can maximize the acceptance of sunshine, create ideal vision, and reduce the blocking to adjacent building at the same time. Unique open top lighting design also makes the building have plenty of sunshine. The atrium and the air garden make the building meet the needs of internal ventilation and lighting only through natural ventilation and natural lighting at most of the work time throughout a year.

4.4. Landscape Analysis

This kind of closing to nature design creates a natural plant community in the primness building and overcomes the primness and depression of the former office. This Atrium space is a lively green environment full of sunshine, which can effectively regulate people's mental state and play a positive role in alleviating fatigue. This green office building gives people not only the visual comfort, but also space and nice mood to pursue health.

The comprehensive application of all these above-natural lighting, ventilation method and intelligent control technology makes the rate of the building natural ventilation reach 60%, which is rare in high-rise buildings.

5. Conclusions

Building energy saving and emission reduction is a highly technical and very complex process. These two aspects have both connections and differences. Generally, energy saving will lead to emission reduction, while emission reduction is not necessary energy saving. Thus the emission reduction projects must strengthen the application of energy-saving technology to avoid the result of one-sided pursuit of the energy consumption and focus on the balance of social benefits and environmental benefits. Although the initial investment of many design of energy-saving and emission reduction is large, technology is difficult and efficiency is low, energy-saving and emission reduction are the direction of new building design. This will benefit sustainable development of urban and rural areas. This is a long-term benefit, so the exploration of energy-saving and emission reduction building design should be continued. At the same time, we should combine the construction function with art and technology, and strive to control energy growth and protect environment with the continuous growth of total construction in order to reduce the running cost of buildings and to develop on the road of intensive, intelligent, green and low carbon. It is significant to achieve self-sufficiency in terms of building energy and to construct "zero energy building" to realize ecological sustainable development of building industry.

Acknowledgements

This work was financially supported by the National Natural Science Foundation of China (51308451).

References

[1] Wenjuan Tang: The Current Situation of Building Energy Consumption in China. Urban and Rural, vol. 25 (2010), pp. 47-49, in Chinese

[2] Linde Liu: The Present Situation of Building Energy Consumption, Energy Saving and Emission Reduction Design and Renewable Energy Utilizationin in China. Hydropower Station Design, vol. 25 (2009), pp. 107-118, in Chinese

[3] Yang Y H: Green Building Development Features in China. Applied Mechanics & Materials, vol. 9 (2014), pp. 725-730, in English

[4] Chunjing Shang, Chenglong Chu, Zhihui Zhang: Comparison of Carbon Emissions in Different Structural Buildings' Life Cycle. Building Science, vol. 27 (2011), pp. 66-70, in Chinese

[5] Guang Yang: Thinking of Building Energy Consumption. Care for Tomorrow, vol. 1 (2015), pp. 18-19, in Chinese

[6] Pengju Yuan: Exploration of Energy Saving Measures in Architectural Design. Construction Engineering Technology and Design, vol. 27 (2015), pp. 13-16, in Chinese

[7] Baoxing Qiu: Six Major Fields of Building Energy Saving and Emission Reduction and Its Prospect. Heating and Cooling, vol. 10 (2011), pp. 26-28, in Chinese

[8] Shasha Wei, Huijiao Wei, Chunling Wu: Study on the Influence of Different Side Window Lighting Methods on the Lighting Energy Saving in Green Office Buildings. Refrigeration and Air Conditioning: Sichuan, vol. 4 (2015), pp. 368-373, in Chinese

[9] Yun Ding: Discussion on External Shading Application of Buildings in the Hot-Summer and Cold-Winter Zone. Huazhong Architecture, vol. 1 (2016), pp. 70-73, in Chinese

[10] Aijuan Chen, Yuhui Di: Research Status and Development Trend of Roof Greening. Clean and Air Conditioning Technology, vol. 2 (2013), pp. 46-49, in Chinese

3

Identification and Management of Risks in Construction Projects

Mohammad Ali Rezvani Befrouei[1, *], Mohammad Taghipour[2]

[1]Department of Civil Engineering, Non-profit Institution of Higher Education, Aba - Abyek, Qazvin, Iran
[2]Research Assistant of Non-profit Institution of Higher Education, Aba - Abyek, Qazvin, Iran

Email address:
Rezvanibefrouei@yahoo.com (M. A. R. Befrouei), mohamad.taghipour@srbiau.ac.ir (M. Taghipour)

Abstract: Today, risk management in construction projects is considered to be a very important managerial process for achievement of project's objectives in terms of time, costs, quality, safety, and environmental sustainability. Instead of employing a systematic approach for identification of risks, their probability and their effects, most of the studies conducted in this area have focused only on a few aspects of risk management in construction project. However, the present study aims to identify and analyze the risks associated with development of construction in the greater city of Tehran, employing a comprehensive approach that is consisted of five aspects. In order to conduct this descriptive-analytical research, after studying, identification and categorization of different types of risks and preparing a relevant questionnaire, a number of 52 active construction companies from different areas have been randomly selected to be questioned about the impact of those risks on construction projects. After the collection and observation of the data, the output was examined by Pearson correlation also, using charts and tables. The results indicated that "tight project schedule" – present in all five categories- imposed the maximum risk. Also "design variations", "excessive approval procedures in administrative government departments" and "unsuitable construction program planning" were identified as next high risk factors.

Keywords: Risk, Risk Management, Construction Projects

1. Introduction

Risk management is one of the most important duties of project managers. If it's not recognized as a crucial responsibility, numerous problems will follow. Effective risk management requires the application of a systematic methodology, and more importantly, it requires knowledge and experience [Serpella, 2014].The main purpose of risk management in construction projects is to provide a safe work environment for the workers, and better achievement of goals. Findings reveal that risk management in construction projects has always faced defects and shortcomings which have had negative effects on the projects' efficiency [Cesarini, Hall, Kupiec, 2013].

For years, undermining risk management in construction projects has led to undesirable results and poor quality. For instance, management inefficiency and incorrect analysis of two important project factors, cost and time, which cannot be accurately determined, may lead to delay in the project and impose further costs. Today, the studies reveal the lack of systematic risk management methods in most of construction projects. The reason is that it has always been the people who adapted themselves to their work environment, and the adaptation of the work environment to the people has been undermined. This in turn will lead to negative consequences in implementation of projects. The statistics show that the vocational risk is an important factor preventing the realization of project goals in terms of time, costs, and quality. The purpose of this study is to identify different types of risks in construction projects, to analyze those risks and to improve the risk management by raising awareness and knowledge in all vocational levels involving the projects, and also to develop an efficient system of management based on the relevant knowledge, and consequently reducing the negative effects of the risks to come.

This approach is part of an attempt by an investigative research project whose primary results indicated that risk management in construction project is still suffering inadequacies; and the main reason for that is the contractors' and workers' lack of knowledge and awareness. It is expected that this study helps overcome the existing flaws and shortcomings and also improve the performance of project risk managers.

2. Risk

Risk is defined as a deviation from possible future events. If there is only one possible future event, then there is less chance of deviation and risk. We live in a world of risks and misadventures. Therefore the risks must be analyzed, and if encountered, must be identified and their consequences must be evaluated. Risk management is the process of documentation of the final decisions and implementation of the criteria that can minimize the risk to an acceptable level. In other words, it is the process of identification, evaluation, taking control measures and course corrections for the risks that would otherwise lead to further loss, or at least no change in the existing status.

Risk management is a systematic method of recognizing dangers and evaluating risks in order to provide required information to decide on appropriate risk reduction measures [Sehat&Alavi, 2010]. Risk management enables a system to make better management of usual risks in its daily activity, and continue its activities more effectively and comprehensively in a stress free environment. It also enables us to reach acceptable results with minimal costs. Risk identification requires the recognition of links in a chain that include srisk factors, resources exposed to risk, and the effect of the first one on the latter. The purpose of risk identification is to collect information on the kind of damage to which the project at hand is exposed [Sehat & Alavi, 2010]. This science presents methods and protocols that equip companies, organizations and governments with foresight and provisions to evaluate, control, and finance for future losses. On this basis, risk management arranges a systematic encounter with risks ahead. And for that purpose it is always faced with two fundamental questions about the future events. The first one is "what will happen?" and the second one is "what should be done?" Risk management is constantly planning ahead to encounter the possible future events.

Picture 1. *Risk management cycle.*

According to the protocols presented by risk management, protecting organizations and investments from dangers and losses requires a system of thought and practice which integrates policies and regulations against risks. Such system is continuously investigating problems of the institutions at stake, and finding appropriate and effective resolutions. On this basis, the mentioned system must distinguish and define the existing problems and structurally analyze them; and by collection and classification of the relevant information, it must produce the best possible method for risk prevention and control.

Risk management operational cycle:

- Identification of problems (risks)
- Identification and explanation of the needs
- Structure analysis of the problems (risks)
- Collection and classification of information
- Devising a strategy for encountering the risks
- Supervision and follow-up

Main stages of risk management process:

- Defining the risk management objective
- Risk identification
- Risk assessment
- Preparing risk treatment plans
- Risk sharing/ transfer to another party such as insurance company
- Risk avoidance, and if necessary, risk reduction/prevention (Iran's I.T professional forum)

3. Data Collection

Risks in construction projects can be divided into five categories: cost risks, time risks, quality risks, environmental risks, and security risks. Each of the categories is consisted of 8 important risk factors, some of which, such as tight programming schedule and unsuitable construction program planning, are shared by all five categories; while some other risk factors are only shared by two or three of them. This study indicates to what extent each of these risk factors affect the execution of a project and achievement of its goals. In order to undertake this analytical-descriptive study, among different construction companies and contractors in Tehran, a number of 52 active companies were selected randomly; and using a questionnaire, the risk managers, employers, advisors and contractors were asked about the impact of the risks on construction projects. After collection and observation of the answers, the frequency of results was obtained by the use of charts and tables.

3.1. Cost Factor

Today, management of construction is faced with unpredicted situations in programming and construction of private and public accommodations to meet the demands of the society. Environmental control and other legal controls and regulations, energy consumption costs and energy accessibility, constant change in public taste and economic state, are among such problems. Some causes of delay and slow pace in projects are incorrect understanding of the project's size and volume, incorrect estimation of required resources, weak cost programming and budget assessment, incapability in control of project's performance costs, and in general, weakness in project's cost management. It is obvious that keeping up with the times, and employment of modern science and techniques, will ensure better management of project costs. Table 1 demonstrates the correlation between risk and the factors related to cost.

Table 1. *The results of the Pearson correlation coefficient between risk and the factors related to cost.*

Factors related to costs	correlation coefficient	Significance level
Project's tight programming schedule	67%	0.01
Design variations and instability	48%	0.01
Inappropriate construction programming	52%	0.01
Emergence of conflicts and disagreement	65%	0.01
Inflation and increase of construction material prices	66%	0.01
excessive approval procedures in administrative government departments and municipality	61%	0.01
Incomplete or inappropriate cost assessment	64%	0.01
Incorrect estimation of timing schedule and project completion	53%	0.01

As the data suggests, the risk of tight programming schedule with 67% has the most correlated and greatest impact, and after that, the risk of inflation and increase in construction material prices with 66%, emergence of conflict and disagreement with 65%, and excessive approval procedures in administrative government departments and municipality with 61% are next factors having great effects.

Budget imposes the primary limitation on all small or large-scale projects. Even in cultural projects, where people participate without receiving any payment, lack of human resources and lack of credit for preparation of requirements pose limitations. Project's budget planning is one the most important duties of project programmers and determining how to provide the required credit for development of project objectives is one of the project managers' difficult tasks, especially in large-scale projects where in addition to providing credits, the orderly arrangement of activities and determining their prerequisites play an important role in effective implementation of the project regarding the financial credits time tables. For instance, it's not wise to finish the base-course layout of a highway without providing credits for the asphalt layer and wait for the next year's budget. Because falling of rain and snow would damage the infrastructure and then a part of budget must be spent on base-course repairs. In scenarios like this, construction of the highway can be completed in separate segments, and those segments can eventually be interconnected. Budget and budget provision is an important element in project phasing and determination of the optimal size for each phase. Managers should obtain accurate information on the budget and project funders' financial commitment and share it with programmers; they should also accurately supervise budget receptions in order to prevent the costs increase in construction projects.

Therefore, cost estimation will be needed in to stages: first before the contract, in order to determine the suggested price, and secondly after the contract and during project execution for budgeting and cost control.

In fact, costs in construction projects can be controlled by taking certain courses of action mentioned below:

- Accurate estimations and special designs
- Application of project execution time control factor, which is directly related to cost control
- Correct and suitable financial management in construction projects
- Seizing opportunities to acquire financial resources
- Providing grounds for private section investment in project execution
- Participation of private section in non-financial investment in projects, in areas such as providing lands, materials, and other investments.

3.2. Time Factor

Another factor that plays the main role in achievement of goals in construction projects and in most of other projects as well is time factor. In fact, time management is the main principle of construction projects execution; and if neglected the project will fall into chaos and further costs will be imposed on the project. The greatest dilemma facing large-scale projects is the delay in completion of its various phases and consequently the delay in completion of the project itself. Delay is an action or occurrence that decelerates the execution of project and increases the pre-arranged duration of a certain action, which manifests as increased time-length of an activity or postponement of its start [Schumacher, 1995]. If a project's operation is delayed, national resources would be wasted and in some cases the project may lose its technical and economical plausibility [Vatankhah, 2003].

The discussion of construction projects' delay in economy council of Iran's organization of management and programming dates back to 1970, and it continues until present time; however it has not yet led to preparation of a legalized solution to this problem [Eshteahradian et al. 2010]. In a report published in 1971, organization of budget and programming divides the causes of delay into 4 categories:

Inaccuracy and mistakes in preliminary estimations

- Increase in task items or task volume, and execution technique
- Executive defects, problems and methods
- Other problems

It is apparent that many of the above mentioned problems are similar, if not identical, to those of present day construction industry in our country. This is a sign of lack of dynamism in the solution of executive problems of construction industry. It is interesting that according to the organization of budget and programming, 90% of construction projects have suffered from increase in duration and costs; and 60% of the unfinished projects need 15 years to complete [Shakeri&Ghorbani, 2005]. It is necessary to mentioned that currently, the average time length of construction projects in Iran is 9 years, that is far off the standard time-length [Shakeri& Ghorbani, 2005]. After the study of questionnaires in this research, the effect of risk factors related to time was evaluated. Table 2 demonstrates the correlation between risk and the factors related to time.

Table 2. *The results of the Pearson correlation coefficient between risk and the factors related to time.*

Factors related to time	Correlation Coefficient	Significance level
Project's tight programming schedule	57%	0.01
Design variations and instability	46%	0.01
Clients' demand to make alterations	45%	0.01
Excessive approval procedures in government departments and municipality	46%	0.01
Unsuitable construction programming	44%	0.01
Government bureaucracy	36%	0.01
High expectations of quality and performance	34%	0.01
Unsuitable construction programming	34%	0.01

As the data suggests, tight programming schedule risk ranks first in terms of effect, and after that, risks of design variations and instability and excessive approval procedures in government departments and municipality with 46%, have the highest ranking.

Several studies on the project's delay have been conducted some of which have tried to reduce the delay by use of different methods [Schumache, 1995]. Some other studies have focused on the project delays in a certain industry or in a certain country. Among such studies the following can be mentioned: "studying the causes of delay and cost increase in Nigeria's construction projects (Mansfield & Doran, 1994), identifying the causes of delay in Lebanon's construction projects (Mezher, 2002), a study of large-scale projects in Saudi Arabia, development report of 164 construction projects and 28 highway construction projects in Jordan between 1996 to 1999, studying the causes of execution delay mentioned in organization of budget and programming report, effective factors of delay in airport construction projects in Iran."

Specialty, commitment and healthy management are the characteristics of successful project managers. A weak project manager does not think of a contractor as his executive assistant, but as a claimant who deserve a great deal of wrong-doing. Such managers believe it to be one of their positive capabilities to postpone project manifests and statements and creating havoc in the process of responding to contractor's demands. They cause the deceleration of project execution and its increasing costs [Ahmadijazzi, 2006]. After the observation of these studies it can be concluded that since projects main elements (employer, consultant, and contractor) play an important role in delaying the project, it is advised the this three element try to systematize project management and develop a procedural approach to management science in large-scale in order to achieve that goal.

3.3. Quality Factor

After time factor, the most important principle in of construction projects is the concern with quality, since low quality execution of construction projects causes the investments to go to waste; on the other hand increase in quality will lead to project durability, and prevents investment and financial waste. Also management of project safety will prevent possible damages.

Unfortunately even today there are some construction projects wherein technical principles are not observed and that leads to technical weak points in project execution. Also lack of continuous inspection by executive managers and supervisors allow the technicalities and construction principle to be overlooked, hence the industrial quality of the outcome is reduced. Table 3 demonstrates the correlation between risk and the factors related to quality.

Table 3. *The results of the Pearson correlation coefficient between risk and the factors related to quality.*

Risk factors related to quality	Correlation Coefficient	Significance level
Tight programming schedule	56%	0.01
Incorrect assessment of project's timetable	41%	0.01
Incorrect estimation of costs	39%	0.01
Unqualified management of subsidiary contractors	36%	0.01
High expectations of quality and performance	34%	0.01
Lack of coordination between project participants	46%	0.01
Lack of skilled workforce	42%	0.01
Unsuitable construction programming	35%	0.01

As demonstrated, the risk of tight programming schedule with 56% ranks first in terms of impact. And lack of coordination between project participants and lack of skilled workforce with 46% and 42% respectively, are the next risk factors in line in terms of impact. Also high expectation of quality and performance with 34% is an effective risk factor when it comes to construction project's quality, as well as an effective time risk factor.

Modern situation of organizations become more dynamic every day and present managers with new challenges. Universal competition and expectations of an evolving community always bring up new management requirements. Concepts such as participation, enabling, teamwork, flexibility and etc. are considered as an organization management's everyday concerns. Additionally, through efficient operation, employment of modern technology, and by commitment to produce high quality products and services for better satisfaction of customers and clients, new demands for efficiency and competitive advantages have emerged. Considering that, comprehensive quality management is one the most effective methods used in management systems to cover all issues related to quality.

Problems in construction quality, such as low quality of execution, lack of attention to workforce and human resources and costs of redoing urged the creation of a quality management system in this industry. Construction as an important industry, receive a great amount of states' budget, therefor, quality management in construction in order to ensure its continuance, supply the demands, avoid redoing, increase the employees' participation and encourage teamwork brings great profits [Harischian et al. 2007]

However it's not enough just to provide a high quality product and service that satisfies the taste of the client; costs of achieving this purpose must also be considered carefully. If the long-term effects of the costs for quality work on the company or organization are rendered as positive, such expenses are acceptable. One of the most important criteria in implementation of quality management systems is the correct understanding of quality costs (i.e. further costs that are imposed on the company because of lack of quality, or too high a quality). Considering its importance, identification of required resources for the project and accurate control and management of costs and analysis of quality costs seem crucial [Abbasnia&Moradisangi, 2010].

Some criteria that can help improve the quality of construction projects are mentioned below:

- Implementation of a quality control team at worksites of major construction projects to examine the quality of concrete in terms of strength
- Making drastic changes in the structure and different aspects of project execution, including management, execution deployment, and contractor
- Direct connection of project design with policies that are sometimes made simultaneous to the execution of the project
- Avoiding wrong and experimental methods and employment of engineering science and improvement of project's scientific standards
- Employment of engineering wisdom and general knowledge along with experiences from previous construction projects
- Establishment of special departments inside civil construction staff, such as department of urban buildings and streets maintenance, project control unit, research and education unit etc.

3.4. Environmental Factor

Workplace is an environment that is prone to risks and dangers. It is necessary for the safety of workers in a construction area to establish order in workshops, collect rubbles and construction waste, and take precautionary measures to avoid falling down. Also adjacent buildings and urban areas must be taken into consideration.

Underground projects are closely related to geo-technique issues. For instance structure collapse must be taken seriously. Preventive measures in such project are taken in three stages:

- At the beginning of the project; short term
- During the operation of vibrating machines; medium term
- After the project and during revenue operation; long term

On this basis, a comprehensive project safety operation must be laid out and regulations must be established for project execution, repairs and maintenance, and periodical examinations.

According to the statistics, safety measures such as scaffolding to prevent people and material from falling are seriously significant in projects conducted in heights, such as high buildings, communication towers, television towers and etc.

Nocturnal jobs or working in dim lights, even in simple urban projects can lead to accidents. Hiring well-trained personnel does not increase costs as much as it prevents accidents and project stoppage or delays. Therefore in night time urban projects some individuals must be hired for the purpose of warning and protecting others. Table 4 demonstrates the correlation between risk and the factors related to environment.

Table 4. *The results of the Pearson correlation coefficient between risk and the factors related to Environment.*

Environmental risk factors	High	Significance level
Tight programming schedule	39%	0.01
Unqualified subsidiary contractor's manager	24%	0.01
Client's demand for making alterations	25%	0.01
Incorrect and inadequate information about worksite (soil examination and analysis report)	35%	0.01
High expectation of quality and performance	24%	0.01
Unsuitable construction programming	22%	0.01
Construction's severe noise pollution	38%	0.01
Lack of professional staff and managers	27%	0.01

As demonstrated, tight programming schedule risk with 39% has the highest impact. And sever noise pollution and inadequate information about the worksite (soil analysis and report) with 38% and 35% respectively, are next risk factors in terms of effectiveness. Also unsuitable construction programming and unqualified subsidiary contractor's management with 22% and 24% respectively, have the least impact between risk factors in construction projects.

3.5. Safety Factor

Before continuing the discussion, three elements of safety, accident, and danger need to be defined.

Safety: in the dictionaries, safety defined and security, peace, and health. It is defined and the level of being secure from risk or danger. In fact, the safety in construction projects is consisted of:

- Security and health of all the labor force and individuals related to construction projects environment
- Security and health of all the individuals who live or pass through adjacent areas of construction project environment (effective radius)
- Security and protection and monitoring of buildings,

vehicles, structures and equipment inside the project's worksite or in adjacent areas

- Protection of living environment inside project's worksite and its adjacent areas

Accident is an unpredicted and mostly harmful incident caused by an unsafe action or unsafe situation –or a combination of both- which has negative influence on completion or natural continuation of a task or activity.

Danger is a condition which has the potential to harm employees, damage tools and equipment and buildings, destroy materials, or reduce the efficiency of a pre-organized action. It is the source or the potential cause of harm to individuals or organizational assets. Studies identify two major dangers in construction projects: danger of physical damage, and danger of diseases or health related issues [Savadkoohi, 2007].

Today, a lot of construction projects are being carried out in our country and a lot of people participate in those projects. It is obvious that this great population of human resources is exposed to a lot of dangers. Considering the sites in which these projects are being carried out, people and civilians of the neighboring areas are also threatened by these dangers. Most of these threats are caused by lack of safety knowledge or oversimplification of safety rules. The overwhelming amount of threats that would lead to damage and losses of lives and assets makes the evaluation of the impact of safety risk factors a crucial issue. Table 5 demonstrates the correlation between risk and the factors related to safety.

Table 5. *The results of the Pearson correlation coefficient between risk and the factors related to Safety.*

Safety risk factors	High	Significance level
Tight programming schedule	45%	0.01
Unqualified subsidiary contractors' management	39%	0.01
Unsuitable construction management	33%	0.01
Accidents and lack of safety regulations observation	42%	0.01
Lack of coordination between project participants	26%	0.01
High expectation of quality and performance	28%	0.01
Lack of skilled workforce	25%	0.01
Lack professional managers and individuals	25%	0.01

As demonstrated, risk of tight programming schedule has the greatest effect, and lack of safety regulation observation and unqualified subsidiary contractors' management with 42% and 39% respectively, are next risk factor in line in terms of effect. Also the risks of unsuitable construction programming and lack of skilled workforce with 33% and 25% respectively, have the least impact in safety of construction projects. Every year, a lot of accidents with irreparable consequences take place in construction projects and worksites only because of indifference toward safety rules. For that matter, implementation of safety regulation as a significant duty of different managerial and executive positions of construction projects is the only way to reduce accidents and losses. Although the chapter 12 of national construction regulations for safety and security is legally binding, still many safety rules are not observed and unfortunately numerous accidents still take place during construction projects.

To achieve desirable standards in construction safety, first social and cultural backgrounds are needed to be established in the community, and then, technical and executive infrastructures should be implemented in companies and organizations that are active in the field of construction. And finally, all levels and positions of construction projects, including employers, project managers, consultant engineers, and contractors must be educated and the technical knowledge must be shared with them. Apparently, this task requires special attention from high ranking managers of the industry, and a purposeful and objectified programming.

4. Suggestions

The first step toward workplace safety must be taken by executive positions; for the workers to take safety seriously, it must become an organizational virtue and value. In all levels of an organization, managers must establish the idea that safety is a responsibility embedded in their job description. In many cases, this responsibility is put upon a safety manager in worksite, or a security manager in organization. Executive project managers and supervisors are required to receive necessary educations in order to be able to encounter unpredicted conflicts, problems and accidents.

Active and credible companies usually use risk avoidance treatment; high level managers, safety managers and operatives create safety committees, discussing an analyzing performance safety to reduce –if not eliminate- risks and risk exposures. These committees must include operative employees, since through their field experience they can provide useful information and ideas on the appropriate ways to implement safety methods and better evaluation and improvement of safety. Although companies usually tend to employ safety managers only in large-scale projects, they need to do the same in small projects as well. Because workplace safety (whether in large or small projects) can prevent potential losses and damages, hence minimizing expenses. However, efforts for safety should not be limited to assignment of a safety manager. Safety managers should constantly supervise and evaluate performance of each individual, because, all the individuals are responsible for the safety of projects. When safety programs, procedures and routines are implemented from above, laborers will also take safety issues seriously. Safety and risk management programs must be assessed and approved on an annual basis, and managers' achievement in implementation of those programs must be evaluated. Companies also need to evaluate their success, by defining criteria such as "hours without accident" or achievement of a sensible annual rate of accidents. Only then they can claim that they have reached an optimal level of risk and crisis management [Alfredo, 2014]. Since a project is first programmed on a paper and on a theoretical basis, its safety and security must also be pre-programmed. A successful pre-program minimizes the threats that would delay a project, or reduce its efficiency, and allows the project to run smoothly.

5. Conclusion

In this research, after clear definition of risk, construction projects' risk factors have been analyzed using questionnaires distributed among organizations and institutions related to construction; and suggestions for better risk management in construction sites have been put forward. It must be noted that knowledge is the key element in successful risk management. A successful risk management requires application of appropriate methods, knowledge and experience such as prediction of accidents that might happen during project execution. Even when the project is running smoothly, lack of preventive measures could have negative consequences. The main issue concerning risk management is the avoidance of risks altogether; and preparation to overcome risks and problems in case they occur. However many companies refuse taking risk management measures because of their financial limitation, while their refusal would lead them to further costs and losses [Marras&Karwowski, 2006]. It is fair to say risk management is an element of success in construction projects.

The project's success depends on an appropriate balance between three important issues of project's timetable, resources and results to present the client with a desirable service. The project serves to satisfy the client, therefore, these three elements must be balanced appropriately. Change in one element will lead to changes in the other two. For instance achievement of higher quality calls for increase in costs (and sometimes duration) of a project. To shorten the duration of project, either the final quality needs to be reduced, or the further costs need to be expended (i.e. more resources will be exhausted). Also with reduction of costs, the final quality will decrease and the project's timetable will become longer. Considering the importance of risks in urban projects, establishment of standards and protocols in such projects seem crucial. Such protocols must be established for all kinds of projects in general, and edited for specific projects in particular; and more importantly, the execution of these protocols must be accurately monitored.

Since procedural elements play an important role in risk preventions, they must be taken more seriously. Methods of project implementation must be determined with concerns for safety and security. Lack of professional considerations in compilation of executive methods is one of the main causes of accidents in construction projects. Education and justification of designers and executive engineers of the project is influential to project safety. Specifically in urban projects, special educational sessions need to be held. Employees' motivation and their demands have an important role in risk prevention and human resource efficiency therefor it must be taken into consideration by managers and engineers. In the end it must be noted that ever-increasing number of construction workshops call for serious supervisions by regulatory systems and organizations. Also emphasis on knowledge and awareness means education of employees and laborers before the start of project. As project programming should be carried out before the start of project, so is the education of employees for better risk management. Risk managers need to constantly evaluate employees and their performance. Also managers' participation will lead to individual commitment to safety. In the way to success, everyone is responsible; and responsibility is the main component in the culture of safety. Laborers, foremen, supervisors and managers are all responsible for the safety of project. The present study suggests that several elements, such as employees' work environment, project timing, tight programming, client's expectations, skilled workforce, and professional managers are taken into consideration in order to improve risk management in construction projects. Although it is still a matter of long debate, it is crucial to note that risk control and risk management is the best way to prevent possible losses and damages to qualitative and quantitative process of construction projects.

References

[1] Eshteharadian E, Nasrazadani S.M, Safavi S.A, Alikhani A. *"studying the causes of delay in urban projects considering project's elements"*, 6th international conference of project management, 2010.

[2] Ahmadi-jazzi R, *"a study of the causes of delay in construction projects in Iran and its solutions"*, M.A thesis, Iran's science and technology university, 2006.

[3] *"What is risk management"*, website of Iran's I.T. professional forum www.itpro.ir/.

[4] *"Analytical report on the causes and solutions of construction projects' costs increase"*, organization of budget and programming.

[5] Savadkoohi F, *"crisis management in urban-construction projects"*, 2007.

[6] Shakeri E, Ghorbani A, *"value engineering and management of conflicts and dispute in construction projects"*,2n international conference of value engineering, 2005.

[7] Shakeri E, Ghorbani A, *"project management and identification of the main claims and disputes of contractors"*, 2n international conference of value engineering, 2005.

[8] Sehat S, Alavi S.S, *"the necessity of application of risk management to third party insurance, and the impact of new third party insurance laws on the related risks"*, Journal of Insurance world news No.144 & 145, 2010.

[9] Abbasnia R, Moradisangi V, *"quality costs in construction projects"*, 6th international conference of project management, 2010

[10] Vatankhah R, *"a study of the causes of delay in construction projects of schools renewal organization"*, M.A thesis, Tarbiat Modarres university, 2003.

[11] Harischian M, Moshtaridoost T, Rahimi-Aloughare I, *"application of comprehensive quality management in improvement of construction projects' performance quality"*, 3rd international conference of project management, 2007.

[12] Alfredo Federico Serpella, *"Risk management in construction projects: a knowledge-basedapproach"*, 27th IPMA World Congress", 2014: 654.

[13] Cesarini, George; Hall, Geoffrey and Kupiec, Matthew; *"Building a proactive safety culture in the construction industry"*, ACE USA, 2013.

[14] Kai Way LI and Cheng-Lung LEE, *"Postural Analysis of Four Jobs on Two Building Construction Sites: an Experience of Using the OWAS Method in Taiwan"*, Journal of Occupational Health", 1999.

[15] N. Jaffar*, A. H. Abdul-Tharim, I. F. Mohd-Kamar, N. S. Lop, *"A Literature Review of Ergonomics Risk Factors inConstruction Industry"*,Universiti Teknologi MARA Perak, Seri Iskandar, 32610 Perak, Malaysia, 2011.

[16] Marras W.S & Karwowski W. Interventions, Controls, and Applications in Occupational Ergonomics. *"The Occupational Ergonomics Handbook (2nd Edition) Taylor&Francis"*, London, 2006.

[17] Mansfield, N. /Ugwu,O./Doran, T.," *Causes of delay and cost overruns in Nigerian constructionptojects* "International Journal of project management, 12 p. 254-260, nov1994.

[18] Mezher, Toufic / Jamali, Dima / Zreik, Carine, article, " *The role of financial institutions in thesustainable development of Lebanon* " May 2002.

4

Study on Roller Compacted Concrete's Nominal Fracture Toughness Based on Crack-Depth Ratio

Ren Li-li[1], Guo Rui-jin[2]

[1]Department of Railway Engineering, Liaoning Railway Vacational and Technical College, Jinzhou, China
[2]College of Civil Engineering, Liaoning University of Technology, Jinzhou, China

Email address:
ly181145@163.com (Ren Li-li)

Abstract: The roller compacted concrete (RCC) has been widely used, and the practical concrete crack usually has different crack-depth ratios, but the researches about the fracture toughness formula for RCC rarely consider crack-depth ratio. Here the wedge-splitting fracture test was carried on the RCC specimens with different crack-depth ratios(0.4, 0.5, 0.6). At the same time, the theoretical nominal fracture toughness formula based on the crack-depth ratio was also studied. At the end, the modified and suitable RCC' formula based on crack-depth ratio was gotten by combining the RCC test results with theoretical formula.

Keywords: Roller Compacted Concrete, Fracture Toughness, Crack-Depth Ratio, Experiment

1. Introduction

Roller Compacted Concrete (RCC) dam construction technology has been widely used with its advantages such as saving cement, fast construction speed and low engineering cost [1]. But it is worth noting that RCC dam still exists temperature cracks what exist in the normal concrete dam. It is necessary to study the fracture properties of RCC. We have carried on many researches about the RCC and made some achievements in China in the 90 's of the twentieth century[1, 2]. Although there are many aspects researches about the fracture toughness of RCC, and there are also many research about the common concrete fracture toughness considered the crack-depth ratio till now[3-6],but the research about fracture toughness of RCC with different crack-depth ratio is rarely. Here the wedge-splitting fracture test for RCC specimens with different crack-depth ratios is carried out, and analysis the results of the wedge-splitting fracture test. We obtain the fracture toughness formula of RCC considering crack-depth ratio, by combining the experimental results with theoretical formula.

2. The Study on Experimental Fracture Toughness of RCC Specimens

2.1. Experiment Design

2.1.1. Material Selection and Mix Ratio

The experiment use common water from life for test water, P.O 32.5 cement, II grade fly ash, and nature medium sand which requires to particle greater size than 5 mm. Coarse aggregate use the maximum particle size of 20mm. Admixture uses high performance water reducer and the mix ratio of the concrete is shown in the Table 1.

Table 1. *Mix ratio of RCC.*

Water	Cement	Fly ash	Sand	Gravel	Admixture
/(Kg/m$^{3)}$	/(kg/m^3)	/(kg/m^3)	/(kg/m^3)	/(kg/m^3)	/%
130	134	90	805	1280	0.75

2.1.2. Parameters of Specimen and Test Device

The wedge splitting specimen is used. The RCC specimen's parameters and quantities are shown in table 2. The wedge splitting test is carried out with the 200 kN electro-hydraulic servo testing machine and the constant displacement loading method is used during the test.

Table 2. The parameters of the wedge splitting specimen.

Specimen type	quantity	Dimension /mm	crack-depth ratio
	14	150×150×150	0.4
RCC150	6	150×150×150	0.5
	6	150×150×150	0.6

2.2. Test Results and Analysis

2.2.1. Selection of Calculation Formula of Fracture Toughness

We use the method in [7] to analysis. The coordinate system shown in figure 1 (a) is established with the center of the specimen as origin. When left and right symmetrical on both sides of the force system were moved to the B_1, B_2 points, the stress system (solid line) of figure1(a) is equivalent as the force system of in figure1(b) of (solid line shows), where M 4 $=1/2P_h$ D-$1/2P_v$ e according to the force balance shows F=1/2mg+$1/2P_v$. Using the superposition principle of force, figure1 (b) can be decomposed into figure1 (c) and figure1 (d) cases. According to Saint Venant principle, figure1 (c) can be further equivalent to figure1 (e).The stress situation of the wedge splitting geometry test is transformed for solutions of problems of uniform tensile force P_x and pure bending with unilateral openings.

Unilateral opening specimens are subject to axial pull action as shown in figure1 (e) and the stress intensity factor can be calculated by the use of the following formula[8]:

$$K_I^1 = \sigma\sqrt{\pi a} f(a/D) \quad (1)$$

in which,

$$\sigma = P_h/(D\times t),$$

$f(a/D)=1.122-0.231(a/D)+10.550(a/D)^2-21.710(a/D)^3+30.382(a/D)^4$.

Unilateral opening specimens are subject to pure bending moment action as shown in Figure 3 (e) and the stress intensity factor can be calculated by the use of the following formula[8]:

$$K_I^2 = \frac{6M}{tD^2}\sqrt{\pi a} g(a/D) \quad (2)$$

in which,
$M=1/2P_hD-1/2P_ve$, $g(a/D)=1.122-1.40(a/D)+7.33(a/D)^2-13.08(a/D)^3+14.0(a/D)^4$.

Superposition type (1) and (2), when wedge splitting pull geometry is loading at any moment, the expressions of stress intensity factor of crack tip can be expressed as follows:

$$K_I = K_I^1 + K_I^2 \quad (3)$$

2.2.2. Test Results and Analysis

We can get each test value of nominal fracture toughness , and take the average value are shown in table 3 by using the maximum load of each specimen and original parameter to substitute type (3).

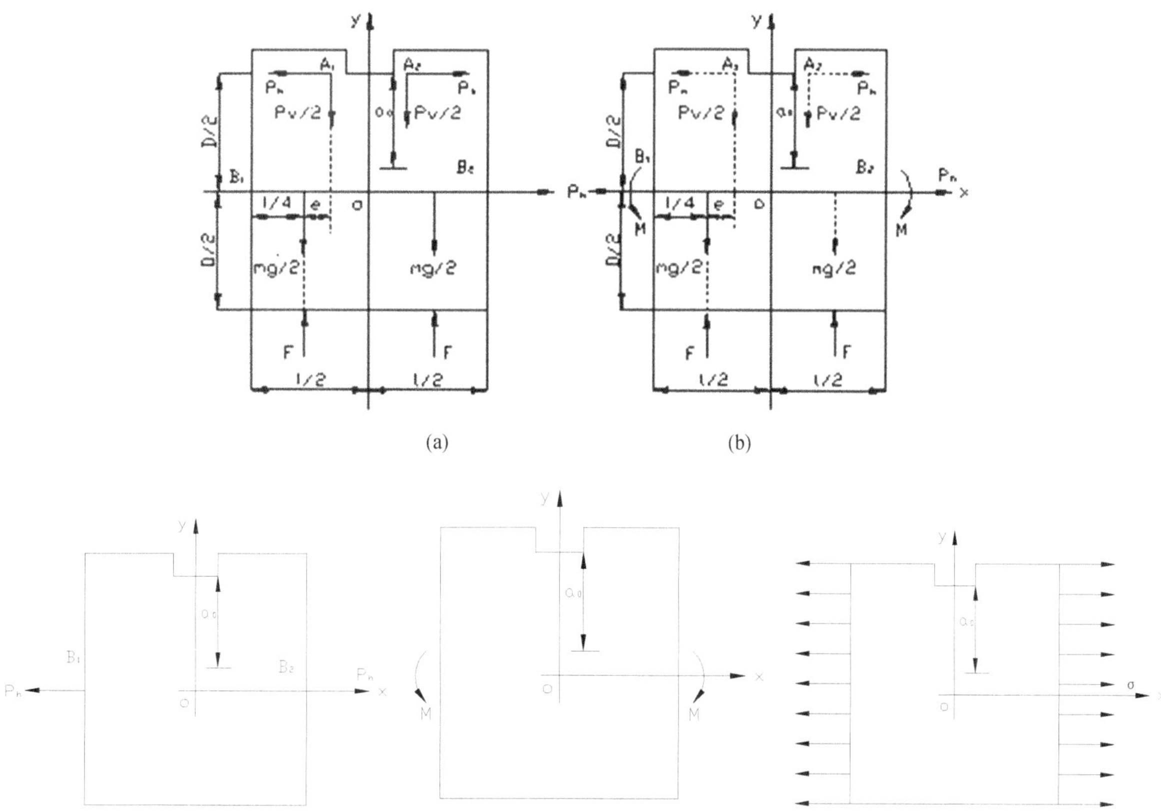

Fig. 1. Diagram of force on the specimen.

Table 3. Test value.

Specimen type	crack-depth ratio	average test value / $MPa.m^{1/2}$
	0.4	0.87
RCC150	0.5	0.76
	0.6	0.89

It can be seen that the average value of a high degree of 150 mm of the nominal fracture toughness of the specimen is 0.75~0.9 $MPa.m^{1/2}$ .m from Figure 4.We can observe the change of fracture toughness with the change of crack-depth ratio, and the fracture toughness of RCC is changed with the change of the crack-depth ratio, but the change rule is not monotonous.

3. Selection of the Formula for the Fracture Toughness of RCC

RCC as a kind of concrete is different from ordinary concrete in the construction process. But studies show that RCC's size effect rules of fracture toughness are same with ordinary concrete's. In addition, the RCC is an anisotropic material, the direction of the roller compacted layer is perpendicular to the direction of the crack. The specimens' force of wedge splitting tensile test is completely symmetrical completely on both sides of the cracks. RCC can be regarded as homogeneous materials along the height direction of each layer, and has same properties with ordinary concrete in this direction. So the RCC maybe have the same nature as the ordinary concrete in the isotropic condition, at the same time due to the different construction process they also have a certain difference. Now the theoretical research on the concrete's formula of size effect has a lot of, such as Xu Shilang [7], Bazant[9], Carpinteri [10], Hu [11]and Wang[12].

3.1. Size Effect Formula

3.1.1. Hu's Size Effect Formula for Fracture Toughness of Concrete [11]

On the basis of nominal tensile strength size effect formula of concrete, Hu obtained the size effect formula for fracture toughness of concrete:

$$K_{\mathrm{Ic}} = \frac{A(\alpha_0)\cdot f_t \cdot Y(\alpha_0)\cdot\sqrt{\pi a}}{\sqrt{1+B(\alpha_0)\cdot a/a_\infty^*}} \tag{4}$$

in which: a is crack length; α_0 is crack-depth ratio, $\alpha_0 = a/h$; K_{Ic} is nominal fracture toughness; ft is tensile strength;

For three point bending specimen:

$$B(\alpha_0) = [A(\alpha_0)\cdot Y(\alpha_0)/1.12]^2;$$

$$A(\alpha_0) = (1-\alpha_0)^2;$$

$$Y(\alpha_0) = f(\alpha_0)/\sqrt{\pi}$$

$$f(\alpha_0) = \frac{1.99-\alpha_0\cdot(1-\alpha_0)\cdot(2.15-3.93\alpha_0+2.7\alpha_0^{\,2})}{(1+2\alpha_0)\cdot(1-\alpha_0)^{3/2}};$$

$$a_\infty^* = \frac{1}{1.12^2\pi}\cdot\left(\frac{K_{IC}}{f_t}\right)^2$$

in which: K_{Ic} is no size effect fracture toughness;

The size effect formula of fracture toughness of the three point bending beam is obtained.

$$K_{\mathrm{Ic}} = \sqrt{\frac{A_1\cdot\dfrac{\alpha\cdot(1-\alpha)^4\cdot f^2(\alpha)}{\alpha_1\cdot(1-\alpha_1)^4\cdot f^2(\alpha_1)}\cdot W}{1+B_1\cdot\dfrac{\alpha\cdot(1-\alpha)^4\cdot f^2(\alpha)}{\alpha_1\cdot(1-\alpha_1)^4\cdot f^2(\alpha_1)}\cdot W}} \tag{5}$$

in which: W is specimen height; A_1andB_1 are the regression parameters are obtained by the test values for a given parameter,

3.1.2. The Modified Formula Based on Hu's Nominal Size Effect Formula [12]

Formula (4) is a function of two variables, which are crack-depth ratio and crack length. At the same time, it is also related to the fracture toughness and tensile strength of concrete with no size effect. When the height of the concrete specimen is large enough, the test of fracture toughness value is no longer the size effect. Based on the fracture toughness of concrete without size effect, Wang study the size effect of nominal fracture toughness and get the modified formula:

$$K_{IcL} = A(\alpha_L)\cdot f_t\cdot Y(\alpha_L)\cdot\sqrt{\pi a_L}\left(1+\frac{B(\alpha_L)a_L}{B(\alpha_s)a_s}\cdot\left(\left(\frac{A(\alpha_s)\cdot f_t\cdot Y(\alpha_s)\cdot\sqrt{\pi a_s}}{K_{Ics}}\right)^2-1\right)\right)^{-1/2} \tag{6}$$

in which: each symbols has the same meaning of various symbols in front.

Compared with formula (5), formula(6) is also considered the effect of crack-depth ratio and crack length. At the same time, it allows the size of the specimens with different crack-depth ratio, has less constraint conditions, less constraint conditions, and has a wide applicable scope. The nominal fracture toughness of the small specimens can be measured easily by the test, and the parameters of the model are not required.

3.1.3. The Nominal Fracture Toughness Size Effect Formula Based on Hu's Formula

When the formula (5) is simplified further, we can obtain the size effect formula of fracture toughness of concrete containing two coefficients:

$$K_{Ic} = \sqrt{\frac{A_1 W}{1 + B_1 \cdot W}} \tag{7}$$

in which: A_1andB_1 are the coefficient, which are related with α_1 、B(α_1)、K_{Ic} and a^*_∞ It can be determined using the test data of two points.

For formula (7), the fracture toughness is obtained by the two known data points and height of the specimen of the third point, and it is different from conventional size effect formula of the fracture toughness which is obtain from the fracture toughness of known specimens to infer the fracture toughness of unknown specimens, and the test of the small size will get the corresponding fracture toughness usually. In order to obtain the formula of common form, this will change the formula (8)、(9).

$$K^y_{Ic} = \sqrt{\frac{A_1 W_y}{1 + B_1 \cdot W_y}} \tag{8}$$

$$K^w_{Ic} = \sqrt{\frac{A_1 W_w}{1 + B_1 \cdot W_w}} \tag{9}$$

in which: K^w_{Ic} is nominal fracture toughness of unknown specimen; K^y_{Ic} is nominal fracture toughness of known specimen, fracture toughness of standard specimen; B_1 is related coefficient withα_1 ,B(α1)and a^*_∞, it can be determined by the test data of two points; W^y is the height of a known specimen; W^w is the height of an unknown specimen;

Further, we can obtain the fracture toughness formula which contain the crack-depth ratio and the height.

$$K^w_{Ic} = K^y_{Ic}\sqrt{\frac{\frac{\alpha \cdot A(\alpha)^2 \cdot f^2(\alpha)}{\alpha_1 \cdot A(\alpha_1)^2 \cdot f^2(\alpha_1)}(1 + B_1 \cdot W_y)}{1 + B_1 \cdot \frac{\alpha \cdot A(\alpha)^2 . f^2(\alpha)}{\alpha_1 \cdot A(\alpha_1)^2 . f^2(\alpha_1)} \cdot W_w} \cdot \frac{W_w}{W_y}} \tag{10}$$

When the height of the known and the unknown specimen is same， we can obtain Formula (11) only considering the crack-depth ratio.

$$K^w_{Ic} = K^y_{Ic}\sqrt{\frac{\frac{\alpha \cdot A(\alpha)^2 \cdot f^2(\alpha)}{\alpha_1 \cdot A(\alpha_1)^2 \cdot f^2(\alpha_1)}(1 + B_1 \cdot W_y)}{1 + B_1 \cdot \frac{\alpha \cdot A(\alpha)^2 . f^2(\alpha)}{\alpha_1 \cdot A(\alpha_1)^2 . f^2(\alpha_1)} \cdot W_w}} \tag{11}$$

3.2. Experimental Verification on Fracture Toughness of RCC Considering the Crack-Depth Ratio

We would removal the large deviation value of the test results obtained earlier and use remaining into formula (11). We obtain the theoretical value of the fracture toughness of RCC. As shown in Table 4.

Table 4. Theoretical value of RCC fracture toughness.

Specimen type	crack-depth ratio	average test value / MPa.m$^{1/2}$	theoretical value / MPa.m$^{1/2}$
	0.4	0.87	0.87
RCC150	0.5	0.78	0.85
	0.6	0.89	0.84

We can see from table 4 that the theoretical values are in good agreement with the experimental values. It can explain that previous theoretical formula has a certain applicability, but its further law also need to be further explored.

4. Conclusions

We get the following conclusion:

(1) The nominal fracture toughness of the specimen with 150mm height is between 0.75Mpa.m and 0.9Mpa.m$^{1/2}$.

(2) Under the influence of the crack depth factor, the fracture toughness of RCC is changed, but its change is not monotonous.

(3) We can obtain the fracture toughness formula of RCC , which is usually used to consider the crack depth ratio, by the Hu's size effect formula:

$$K^w_{Ic} = K^y_{Ic}\sqrt{\frac{\frac{\alpha \cdot A(\alpha)^2 \cdot f^2(\alpha)}{\alpha_1 \cdot A(\alpha_1)^2 \cdot f^2(\alpha_1)}(1 + B_1 \cdot W_y)}{1 + B_1 \cdot \frac{\alpha \cdot A(\alpha)^2 . f^2(\alpha)}{\alpha_1 \cdot A(\alpha_1)^2 . f^2(\alpha_1)} \cdot W_w}}$$

(4) We can get the theoretical values are in good agreement with the experimental values. It can explain that previous theoretical formula has a certain applicability, but its further law also need to be further explored.

Acknowledgements

This work was financially supported by Project of Education Department of Liaoning Province (L2013535).

References

[1] Yang Huaquan, Li Wenwei. Research and application of hydraulic concrete[M]. Beijing: China Water Conservancy and Hydropower Press, 2005.

[2] Zhang Lin, Xu Jin, Chen Xin, et al.Experimental study on fracture of RCC[J]. Journal of water conservancy. 2001, (5): 45-49.

[3] HU Shao-we, XIE Jian-feng, YU Jiang. Frac ture Toughness of specimen with different initial seam-height ratio under wedge splitting test[J]. Journal of Yangtze River Scientific Research Institute, 2015, 32(2): 114-118.

[4] RONG Hua, DONG Wei, WU Zhi-min, FAN Xing-lang. Experimental investigation on double K fracture parameters for large initial crack-depth ratio in concrete[J]. Engineering Mechanics, 2012, 29(1): 162-167.

[5] HU Shaowei, MI Zhengxiang. Experimental study on double-K fracture characteristics of standard reinforced concrete three-point beam[J]. Journal of Building Structures, 2013, 34(3): 152-157.

[6] HU Xiao-wei1, ZHANG Xiu-fang1, XU Shi-lang. Determination of double-K fracture parameters of concrete using split-tension cubes[J]. Journal of Hydraulic Engineering, 2012(S1): 98-109.

[7] Xu Shilang, Zhang Xiufang, Gao Hongbo. Concrete wedge splitting specimens of double K fracture parameters superposition calculation and boundary effect[J]. Journal of Dalian University of Technology, 2006, 46 (6): 867-874.

[8] TADA H, PARIS P, IRWIN G. The stress analysis of cracks handbook, (2nd, edn) [M]. St. Louis: Paris Productions Incorporated, 1985.

[9] BAZANT Z P. Size effect in blunt fracture: concrete, rock, metal[J]. Journal of Engineering Mechanics, 1984, 110 (4): 518-535.

[10] CARPINTERI A, CHIAIA B, FERRO G. Size effects on nominal tensile strength of concrete structures: multifractality of material ligaments and dimensional transition from order to disorder[J]. Materials and Structures, 1995, 28 (180): 311-317.

[11] DUAN K, HU X Z, WITTMANN F H. Size effect on fracture resistance and fracture energy of concrete[J]. Materials and Structures, 2003, 36: 74-80.

[12] Wang Xuezhi, BiZhong, Zhang Xiaogang, et al. Experimental study on the size effect of the fracture toughness of RCC[J]. Journal of Wuhan University of Technology, 2007, 29(6): 54-57.

Use of Debris Program of Civil Construction Waste Management in Containment Residential Erosion Geovanni Braga in Anapolis in the State of Goiás

Welvis Furtado da Silva[1], Ernane Rosa Martins[2]

[1]Department of Environmental Engineering, Metropolitan Institute of Education and Culture, Anápolis, Brazil
[2]Coordination of Information Technology Area, Federal Institute of Education Science and Technology Goiás, Luziânia, Brazil

Email address:
welvisfurtado@anapolis.go.gov.br (W. F. da Silva), ernane.martins@ifg.edu.br (E. R. Martins)

Abstract: The gullies can cause great damage to the environment and also the very human environment, as one of the main consequences of this erosion is the major loss of soil mass. This work aims to present the use of debris originating from demolition in construction to contain the degraded area that causes environmental impact at Residential Geovanni Braga by the Municipal Works, Urban Services and Property Annapolis City Hall in the State of Goiás in the year of 2015. The methodology of the proposed study is theoretical, / literature, practical / field work / case study and qualitative and quantitative descriptive. Adopted the use of discarded materials such as: debris of construction material, tires, wood and shackles for sewage or storm water. As a result it was found that the techniques used in the treatment of erosion was rapid, extremely efficient and extremely cost-effective when compared to traditional methods, pleasing the residents' association and improving the aesthetics of the place before degraded.

Keywords: Construction Waste Destination, Erosion Containment, Degraded Areas, Environmental Impact

1. Introduction

Erosion is a ground displacement process or rocks from a surface. Erosion can occur per share of natural phenomena or human. With regard to acts of nature, we can mention the rains as the major agent of erosion. Upon reaching the ground, in large quantities, causing landslides, seepage and changes in terrain consistency. Thus, it causes the earth displacement. The wind and temperature change are also important cause erosion.

The implementation of vegetables in areas with natural kicking erosion is to minimize or eliminate the adverse effects of interventions and environmental changes inherent to drag the soil at a particular location.

In urban areas these spaces have the potential to function as temperature amenizadores (climate control), lower noise and carbon dioxide levels (improving air quality) promote the medium disorders balance (protection against floods and droughts) protect watersheds to supply clean water (control and water supply), provide shelter for wildlife (fauna and biological control of refuge) and contribute to the aesthetic improvement of the landscape.

To work around this urban problem, the development of normative solutions and appropriate works projects to each situation is needed. These solutions should be simple and less costly but efficient and to justify public investment, resulting in security for the population and for public facilities in order to return the area to its urban use.

Thus, this research aims to present the use of debris Construction in curbing the environmental impact of erosion Residential Geovanni Braga by the Municipal Works, Urban Services and Property Annapolis City Hall in the State of Goiás in 2015.

The methodology of the proposed study is theoretical, / literature, practical / field work / case study and qualitative and quantitative descriptive.

It adopted the use of discarded materials such as: urban cleaning debris, construction material, tires, wood and shackles for sewage or storm water.

This research is justified by the limited literature on the use of rubble from civil construction and / or street cleaning in curbing the environmental impact of erosion.

So based on the above, this article is divided into five sections, in this present section presents besides the introduction, the definition of the research problem, the goal, the rationale and importance of the study and structure of this research. Section two (02) brings the theoretical framework, with the formation of a conceptual and theoretical base, providing subsidies for the development of this study. In section three (03) presents the method used and the technical and methodological procedures used for the study. In section four (04) presents the analysis and discussion of the results obtained in the research. Finally, in section five (05) conclusions are drawn.

2. Literature Review

The environmental protection has been instrumental for establishing and maintaining urban developments, industrial, agricultural, commercial and road, under the legislation and also for the surveillance of society thus avoiding degradation and environmental destruction. The Brazil, a tropical country with strong intemperate conditions such as heat stroke and heavy rains have contributed to accelerate the erosive processes that normally there is no preventive treatment, but in some cases only the concealer, bringing recovery of costs [1].

The removal of vegetation from an area leaves it exposed to erosion caused by falling water droplets, mainly from rainfall, which ends up resulting in a mass movement in the soil. "The process responsible for the breakdown of the soil, after removal of the vegetal layer on its surface, is the impact of rain water droplets, thus the pellets are transported from one location to another" [2].

After a long rainy season, these water impacts with the ground end up generating a stream of sediments that can lead ravines, and if the process is continued cause incessant deepening of the soil, being able to reach the level of a gully.

The gullies are the most advanced stage of erosion and thus more difficult to contain. Are products of surface waters, running on the surface, causing wear, landslides, soil hauls etc., can also be caused by deep waters, infiltrating the ground, walk through into the profile, more or less vertically, to find a impermeable or less permeable layer where they accumulate and moving horizontally, causing mudslides and landslides [3].

Erosion and other factors is caused by a group of geodynamic processes, driven by energy and gravitational potential energy associated with the uneven terrain [4].

Ferreira et al. (2007), states that "the gullies are considered one of the worst environmental problems in areas of crystalline rocks in mountainous tropical regions where they are frequent and can reach large" [5].

To Ab'Saber (1968), the process of formation of gullies is associated to landscapes from which it was withdrawn their vegetation. In these landscapes, water runoff to linearly move through the soil and reach the water table, compromises the stability of the area and lead to the formation of gullies [6].

According to Lima (2003), soil erosion affects life in many ways, there is an issue that involves professionals from various fields of knowledge, sometimes with different views on the same issue. There are lines of research on erosion nationwide in agronomy, geography, geology, geotechnical and other areas of knowledge [7].

According Iwasa & Fendrich (1998), most municipalities is affected by linear erosion, mainly caused by the concentration of water runoff [8].

The great detonator of urban erosion is concentrated runoff on impermeable areas with weak or non-existent drainage system.

Camapum de Carvalho et al. (2006) point out that, in engineering works such as dams, transmission and road lines, the inappropriate release of water and the loan abandonment of areas explored without adequate technical care may also be related to the genesis of erosion [9].

Landslides are fast-moving, large amounts of material, which typically occur as a result of the existence of discontinuity in the soil profile, imperfect geotechnical works (embankments) and the very dynamic areas with rugged terrain. Are also considered within the scope of erosion, although they are usually neglected in the context of soil erosion study, which can also lead to the formation of gullies [10].

According to Embrapa soils (2002), the correction voçorocamento areas can be made in order to "control erosion in the area upstream or the slope bedside, retention of sediments inside the gully, reforestation of catchment areas (bedside) and inside the gully with plant species that are able to develop properly in these locations." [11].

Many are the costs involved for the recovery of degraded areas by gullies, such as hand labor used, inputs, cost of seedlings and transport of the same, etc. The cost of recovering such an area will depend mainly on the size (length, width and depth) of the gully one wants to retrieve, thereby evaluating whether an intervention is economically viable in voçorocada area.

In this context, this study aimed to demonstrate the areas recovery method eroded using rubble and see if the technique used to stem the erosion is valid or if there are flaws that can be corrected.

It is understood by the rubble fragments together from the waste in the construction, renovation and demolition of several buildings structures such as buildings, houses and bridges.

Among the projects that used rubble and succeeded in erosions of contention are:

The project carried out by Carvalho (2004), which was used from the rubble generated in the city to contain about 10 gullies as the rehabilitation of the Villa St. Matthew, in the neighborhood Estrela Dalva in Contagem (MG), which was transformed into terrace (flat area) for disposal of inert waste (debris) and will eventually be deployed a linear park with funds from the Brazilian Institute of Environment and Renewable Natural Resources (IBAMA). In addition to this, the same method was used in Piraquara neighborhood near the center of the city, and the region of Wholesale Markets (CEASA) [12].

Cappi (2004) also conducted an environmental recovery work of eroded areas as the destination alternate ending of scrap tires. According to him, this technology presents operating low cost characteristics and the advantage of providing indirect control the dengue mosquito, and recover degraded areas by large erosion (gullies) or build them siltation barriers as part of the landscape recovery strategy eroded, which will then be revegetated. For this, we used two different techniques: the construction of embankments to sediment through tire barrier and complete burial of tires in the gully bottom. The second technique used more tires and the restoration of landscape after the end was faster [13].

It is known that the construction industry is the major generator of rubble and it is known, too, that this practice is inevitable. However, one can adopt efficient measures that do not degrade the environment and, in turn, will bring some social benefit.

It is also known that the use of whole tires associated with large plant roots can be used to assist in the containment of soil erosion [14]. As the filling of landfill using shredded tires to replace part or whole of aggregate with low cost and keep the soil with good drainage [15].

Another technique that can be used is composting, which is the aerobic decomposition process of organic matter under appropriate conditions, which aims to turn the waste into a product of good quality, which has several benefits for agricultural products besides being a good soil conditioner because of its physical characteristics, chemical and biological.

The composting technique is shown to be of utmost importance to reduce the environmental damage caused by the improper disposal of organic waste. Therefore, the compost should be seen as an environmental biotechnology, as the benefits of your choice as a key strategy to make up the management of organic waste, whether in urban or rural setting, or generators as agro-industrial or industrial complex, constitute the process enabling compliance with the items considered fundamental the concept of sustainable development for the efficient treatment and disposal of solid waste. With the minimization of environmental impacts, minimizing waste and maximizing recycling.

Haug (1993) means that this process (composting) occurs on conditions allowing high temperatures, and a result of the aerobic biological decomposition of organic substrates, achieving a stable end-product for storing and applying to the soil, further highlighting the lack of environmental effects Undesirable [16].

Note also, that composting is essential for sustainable development, and that literature poses many benefits from composting technology, "compound (...) is a valuable soil conditioner, source of humic substances, and range of nutrients that provide growth of plants in horticulture, gardening and horticulture" [17].

Thus, bioengineering is a combination of alternatives involving biodegradable structures such as vegetable fibers, live cuttings, timber and rigid structures such as stone, concrete, iron, among others.

According to Gray and Sotir (1995), all erosion has a negative visual appearance showing signs of neglect, danger and environmental degradation, and most often silted sediment to waterways, contributing to assoreá them and washing / removing the layer fertile soil [18].

According to Goldman (1986) almost all erosion is possible to control and recovery, so the techniques used here will be useful in the control of existing erosion [19].

3. Research Methodology

The methodology of work is theoretical / bibliographic nature, practical / field work / case study and qualitative and quantitative descriptive.

To carry out this study a literature review was performed, since this theme recurs in several studies conducted in various regions of Brazil.

They used photos and satellite images, to study the area and understanding of the problems caused by erosion.

This study is characterized as descriptive of the nature of the objective, as it has in order to describe the characteristics involved in the project undertaken [20].

The case study was chosen because it is a study of empirical investigating a particular phenomenon, contemporary, within a real life context, in the case of an in-depth analysis of an object (case), allowing its broad and detailed knowledge [21]. Its aim is to deepen the knowledge about a problem not sufficiently defined [22] to stimulate the understanding, suggest hypotheses and questions or simply develop the theory. This attempts to explain why a decision or set of decisions were taken, how they were implemented and what results have been achieved [23].

We opted for the recovery of degraded areas by reusing discarded materials, such as debris materials from construction and urban cleaning, tires, wood and shackles for sewage or storm water, in order to decrease public investments. Since the spoils were donated by gathering companies such materials of construction of the city.

We used field research to collect data through semi-structured questionnaire to identify the level of satisfaction of the neighboring population to work, comprising an objective question (approve or not approve) and four essay questions, whose answers were recorded by the researcher and later transcribed.

Respondents were invited to participate in visit "in loco", being chosen the residents of the neighborhood where it is located erosion. It stipulated a distance of 200 meters from any axis thereof to denote the sharp contours.

The method of analysis used data primarily consisted of using Microsoft Excel 2013® program, where he designed a database that has been fed with those obtained questionnaires. Then there was the descriptive analysis of the answers given by local residents.

It was used content analysis to analyze strategy of research data.

4. Analysis and Discussion of the Results

4.1. Presentation Area

The gully Geovane Braga is located in the eponymous district and is bordered upstream by Street May 13 and its left bank by the Ombudsman Street and 4th Street, at coordinates 717928 E / 8189088 N. This process voçorocamento settled perpendicular to the slope Goes Stream, a tributary of the Antas River, as shown in Figure 1 Goggle Earth date of 05/31/2015.

Figure 1. *Local intended to edáfico Method.*

The erosion was arranged in the direction North West / East, it has a linear shape with a length of approximately 550 meters, the average width of approximately 25 meters and an average depth of approximately 12 meters.

Its banks were still expanding, because of the development of the mass movements of the type slipping and his head had large amount of landfill with debris and trash.

The head of the gully position of a concave side, the upstream soil sealing and inefficient drainage system and extensive street in the slope favored the concentration of runoff flow this hillside contributing to the erosion. In addition, the release of water from the rainwater catchment system the hillside was highlighted as the triggering of voçorocamento.

During the conducted field survey it was found that part of the vegetation of the area in natural kicking erosion, especially on hilly region of the right bank of the above address was virtually wiped out by man-made phenomena and the nature of that soil range for several years, without steps were taken to recover said area. However held civil engineering works in the vicinity thereof to contain the problem of excess runoff and disposal of municipal waste, but reflections of the latter offense can still be seen at the site due to the insistence of some community members to discard various waste in the area.

4.2. Area Characterization

According to Nimer (1989), the dominant climate in the savannah region is humid subtropical warm type characterized by two distinct seasons, a drought that matches the autumn-winter period, and another humid with heavy rains corresponding to the period of spring-summer, where 70% of the total accumulated rainfall during the year is concentrated between November to March, usually the wettest quarter from January to March, when it rains on average 45-55% of the annual total [24].

Though quite antropisada, the area under review presents an excellent kicker processes in the erosion problem and disposal of municipal solid waste that were frequently on site.

The area occurs in topography in sharp gap not being semi flat whose adjacent drainages belonging to the Rio das Antas basin.

From the point of view of geology, the city of Anapolis is in the field of Goiás complex, existing rocks of the granulite complex (Archean), the Araxá Group (Mesoproterozoic) Detritus and-lateritic cover (tertiary and quaternary). In the southwest portion predominance of lateritic yoke in relief flattened and granulites the dissected areas.

The main feature lies in the dominance of deciduous species such as red mimosa (governed Parapiptadenia), mastic (Astronium SP), yellowing or garapa (Apuleia praecox) and others.

In general the massive degradation is more relevant on the surfaces receiving the afternoon sun and / or those who receive the sun all day, because this is a big temperature difference between day and night. This gradient promotes drying of the soil which in turn tends to increase the pressure in the air inducing phase greatest amount of contraction cracks which can be reflected inside the mass unsettling it and favoring the erosive process.

The gully Geovane Braga evolved in the shed on the left bank of the stream Goes, a tributary of the Antas River. The highest peaks are in the watershed tops and smaller close to the watercourses. Since the tops predominate altimetry diversity of Medium / Low to Medium that are accentuated on time in erosive edges of the tops Plans west and east dissecting modeled where stands altimetry diversity predominantly Alta. The High altimetry diversity of sites also have high slopes of 10 to over 45%. The topographic description allows watching even the inclination of the slope steepness marked by greater than 45%. On the left bank of the gully prevailing slopes ranging from 5 to 10% and on the right bank predominant slopes between 15 and 30%. Low 2% lower slopes outline the flattened tops. Immediately upstream of the gully it can check the presence of timely Concave-concave relief transitioning to aspects predominantly Rectilinear-Rectilinear and later Concave-convex to the head of the gully become convex-concave and later Concave-convex, the latter being very concentrators Flows [25].

With regard to the orientation of the slopes west bank of the watershed is predominantly oriented to the west and northwest. On the east bank of the watershed where processes have a direct relationship to the gully described in this item predominate slopes oriented to the east, but the closest banks of the gully have a different orientation, with the left-oriented bank the South and the right to North. Then the right edge gets higher solar radiation, and this interaction soil / atmosphere may be observed in the field, since the slope of that edge presents less humid than the other side. Evolutionarily, the right bank that receives more sunlight tends to evolve laterally more sharply [25].

Moreover, it is observed that this enlargement diminishes the sharp edge is present in the portions where the slope is flanked by a fragment gallery forest. In this, the vegetation may be acting as a mitigating action of solar radiation. As for drainage away from the Modeled grooming remains distant 100-1000 meters from from water courses while on patterned dista dissecting 0-100 meters, and the interception of the water table caused by the great depth of the incision [25].

4.3. Recovery Procedure

The recovery of the area was carried out by edáfico method, which includes the recovery of all battered area in the primitive land containing approximately 8214 m², the more the soil profile (slope) on the edge of the forest, as shown in Figure 2.

Figure 2. *Erosion.*

The following describes the techniques and methods used in operating result in the stabilization and control of erosion.

4.3.1. Removal of Loose Sediment Segregation of Waste

This operation consists of removing all loose sediment and also those who were in unstable conditions. The removal was done manually and with the use of machines as shown in Figure 3.

Figure 3. *Use of machinery.*

It was allowed to waste pickers perform the collection, as shown in Figure 4.

Figure 4. *Waste Segregation manually.*

The local government allowed the population had free access to the site, however, was allowed only the disposal of rubble. Companies (disk-rubble) responsible for collecting this type of material were asked to sift through and gather their buckets, any other kind of material between them, such as soda cans, water bottles, candy wrappers and ice cream, as shown in Figure 5.

Figure 5. *Buckets with segregated rubbish.*

The method used by the city is second to no other known.

4.3.2. Protection of the Headwaters of Erosions

The protection of headwater erosion was made with the construction of drainage channels and protective barriers, in order to reduce the speed of water that could cause erosion causing all the work performed was damaged.

Channels are designed so as to meet local needs, due to rainfall.

4.3.3. Anchoring Sediments

The first rains could compromise the work if the sediments were not anchored. These sediments should be retained until the vegetation is established.

The anchoring sediment was performed with the use of wood palisades.

This method was chosen because it proved adequate to prevent failure in the work and cost reduction.

The construction of the palisades was taken downstream of the gully with sewage or rainwater shackles no longer usable available in town, as shown in Figure 6.

Figure 6. *Anchoring.*

To fix the palisades were held drainage lines of the gully. They were then accommodated tire to absorb the impact of the water passing through the palisade. The use of tires prevents water make a drag (hole) after the stockade, compromising its structure, as shown in Figure 7.

Figure 7. *Tire use.*

4.3.4. Reconformation Ravines and Slopes

The area voçorocamento presents part with ravines and unstable slopes, conducting reshaping is necessary to ensure stability and allow the planting and establishment of vegetation. This work was carried out with the use of machines in the dry season in order to have the area ready to accommodate planting at the beginning of the rainy season, as shown in Figure 8.

Figure 8. *Use of machines for conformation.*

4.3.5. Initial Coverage Area

After performing the mechanical practices, construction of palisades and the reshaping of slopes and ravines, passed to the application of plant remains for the formation of mulch on the eroded surface until the herbaceous, shrub and tree implanted to establish and ensure good protection to the soil. For this, we used the residue of vegetable origin such as straw, bark, sawdust, wood and other materials available in town, as shown in Figure 9.

Figure 9. *Used woods.*

4.3.6. Application of Biodegradable Products

It was directly applied to the right surface completely biodegradable product (Embiotic - Throttle composting) to speed up the composting of organic materials that contribute to the recovery of the biota in the soil, control erosion and stabilization of slopes and embankments, Figure 10 shows the result during rain after application of this product.

Figure 10. *Result of the composting of organic matter.*

4.3.7. Revegetation of the Area

For restoration of vegetation were used native species adapted to the region, such as legumes, rapid development of grasses and planted fruit still aiming to attract small animals. Deploying the process and accepting the assumption that all species regenerate naturally, it is expected that the same occurs with the production of seeds (flowering, pollination and maturation) and their consequent dispersion by the wind, birds, rodents, etc. ants in degraded points the old aggressive action.

The beginning of the planting process coincided with the beginning of the rainy season (fall / November). For the implementation of the measures in question, we used a least possible preparation of the soil, so as not to interfere with the physical structure of the same, because of the shallow. Thus, soil preparation occurred with small machines for plowing service and soil disking and planting species, forming a containment range in level, and also used the green mass of the corporation legumes, promoting Nitrogen fixation in the soil, improving soil fertility, increasing microbial activity and minimizing erosion. It was also used revegetation technology of slopes using material from other sites (organic fertilizer + soil + seed) packed in jute bags, set in inclined slope (of exposed Profile ravine) with wooden stakes.

The management made in floristic reconstruction of the area was performed using the intercalated planting of native species in Table 1.

Table 1. List, spacing, timing and method of planting species.

Legumes / Native	Time	plantation	Spacing	Planting method
Crotalaria juncea leucena Leucenaleucocephaka Calopogonio Callopogomim Muconoides Guandu Cajanus cajan Dolichos lab lab Mucuna black Stizolobium atenium	October	January	1,0 m x 1,0m	The haul (taking care to make the coating of seeds with little overburden)
Embaúba, yellow Ipe, Tabebuia sp, Garapa, leiocarpa Apuleia, Angico, Gameleira, Ficus SP, purple Ipe, Tabeuia SP, Inga, Inga SP, Jatoba.	November	January	3,0 m x 2,0m	Opening holes (40cmx40cm), mix the soil removed from the hole with tanned manure at the rate of + 100 g of limestone.

In the range of 53 meters from the shore Street 04 into the attacked area was maintained at still existing vegetation and was not done cleaning so that not only create a protective strip, but the total coverage area, following the principle of the pioneer species will offer shading and fitting climax for the development of secondary that will arise during the process of natural revegetation.

After executing the planting, the site was irrigated with water until the 3rd (third) month to effective rooting (plant catches) and later made replanting in the event of death of some species. After 45 (forty five) days after the planting was held on topdressing with 100g of NPK 10-10-10 fertilizer per hole and cultural practices (weeding) around the graves. Combating the presence of ants and termites was necessary so that not having the inhibition of catches of the species. To this end, we used ant poison baits. The species were chosen taking into consideration the ease of obtaining seedlings and adaptation to the site to be vegetated.

Periodic inspections were carried out the excavation, erosion containment terraces and picks up the seedlings that were inserted into the area of interest. And carried out periodic maintenance on replanting to prevent the loss of grass layer formed on the surface thus accelerating the retreat of the attacked area. In Figure 11 is shown downstream of the gully recovered and restored vegetation.

***Figure 11.** Downstream of the recovered gully.*

Figure 12 provides a front view of the gully recovered and restored vegetation cover (quarantine area).

***Figure 12.** Front view of the recovered gully.*

4.4. Field Research

The results of the field research were analyzed and represented in tables.

Observing Table 2, we realize that, in relation to the frequency of the means of transport that arrived at the scene to deposit the spoils, it was found that the most frequent day is Tuesday followed by the second and Thursday, after Wed fair and, finally, to Friday.

***Table 2.** Frequency of transport means a day.*

Days of the week	Frequency
Monday	39
Tuesday	43
Wednesday	24
Thursday	39
Friday	18
Total	163

The responsible authorities have taken the initiative to surround the whole area and prohibit your household waste disposal on site, making it clear, using cards that could only be deposited there rubble (construction and demolition debris), twigs and leaves, with accompanying supervisor of works and services of city hall.

Thus, it was found that only been deposited on-site materials listed in Table 3.

Table 3. *Frequent Debris erosion.*

Rubble Types	Frequency	Frequency %
Construction	68	43
Organic (branches and leaves)	64	41
Mixed	25	16
Total	157	100

In the case of the types of means of transport responsible for the unloading of these materials, it was noticed that was common the presence of tractors, disk-dump trucks, carts, private cars and wheelbarrows.

The most frequent means of transport for the day were disk-dump trucks, followed by tractors, carts and finally other means of transport, as shown in Table 4.

Table 4. *Types of discharges of transport.*

Types of transport	Frequency
Truck (disk-dump)	91
Tractor	55
Cart	13
Particular	02
Wheelbarrows	07

4.5. Opinion Poll with Neighboring Residents

An opinion poll in the homes of people with the association of neighborhood residents was held for approval by the recovery of the gully project (erosion) with segregated alternative materials (waste class IIB - inert - according to NBR 10,004 - ABNT 2004a) in the residential Geovanni Braga in Annapolis in the state of Goiás, as shown in Figure 13.

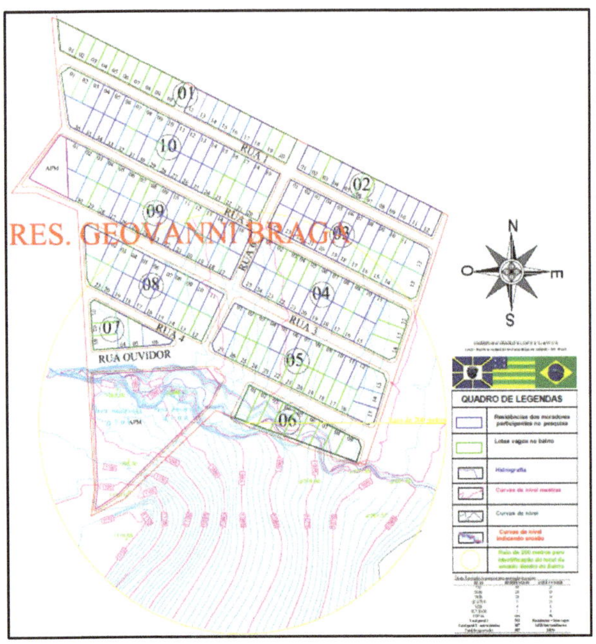

Figure 13. *Survey map.*

The survey was conducted on a total of 104 residences and 33 of One Street, Two 28 Street, 28 Street Three, 9 of Fourth Street, Sixth Street 4 and 2 of the Ombudsman Street. With a total of 367 residents interviewed, where he obtained 100% approval with respect to the objective issue (approve or not approve).

Among the subjective responses of the respondents, the most indicative were chosen according to the objectives of this research, avoiding repetitions.

As to the question "What is your level of satisfaction (dissatisfied, satisfied or very satisfied) with regard to intervention in Residential Erosion Geovanni Braga? Why? "Were reported by respondents residents:

"We are very pleased because this erosion dating back to 1965 and never any government pledged to solve the problem."

"I am pleased because there was never any initiative on solving the problem. People threw garbage at this site, which caused various diseases in people in the region. "

"Yes I am very satisfied because they always played waste improperly on this site, and today over this bad habit."

As for the question "Do you believe that the work performed meets your expectations and wishes of the population? Why? "Were reported by respondents residents:

"Yes I believe. Why now this place will be able to build homes, and the neighborhood will become inhabited. "

"Yes, why the look of the region was something else, much better."

"Yes, why the improvements made were very good."

As for the question "Do you have perceived change in the quality of life for residents of this neighborhood? Why? "Were reported by respondents residents:

"Yes I have, people pass near the site where there was erosion without fear of being attacked by robbers who were hiding inside."

"Yes, I believe that improved in terms of safety and cleanliness."

"Yes, because there was too much waste deposited on this site, and today is a nice place to live."

As for the question "Do you believe are certain the actions taken to contain the erosion of Residential Geovanni Braga? Why? "Were reported by respondents residents:

"Yes I believe, why the professional who has developed and leads the project has assignments in two-class professional advice."

"I believe so, by involving the whole community in the neighborhood in order to solve the problem."

"Yes, the material used to contain the erosion complies with environmental standards."

The answers of respondents residents neighboring the recovered area, showed that they were satisfied with the intervention performed. No cases of dissatisfaction among respondents residents not being identified.

5. Conclusion

This research aimed to present the use of debris

Construction in curbing the environmental impact of erosion Residential Geovanni Braga by the Municipal Works, Urban Services and Property Annapolis City Hall in the State of Goiás in the year 2015.

Based on the results obtained in this study it can be said that the use of the techniques employed proved to be surprising, presenting no erosive focus, and the visual impact of the site has been significantly improved. Since implementation costs were lower than traditional techniques.

It was concluded that the techniques used were adequate, efficient and mainly economic, to use reuse material donated by city collection companies.

The whole area was treated and covered with vegetation, which facilitated the infiltration of rainwater, being held in place only a small part that hit the lowlands and streams, thereby conserving soil moisture and improving the conditions of the same and adjacent.

The application of the composting accelerator immediately protected conserving soil moisture for a long period of time. Besides that after certain period it initiates composting providing nutrients to the soil which is absorbed by the root system of the plants and keeping the freshness of vegetables.

The mix of seeds and seedlings used was very important to have roots of species with different architecture and growth may keep the green all year round.

In synthesis techniques used in the treatment of erosion were rapid, highly efficient and greatly reduced cost compared to traditional methods. The results pleased the neighboring population improving the look, and not presenting that there was human interference there.

To continue this research is suggested to carry out similar studies in other regions with erosions.

References

[1] PEREIRA, A. New Technology in Erosion Control and Revegetation of Degraded Areas. I SYMPOSIUM BRAZILIAN IRON ORE, Ouro Preto, MG - Brazilian Association of Metallurgy and Materials, p 615-619, 1996.

[2] LOPES, S. L; WAR, A. J. T. Gullies Monitoring by GPS satellites in quartz sand areas podzolizada: Praia Mole, Florianópolis-SC. In. VII National Symposium on Erosion Control, Goiânia-GO, V. 1, No. 1, p. 106, 2001.

[3] GALETI, P. A. Practice control erosion. Campinas: Campineiro Institute of Agricultural Education, 278 p., 1984

[4] PEREIRA, A. R. Bioengineering techniques used in soil protection and recovery. WORKSHOP ON ADVANCED TECHNIQUES IN Erosion control. PROCEEDINGS. SME Mining Society of Engineers, 1-8 p., 1997

[5] FERREIRA, R. R. M.; FERREIRA, V. M.; TAVARES FILHO, J.; RALISCH, R. Origin and evolution of gullies in Cambissolos in the basin of the upper Rio Grande, Brazil. In: XXXI Brazilian Congress of Soil Science, 2007, Gramado-RS. Proceedings, 2007.

[6] AB'SABER, A. N. Boçorocas of France. Journal of the School of Philosophy, Sciences and Letters of France, 1 (2): 5-27, France, in 1968.

[7] LIMA, M. C. Physical and chemical degradation and mineralogical massive along the gullies. Doctoral Thesis in Geotechnical Engineering, Department of Civil and Environmental Engineering, FT, UnB, Brasília, DF, 2003.

[8] IWASA, O.Y. & FENDRICH, R. Urban Erosion Control. Oliveira, A. M. S., Brito, S. N. A. (Eds.). Engineering geology. Brazilian Association of Engineering Geology - ABGE, São Paulo, SP, p. 271-282, 1998.

[9] CAMAPUM DE CARVALHO, J.; SALES, M. M.; MORTARI, D.; FAZIO, J. A.; MOTTA, N. O. & FRANCISCO, R. A. Erosive processes. Camapum de Carvalho, J. et al. (Org.). Erosion in the Midwest Brazilian, Finatec, Brasília, DF, 2006.

[10] MORGAN, P. R. C. Establishment of plant cover and parameters for modeling splash detachment. In: EL-SWAIFY, W. C.; MOLDENHAUER, W. C.; LO, A. (Eds.). Soil erosion and conservation. Ankeny, Iowa: Soil Conservation Society of America, p. 377-383, 1985.

[11] EMBRAPA SOIL. Technical report and monitoring plan of Degraded Areas Recovery Project. Rio de Janeiro, in 2002.

[12] CARVALHO, E. T. Project Manuelzão: river basin management course. Committees watershed of Arrudas and Oz. Belo Horizonte, in 2004.

[13] CAPPI, M. D. Environmental restoration of eroded areas as the destination alternate ending of scrap tires. 60 2004. f. Thesis (MS) - Luiz de Queiroz College of Agriculture, University of São Paulo, Piracicaba, 2004.

[14] MARTINS, H. A. F. A. Use of Tire Rubber in Asphalt Pavement. Work Completion of course Universidade Anhembi Morumbi under the Civil Engineering Course with environmental emphasis. São Paulo, 2004.

[15] RAMOS. L. S. N. Reverse logistics scrap tires: The problem of location of collection points. Masters dissertation. Federal University of Santa Catarina, Program Graduate and Production Engineering, Florianópolis, 2005.

[16] HAUG, R. T.; The Practical Handbook of Engineering composting. Boca Raton, Lewis Publisher, Boca Raton, FL, 1993.

[17] IGNATIUS, C. T.; MILLER, P. R. M.; Composting: science and practice applied to waste management. EMBRAPA. 2009.

[18] GRAY, D. H. E. SOTIR, R. B. Biothechnical bioengineering slope and soil stabilization: A Practical guide for erosion control, John Wiley & Sons, Inc., 1996.

[19] GOLDMAN, S. J. Erosion and Sediment Control Handbook, New York. McGraw-Hill, USA. 1986.

[20] GIL, A. C. Methods and techniques of social research. São Paulo: Atlas, 1999.

[21] GIL, A. C. How to develop research projects. 4. ed. São Paulo: Atlas, 2002.

[22] MATTAR, F. N. Marketing Research: Methodology and Planning. São Paulo: Atlas, 1996.

[23] YIN, R. K. Case Study: Planning and methods. Porto Alegre: Bookman, 2005.

[24] NIMER, E. Climate. In: Climatology in Brazil. Rio de Janeiro: IBGE, 1989.

[25] JESUS, A. S. Multidisciplinary research of linear erosion: a case study of the city of Anapolis - GO. Doctoral Thesis, Publication G.T.D. 087/2013, postgraduate program in Geotechnical Engineering, Department of Civil Engineering University of Brasilia, 2013.

Project Risk Management: Methodology Development for Engineering, Procurement and Construction Projects a Case Study in the Oil and Gas Industry

Ashwin Amarshi Maru

L.E.College, Civil engineering department, Saurashtra University, Rajkot, State: Gujarat, India

Email address:
maru.ashvin@gmail.com

Abstract: EPC Project comes with a set of risks that are unique and hence cannot be addressed in the context of general construction risks. This paper researches existing methodologies and presents a customizable framework for project risks assessment with EPC project. A case study specific to the Oil and Gas industry has been used. The developed framework is also extendable for other EPC works.

Keywords: Project Risk Management, Integrated Risk Management, Risks in Procurement, Risks in Construction, construction Management, Oil and Gas Industry, EPC (Engineering, Procurement, Construction)

1. Introduction

THE Construction Scientific management of risks is a growing need for construction projects. This is because of the rising amount of risks that are faced in the construction industry. Current risk management approaches need to be revised. This is primarily because most risk management approaches are applicable for different sections of the project development and there is a lack of an integrated approach in risk management [2]. The advantages and benefits of an integrated risk management from the procurement to the final engineering development would be helpful in handling concerns efficiently [3]. The basic aim of this project is to study about risks, risk management and to critically evaluate it for a case study. The case study is from the engineering, procurement and construction projects of the oil and gas industry.

Although various studies are available for information technology, manufacturing sectors, there has been no specific study for oil and gas industry [6]. This case study intends to benefit the oil and gas industry, and makes recommendations both general and specific that can be applied to meet the wholesome needs of the Oil and gas Industry. The Engineering procurement and Construction EPC background is the focus of research.

2. Research Background

A risk is defined as a deviation from a defined target. This deviation for a target with specifications is normally defined by uncertainty. It is unpredictable (although not always) and will have negative consequences on the defined target [9]. There is a great demand for Risk management solution providers, who systematically solve the issues and the problems to be, and also suggest a solution. Due to this calculated and scientific approach, the risks identified can be managed on a real time basis, without these problems affecting the whole project [5]. As such, this also holds benefits in the context of real time budgetary needs. Where risk management is achieved, it becomes easier to control the project costs, make the project work within the anticipated budget or to avail of fresh funds, and for easy prediction of the tentative project completion dates, as the Risk management solution involves all aspects of the project from Inventory, tender work, order finalization, human resources and the standards and quality aspects [8].

The purpose of this research is to analyze the overall parameters which are involved in uncertainties and can hence result in affectations to project completions during critical times. There are different methodologies towards risk management, a customized methodology that can be applied to Engineering, Procurement and Construction (EPC) projects in the oil and gas industry is discussed here. The case study project has been critically evaluated for risk management opportunities and strategies based on different methodologies.

3. Literature Review

There are many a concepts and studies with varying standards and methodologies to be applied in Risk Management, but none seems to be specific to address the issues of the Oil and Gas Industry. This project will ease that doubt and will try and cater to the issues pertaining to the Oil and gas Industry, and a systematic procedure will be developed to accurately evaluate any Oil and gas Projects.

The risk management study would identify the areas associated or prone for Risks right from the start and address those issues with a valid response or Decision making process [4]. As such valid decision making is essential in finalization of the contracts with the suppliers / vendors or in turnkey projects or any other joint venture or Build own operate processes. In any EPC Project, the Supplier of plant and machinery and his team is responsible for proper erection and execution of the plant and machinery. The key to identify the best realistic terms to be achieved rests with the identification of the inherent risks and the knowhow to solve the same.

Studies of Haghnevis and Sajedi present the gaps inherent in existing literature. Although there are several models, none of them address the EPC aspect comprehensively and every company's aspects on risk is much different from the other. Specificity often makes the research work inapplicable in a general context for the construction field [3]. EPC projects need a more focused framework.

Project risks in EPC will benefit by the application of the PMBOK (project management body of knowledge) standards. According to the standards, a project risk "is an uncertain event or condition that, if it occurs, has an effect on at least one project objective. A risk may have one or more causes and, if it occurs, it may have one or more impacts" [10]. These risks are either dynamic risks or static risks as can be classified according to their characteristics [9].

EPC contractual arrangement related risks can be of either type. EPC is a contractual arrangement in the construction industry where the contractor is made responsible for the engineering, designing and procurement details for construction [7]. There are cost based contracts and work based contracts and for the most part these risks are transferred to the [10].

Specific points of analyzing for the risks in these contracts are by planning, control and facilitating simultaneous activities and by use of risk management suites such as the IRM (Institute of risk management), PMBOK, PRMA (Public risk management association) and more.

4. Methodology

There are many a varied model of risk management theories, concepts, case studies available to suit many industries, but the one which can be related to be close to the Oil and gas industry would be the Construction industry and some of the common methodologies used for this industry are that of the Boehm, Fairley, SEI, Kliem & Ludin, SHAMPU, PRAM, PMBOK, PRMA, Smith& Merritt-all try to find solutions in Risk Management.

The most commonly used profile is that of PMBOK Model which is registered as an ANSI Standard. The PRINCE 2 Model is a solution provider is controlled Environments is widely used the Government Bodies in the United Kingdom, this software is available for free and due to which is used widely throughout the world.

Though these above cited methodologies provide solution in Risk Management, they were not developed to cater to any particular industry. They were developed as a Common application platform only, and not need based [5]. The Current project was assimilated by conducting various sessions of rigorous data Collection for Empirical work and for theoretical framework, data analysis, and associating those data with the selected Oil And Gas Company. The company for which information is collated and studied is the on-going Iranian EPC oil and gas project called "Bidboland II Gas Treating Plant".

For the case study, the experience on both main office and construction site of the project are analyzed and the specific risks associated with that particular industry is studied in depth to evolve a solution.

5. Case Study: Data Analysis and Discussion

Project name: Bidboland II Gas Treating Plant
Client: National Iranian Gas Company
Location: Behbahan, South West of Iran
Contract price: 3.3 Billion USD
Duration of work: 43 months

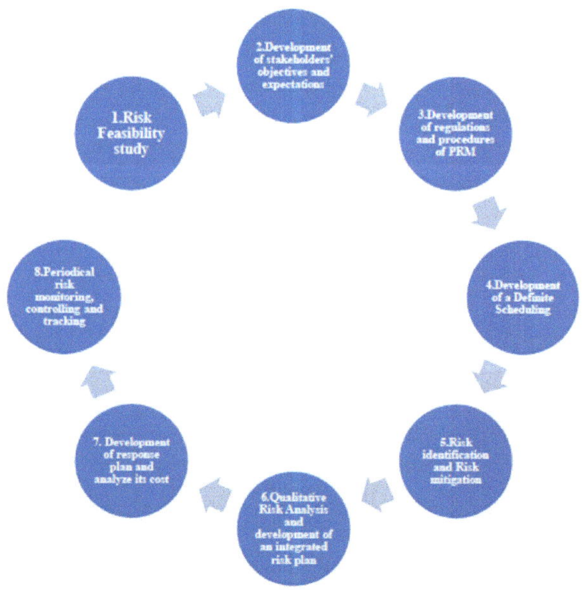

Figure 1. *Risk Management Model for EPC Projects*

The specific risks associated with the Bidboland Gas Treating Plan have been studied and it was identified that the risk are spread between the Oil, Gas and the Petrochemical Industries. The following framework was constructed with existing literature work

Step 1: A Risk Feasibility Study was conducted. The major activities that were involved in this study were that of analyzing the risk management strategic planning, project life cycle study and the project implementation method. The contractor and consultants involved are also analyzed. A risk feasibility package is the main outcome at this step.

Step 2: Stakeholders objectives have to be developed by working with them. Their specifications have to be identified and based on the same; the risk feasibility analysis is conducted. The output of this stage is a risk management package that will be considerate of the stakeholder's requirements.

Step 3: The regulations and the procedures necessary for developing risk management for the EPC is identified. The main inputs to this stage are that of the risks feasibility package, the stakeholder's requirements as collected from step 2, using which the objectives of the bidboland Gas Treating Plant will be determined. The technical tools that will be used, the software packages to be used and more are determined. The risk breakdown structure is considered in this stage also. The output of this stage is the risk management manual. This manual is more custom made to the project at hand.

Step 4: From considering more top level planning, the more specific planning begins at step 4. Here the planning output a definite plan. The main inputs that are considered at this step are that of the time schedule, the outputs of the previous steps, the risk management entities identified and documented till this step and more.

Step 5: This is the risk identification and mitigation stage. At this point, the risk management has moved on to include the plan for the different risks types. Here the input to the step are the work breakdown structure WBS and the Risk Breakdown structure RBS, risk management procedures, the manual prepared from the previous step and the specific time schedule that has been identified from the WBS. The risks that has been identified from the previous state has been refined at this stage. By refinement the meaning here is that the risks are itemized and sub itemized and have been prioritized based on their different levels of impact. Furthermore at this step, the secondary level risks are also included.

Step 6: This s a critical step at this level, a qualitative risk analysis is conducted. The integrated risk plan is developed and this risk plan is applied on the project. Here accosts and time analysis will also be concluded so as to make the risk management plan more real time. The project schedule will be the output at this stage. The project schedule will not reflect the applied risks in the previous steps; here the project schedule has been updated to reflect the applied risks. The main inputs that are used in this step are the outputs of the previous steps which includes the risk management manual, the itemized risks register, the sub procedures, the WBS and RBS applied and more. The risk response plan is developed here and is one of the key output of this stage. A risk response plan makes it easy to analyze and estimate the risk impact before and after (if the risk cannot be mitigated). Also the definitive plan will be applied here, so the key responses can be recorded for. In addition one another significance of having this as a separate step is that all the involved stakeholders and the ownership of the different project entities assigned to them, and the risk response corroborated to the specific entities are mapped. These outputs are then passed onto the risk monitoring system that will work parallel to the entire project unit.

Step 8: Periodical risk monitoring, controlling and tracking is the main emphasis on step 8. Here it is noted that the main inputs of the risk plan are identified. The risk response is taken as the input and the different owners of the plan are also identified as inputs to this subsystem. The risk response plan is executed along with the necessary units for periodical updates. The updates risk response plans for the project are the output at this stage. The benefits of using such a system in the oil and gas industry is that risk responses are constructed real time as well as planned for.

The specific risks such as Economic sanctions, Nonpayment and violation of Contracts, Gap between consumers and producers, technical Complexity and high technology used are addressed. The core area of funding by the developed nations in SEED projects as the Oil gas Project requires huge Investment and the role of Insurance players is analyzed in depth.

Table 1. *List of Contractors connected with the EPC Project*

Company	Related Experience Rate	Main Expertise	Participation rate
Sarah (Iran)	Over 30 years Engineering, Procurement	Engineering, Procurement	High
Jahan Pars(Iran)	Over 30 years Engineering, Procurement	Construction, Procurement	High
SNC Lavalin (Canada)	Over 50 years	Engineering	Moderate
RIPI (Iran	About 20 Years	Engineering	Low
Prosernat (Malaysia)	About 30 years	Engineering	Low

The lists of contractors that are associated with the EPC project are presented above. These are the primary stakeholders associated with the project. A feasibility study will help in presenting the key risks, and the risks can then be linked back to help the stakeholders take up ownership of the risks.

***Table 2.** Feasibility Study: Need for Risk Management in EPC Project*

Item Description	Weight of Item	Need to Risk Management	Score
Project risk Management Strategic Planning	0.16	9	1.44
Project Life Cycle Statutes	0.14	10	1.4
Project Overall Structure	0.06	5	0.3
Project Implementation Method	0.16	9	1.44
Project Contract and Payment Method	0.18	10	1.8
Contractors and Consultants	0.14	4	0.56
Project at the Portfolio Level of organization, ministry, country, region	0.1	3	0.3
Desired level of risk management	0.06	7	0.42
	1		7.66
1-10, Number 10 means Strong need for Risk Management and 1 means low need			

In general the overall risks identified can be listed as to accurate understanding of Bid documents, not undertaking a clear process to enter in the fray, not to anticipate the costs involved in getting the requisite licenses [4], not to select a proper legal person to understand the terminology used in the bidding document, and to quote a higher bid price which is not realistic in the primary area, and in the secondary area, non completion of projects on time, delay in completion of the project, not to sign an cost increase pact, non proper supervision of the Construction work and others. Each of these risk issues in EPC can be connected across to the different stakeholders which in this case are the contractors.

6. Conclusion

The area of project Risk Management is the new emerging field of comprehensive study, as the Mega Projects coming up currently are of highly technical and complex in detail. But there are instances where in the Oil and Gas companies are hesitant in investing on Risk Management Services [1]. They try to tackle the issues as any common problem faced in the manufacturing sector without realizing the real depth of the problem. Yet another problem is that, the project Managers are trying to empower themselves with various tools to prevent the risks, and to identify beforehand the problems which would pose a problem in the project not getting commissioned on time without a specific tool to tackle the problem. Hence, nowadays, the Engineering, Procurement and Construction companies start their risk management schedule right from the day one from floating of the tenders to the bidding of the tenders, and in finalization of the work contracts and so on. Even if the Risk management protocol is observed, the grey area has lack of continuous monitoring and identification, control, and insufficient risk response plan. Such is the situation prevalent in this industry, and as the case study highlights a continuous and constant watch upon its own actions is needed. This means keeping abreast of the new technologies and inputs towards track risk based problems and finding a logical and correct solution.

Adding to the above problems, the risk management procedures that are made use of in the sector are often very generic. They are projected for use for many things and have to be tailored down to meet the demands of a specific industry. This research has pointed out how a customized methodology can be made possible. Through this methodology it will be possible to identify the top level risks at the time of project procurement itself, so the planning of the project can be done around it. It is the top level abstraction of risks that when are not mitigated results in heavy loss in the form of time and money. This situation further gets exacerbated by element such as the project location critical parameters, the environment, the macro-environmental factors and more. Identification and evaluation of risks earlier on will help create mitigation strategies that focus on the risk. Even if the actual risk cannot be averted, the risk will be managed or minimized. Adverse consequences are reduced. Only with a defined risk management model this is possible.

The research specifically presents a risk management model that is applicable for risk analysis in the EPC projects of the oil and gas industry. The risk management framework that is presented is more detailed and internetworked with software technologies and techniques that most other generic models do not present. Some of the techniques that are being used here are the PERT, GERT and/or Mont Carlo. Higher performance and productivity is possible with these techniques. The technical requirements and other aspects of the model were developed keeping in mind the research literature available on risk management in the oil and gas industry. It is noted that there is a general combination of the PMBOK, PRINCE2 and Cooper that has been used in the Gas industry. Each of these techniques comes with its own set of advantages and disadvantages, and care has to be taken to ensure that the application is well thought of. Also most of the models in risk management start off with the risk identification procedure and the management, while this does have its strength, the construction company's entire process structure does not come into analysis. This research has suggested for a risk feasibility study. This is a very comprehensive study. The risk feasibility analysis will be conducted with the a comprehensive overview of the company processes, such as its strategic plans, the entire life cycle for the oil and gas constructions, the implementation, procurement processes and details, payment, client portfolio, contractor level profiles, and more. This form of a comprehensive arrangement that is done for risk study even before the actual project is set up actually offers many advantages. Primarily it can be seen companies might base their decisions to attend the tender meetings based on the foreseen risks. Companies are better advised on how they could handle the risks and how this would affect the tender contracting. There are much legal entities involved once the contract is signed off, and a risk management analysis done after the procurement most often would mean that the constructor is constrained for options to deal with the risks. There will be geo-political issues and additional legislative factors that might constrain them from having adequate support. When a proper feasibility study is submitted before hand, then the companies have better competitive advantage [8] In order to complete a project successfully it is necessary that there is a periodical risk monitoring available. Risk registers have to be developed and additional resources be deployed to handle the identified risks. These and more issue can be negotiated at the tendering time when the research recommendation is followed.

The research work is not a final recommendation however, it is more of groundwork on how the EPC projects will be able to better make use of risk management when tendering, accepting and implementing projects. Bidboland II gas treating plant project is where the methodology has been tested out. It is seen that functional risks are averted better and there is also better documentation procedure in risk handling that will be useful for future commissioning. Furthermore the risk management is spread across, from the pre-tendering stage to the post-commissioning stage. This risk management recommendation hence will be useful for the oil and gas industry to further develop custom methodologies according to their company policies and can also be employed by Government entities when they subcontract out their projects to private contractors, as they will be able to pack much transparency of risks involved when subcontracting.

References

[1] Burke, R., 1999. Project Management: Planning & Control Techniques. 3rd ed. New York: John Wiley and Sons.

[2] Fairley, R., 1994. Risk Management for Software Projects. IEEE Software. 11(3), pp. 57-67.

[3] Haghnevis, M. and Sajedi, H., 2006. A model for assessment of project management risk models. 4th International Management Conference. Tehran, Iran, 19-20 [online] Available at: http://mba.mbairan.ir/Portals/0/388_Full_moeed.pdf, [Accessed on March 18, 2015]

[4] Heldman, K., 2005, Project manager's spotlight on risk management. San Francisco: Harbor Light Press.

[5] Huse, A. J., 2002. Understanding and negotiating turnkey and EPC contracts. 2nd ed. London: Sweet & Maxwell Ltd.

[6] Kangari, R., 1995. Risk management perceptions and trends of US construction. Journal of Construction Engineering and Management. 121(4), pp. 422-429.

[7] Loots, P. and Henchie, N., 2007, Worlds Apart: EPC and EPCM Contracts: Risk issues and allocation. International Construction Law Review. 24(1), pp. 252-270.

[8] March, J.G. and Shapira, Z., 1987. Managerial perspectives on risk and risk taking. Management Science. 33(11), pp. 1404-1418.

[9] Mulcahy, R., 2003. Risk Management: Tricks of the Trade for Project Managers. Minnesota: RMC Publications.

[10] Project Management Institute, 2008. ANSI/PMI 99-001-2008, A Guide to the Project Management Body of Knowledge. Newtown Square, Pennsylvania: PMI (Project Management Institute).

[11] Ramasamy, S., 2005. Total Quality Management, Delhi: Tata McGraw-Hill.

7

Deterioration and Service Life Prediction of Concrete Subjected to Freeze–Thaw Cycles in Na_2SO_4 Solution

Guo Li*, **Dan Wang, Jian-Min Du**

School of Mechanics & Civil Engineering, China University of Mining and Technology, Xuzhou, China

Email address:
guoli@cumt.edu.cn (Guo Li), 1075207208@qq.com (Dan Wang), djm1975@cumt.edu.cn (Jian-min Du)
*Corresponding author

Abstract: To investigate the resistance of normal concrete to sulfate solution frost, concrete specimens with different water/cement (w/c) ratios, mineral admixture types, replacement ratios, and air contents were fabricated. Then, these specimens were frozen and thawed cyclically in 5% concentration Na_2SO_4 solution. As the freezing and thawing cycles proceeding, the appearance morphology, mass loss, and dynamic elastic modulus (DEM) of the specimens were observed. The service life of concrete that is subjected to the freeze–thaw cycles in Na_2SO_4 solution was calculated based on specimens' DEM losses. Results indicated that the appearance damage and mass loss of concrete along with the freeze–thaw cycles were unnoticeable until failure, and DEM losses played a controlling role in determining specimens' failure. In addition, a sudden fracture failure in the middle occurred easily in specimens with low w/c ratio. Decreasing concrete w/c ratio can slightly increase the concrete resistance to the freeze–thaw cycles in Na_2SO_4 solution, whereas incorporating fly ash or slag has almost no effect. Moreover, higher replacement ratio of fly ash or slag increases the adverse effects on concrete. Adding an air-entraining agent to concrete can significantly improve its resistance to the freeze–thaw cycles. Air content at 4.6% and 5.7% can extend the service life of concrete under the freeze–thaw cycles in Na_2SO_4 solution by more than 5 times its ordinary life span.

Keywords: Concrete, Freeze–Thaw Cycle, Deterioration, Na_2SO_4 Solution, Service Life Prediction

1. Introduction

The unique geography and harsh natural climate in the western regions of China pose challenges to concrete bridges in the area, which causes the service life of some of these structures to last for only 3 to 5 years [1-2]. Sulfate attacks from saline soil or salt lakes and freeze–thaw cycles resulting from seasonal temperature changes are the two main causes of structural damage, and sulfate attack and freeze–thaw circulation occur simultaneously most of the time.

Studies on the performance of concrete that is subjected to salt solution freeze–thaw cycles have been conducted and significant conclusions have been presented [3-6]. Mu et al. reported that various salt solutions have different effects on the frost resistance of concrete [7]. They also argued that NaCl solution causes no change in the failure mechanism of concrete freeze–thaw cycles, whereas Na_2SO_4 solution can cause freeze–thaw damage and sulfate attack at the same time. Yu et al. found significant differences in the final number of freeze–thaw cycles that concrete can withstand when subjected to different salt lake solutions [8]. They also demonstrated that some salt lake solutions promote concrete freeze–thaw damage, whereas others restrain such damage. Yuan et al. showed that, with the increase of Na_2SO_4 solution concentration, the freeze–thaw damage on concrete is gradually inhibited rather than increased [9]. Zhang et al. suggested that $MgSO_4$ solution can significantly slow down the freeze–thaw damage and failure of concrete [10], whereas Jiang et al. showed that the freeze–thaw cycles in $MgSO_4$ solution causes more damage to concrete compared with water and Na_2SO_4 solution [11].

In terms of concrete grade and raw materials, low concrete water/cement (w/c) ratio is commonly believed to increase the frost resistance of concrete [12]. However, Mu et al. found that the freeze–thaw damage rate of low-strength concrete is unaffected by the Na_2SO_4 solution, whereas high-strength concrete is damaged in advance because of the joint action of sulfate attack and freeze–thaw cycles in late period [13]. An entraining agent can improve the frost resistance of concrete to water freeze–thaw and can enhance solution freeze–thaw [12] [14]. Du et al. and Ghafoori et al. found that adding fly ash into concrete can improve its sulfate corrosion resistance [15-16]. Other researchers observed that incorporating fly ash, slag, silica fume, and other active mineral admixtures significantly reduces the frost resistance of concrete [8] [17]. Fly ash, slag, or silica fume has been widely used in preparing high-performance concrete. However, further study is needed with regard to the effects of applying mineral admixtures on the performance of concrete resistance under salt solution freeze–thaw cycles. Thus, the present study examines the performance of concrete specimens with different w/c ratios, mineral admixture types, and air contents under 5% Na_2SO_4 solution freeze–thaw cycles. The service lifespan of these specimens are estimated to provide a useful reference for engineers.

2. Experimental

2.1. Raw Materials and Mixture Proportions

Chinese-standard P·O 42.5R ordinary Portland cement produced by Huaihai Zhonglian cement factory was used. Grade I low-calcium fly ash powder and S95 slag powder were used in the test, and their cement replacement ratios (by mass) were set as 10%, 20%, and 30%. The specific chemical composition of each cementitious material is presented in Table 1.

Table 1. *Chemical composition of cement, fly ash, and slag/wt%.*

Item	SiO_2	Al_2O_3	Fe_2O_3	CaO	MgO	Loss on ignition
Cement	22.3	5.05	3.16	64.78	0.92	1.32
Fly ash	57.2	30.3	4.5	2.5	0.6	2.65
Slag	29.3	13.7	0.4	33.2	9.2	2.31

Natural river sand with a fineness modulus of 2.53, crushed stone with particle size of 5–20 mm, and ordinary tap water were chosen as fine aggregate, coarse aggregate, and mixing water, respectively. In addition, an air-entraining agent was used for partial specimens (PYA and PYB) to obtain different concrete air contents. A total of 12 concrete mixture proportions were designed, as shown in Table 2. The measured air contents for fresh concretes of PY0, PYA, and PYB were 1.3%, 4.6%, and 5.7%.

Table 2. *Concrete mixture proportion/kg/m*^{3}.

Item	Water	Cement	Fly ash	Slag	Sand	Gravel
P0.4	190	475	0	0	607	1178
P0.5	200	400	0	0	666	1184
P0.6	210	350	0	0	737	1153
F10	200	360	40	0	666	1184
F20	200	320	80	0	666	1184
F30	200	280	120	0	666	1184
K10	200	360	0	40	666	1184
K20	200	320	0	80	666	1184
K30	200	280	0	120	666	1184
PY0	190	380	0	0	623	1157
PYA	190	380	0	0	623	1157
PYB	190	380	0	0	623	1157

2.2. Test Methods

Prism specimens with a size of 100 mm × 100 mm × 400 mm were used. After demolding, the specimens were kept in a water tank for curing at 20 ± 2°C over a period of 28 days. Then, the specimens were collected and placed in an indoor natural atmosphere for another 28 days. Before the start of freeze–thaw circulation, the specimens were immersed in 5% (by weight) Na_2SO_4 solution for 4 days. In this study, "rapid freezing method" from Chinese standard on "ordinary concrete long-term performance and durability test" (GB/T50082-2009) was adopted. During the test, the specimens were frozen and thawed at 5% mass fraction of the Na_2SO_4 solution. Each freeze–thaw cycle test lasted for approximately 3.5 h, and the thawing time could not be less than one-fourth of the entire freeze–thaw cycle time. The temperature of the transformation between freezing and thawing was set at –15°C and 6°C. During the entire process of freezing and thawing, the lowest temperature of the specimen center was –18°C and the highest temperature was 10°C.

At the beginning of the freeze–thaw circulation and after every 25 freeze–thaw cycles, the specimens were collected, weighed, and tested for ultrasonic velocity using a nonmetal ultrasonic detector. The specimens were then placed back into freeze–thaw cycle boxes. When the mass loss of a specimen exceeded 5% or the dynamic elastic modulus (DEM) dropped to 60%, the specimen was considered to be a failure. Eqs. (1), (2), and (3) were used to calculate specimens' mass loss ratio α, relative dynamic elastic modulus (RDEM) P, and RDEM losses D.

$$\alpha = \frac{m_0 - m_N}{m_0} \times 100\% \tag{1}$$

$$P = \frac{v_N^2}{v_0^2} \times 100\% \tag{2}$$

$$D = 1 - P \tag{3}$$

where m_0, m_N are the mass (g) of concrete specimens before freeze–thaw circulation and after N freeze–thaw cycles, and v_0, v_N, are the longitudinal wave velocity (km/s) of concrete specimens before freeze–thaw circulation and after N freeze–thaw cycles.

3. Results and Discussion

3.1. w/c Ratio on Concrete Resistance to Freeze–Thaw Cycles in Na_2SO_4 Solution

Conventional concrete failure modes in freeze–thaw cycles can be described as follows: as the freeze–thaw cycles proceed, the grout spalls from the concrete surface, and the fine aggregate is exposed; the coarse aggregate becomes exposed and even scaled; finally, the concrete section decreases until complete failure [17]. In this study, with the action of 5% Na_2SO_4 solution freezing and thawing cycle on the specimens, sand and cement mud scaled from concrete surface also appeared. However, damage on the specimens was unnoticeable in general. When the freezing and thawing cycles reached 125 times, the specimens with w/c ratios of 0.4 and 0.5 showed a sudden fracture in the middle part, but such a phenomenon did not occur in the specimens with 0.6 w/c ratio (Fig. 1).

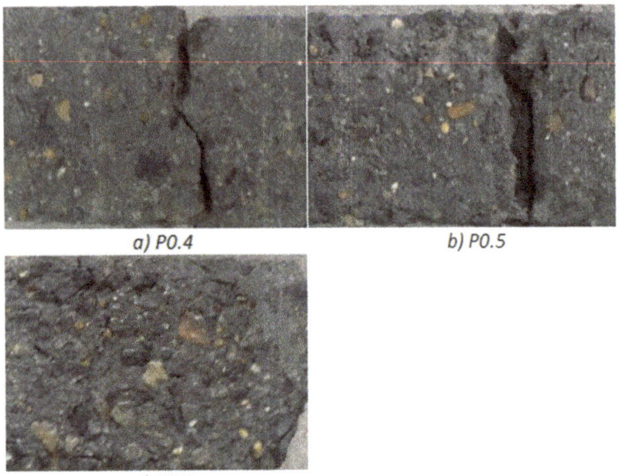

Figure 1. *Appearance of specimens with different w/c ratios after 125 freeze–thaw cycles.*

Sulfate solution has better compressible plasticity than water and chlorine salt solution after freezing [7]; thus, as the freeze–thaw cycles proceed, the phenomena of concrete surface damages are not serious. However, with the continuous accumulation of sulfate solution invasion and salt crystallization [14], specimens subjected to the Na_2SO_4 solution freeze–thaw cycles are more prone to failure characterized by a sudden fracture, especially in the case of concrete specimens with low w/c ratio and high strength. These findings have also been reported by other researchers [13] [18]. The mass loss and RDME development of specimens with three kinds of w/c ratios along with the freeze–thaw cycles are shown in Figs. 2 and 3.

As shown in Fig. 2, the mass of all specimens increased slightly at the beginning, and then decreased thereafter. However, three kinds of specimens did not have serious mass loss until 125 freeze–thaw cycles and did not reach the failure criterion of 5% weight loss. For specimens with 0.4 w/c ratio, the mass loss had a negative value until 100 freeze–thaw cycles. The mass loss and appearance damage of the specimens were consistent, which suggested that concrete specimens do not have serious mass loss until the occurrence of failure.

As shown in Fig. 3, the RDEM of concrete decreased gradually with the freezing and thawing cycles; the decrease was slow at the initial stage (25 to 50 cycles), and then it accelerated. The RDEM of the specimens with 0.6 w/c ratio dropped to 63.2% after 50 freeze–thaw cycles, which was close to the failure criterion. The RDEM values of the specimens with different w/c ratios dropped to 60% or lower after 75 freeze–thaw cycles, which reached the failure criterion. Until this point, the mass losses of the specimens were extremely low. Notably, the RDEM of concrete is the control index used to determine the damage caused by the Na_2SO_4 solution freeze–thaw cycles.

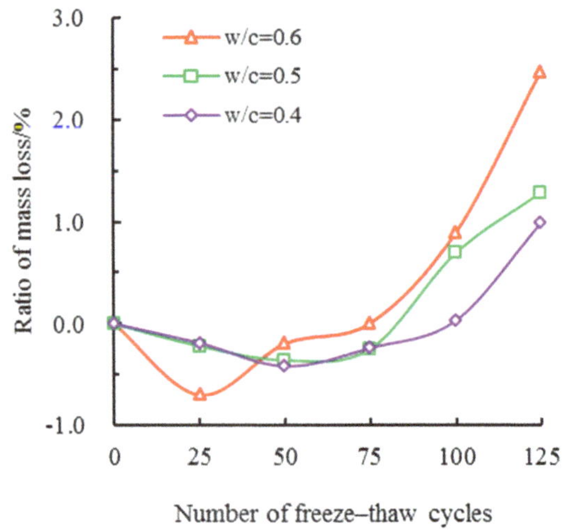

Figure 2. *Mass loss of concrete with different w/c ratios.*

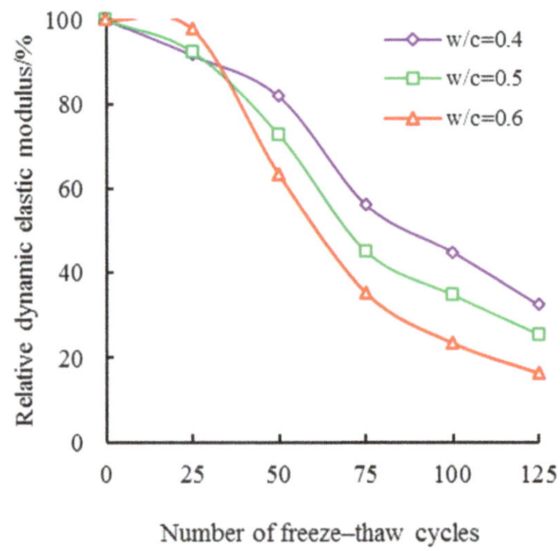

Figure 3. *RDEM of concrete with different w/c ratios.*

At the same time, with the same freeze–thaw cycle, the specimens with higher w/c ratios usually suffered more mass loss and RDEM loss except at the initial stages. This finding was consistent with the results of many previous studies [8] [13] [18]. Lower concrete w/c ratio often corresponds to higher concrete strength and density, which is beneficial in resisting freeze–thaw cycle damage. However, in another aspect, lower concrete w/c ratio causes significant brittleness, which results in sudden fracture failure in the middle part of the specimen.

3.2. Mineral Admixtures on Concrete Resistance to Freeze–Thaw Cycles in Na_2SO_4 Solution

After the addition of fly ash or slag, the surface spalling of the specimens, which results from the freezing and thawing cycle becomes more obvious than that in ordinary specimens (no fly ash or slag), and the specimens with higher replacement ratio usually exhibit more serious surface damage. Figs. 4 and 5 present the mass loss and RDME development of the specimens with two mineral admixture types subjected to the freeze–thaw cycles.

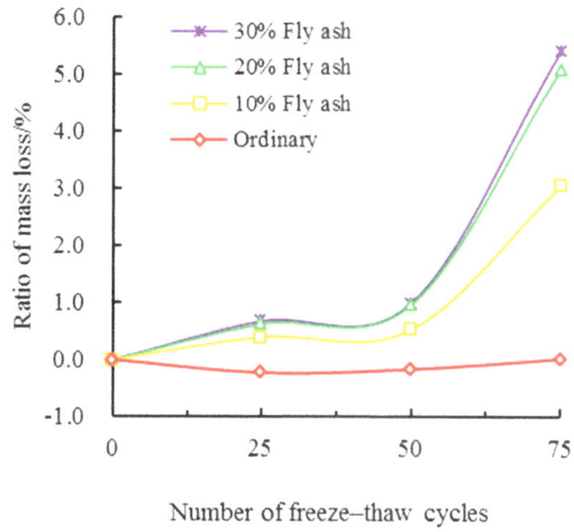

a) Fly ash specimens

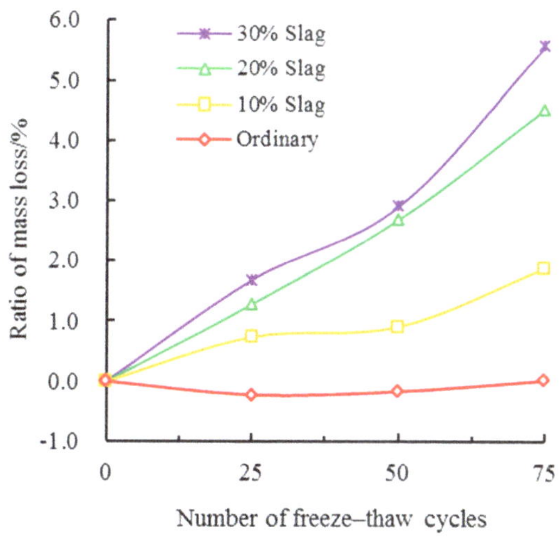

b) Slag specimens

Figure 4. *Development of mass loss of mineral admixture specimens.*

As shown in Fig. 4, the mass losses of either fly ash specimens or slag specimens were significantly higher than those of ordinary specimens at the same freeze–thaw cycle, and the specimens with higher cement replacement ratios usually have greater mass losses. As shown in Fig. 5, the RDME of fly ash and slag specimens showed a gradual decline with the progress of the freeze–thaw cycles. However, some differences were observed between fly ash specimens and slag specimens. With the same freeze–thaw cycle, the RDME values of fly ash specimens were all lower than those of the control specimens, and these values dropped below 60% at the period of 50 freeze–thaw cycles. By contrast, the addition of slag improved the RDEM of concrete. The specimens with slag tended to have higher RDME than the control specimens. In addition, the higher the replacement ratio is, the greater the improvement is.

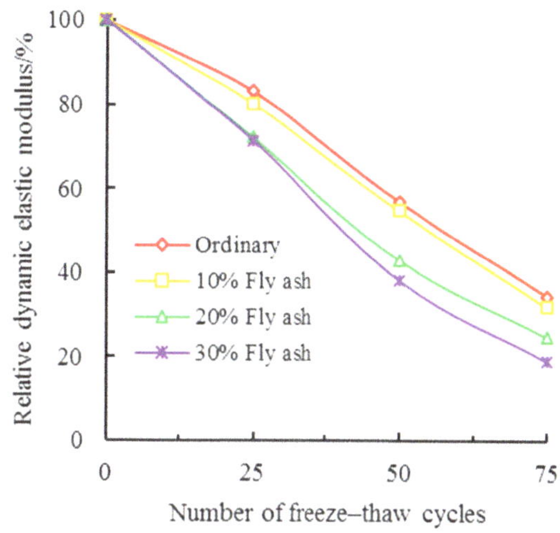

a) Fly ash specimens

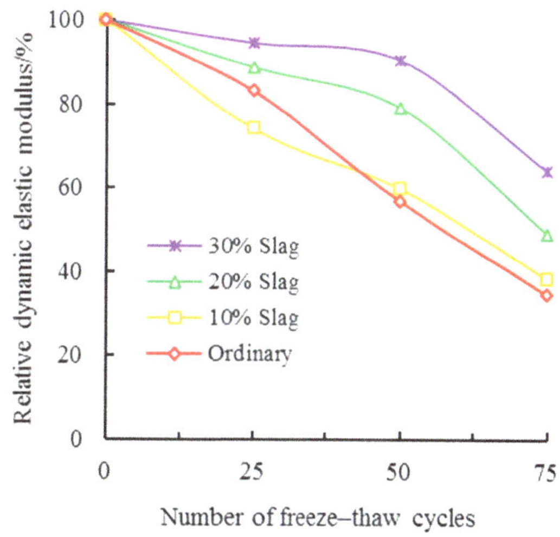

b) Slag specimens

Figure 5. *Development of RDME of mineral admixture specimens.*

The addition of fly ash and slag is believed to reduce the early strength of concrete because of relatively low pozzolanic activity. The pozzolanic hydration reaction of fly ash and slag is a lagging reaction. The failure mechanisms of concrete under Na_2SO_4 solution frost are mainly due to freeze–thaw damage and salt crystallization damage; the resistance to sulfate chemical corrosion of fly ash and slag has no effect. Hossack and Thomas [19] studied concrete with mineral admixtures for long-term immersion experiments in 5% Na_2SO_4 solution at different temperatures. They also found that the addition of fly ash can improve corrosion resistance to sulfate at higher temperatures (10°C, 23°C), whereas nearly no effect is observed at lower temperatures (5°C, 1°C). Therefore, fly ash and slag can reduce the concrete resistance to sulfate freeze–thaw cycles. Meanwhile, the pozzolanic activity of slag is slightly higher than that of fly ash. Accordingly, the addition of slag has slightly lower adverse effects on concrete than that of fly ash. Other researchers have confirmed this phenomenon [8] [18]. In general, fly ash or slag is not recommended for application in concrete members used in the environment of sulfate freeze–thaw cycles.

3.3. Air Content on Concrete Resistance to Freeze–Thaw Cycles in Na_2SO_4 Solution

Compared with the specimens that had no air-entraining agents, which showed surface mortar peeling and stone exposure phenomenon at a very early time, the specimens of PYA and PYB did not show any sign of damage until 200 freeze–thaw cycles. Remarkably, even after 300 freeze–thaw cycles, the surface changes of these specimens remained small, and the specimens did not show serious spalling. The mass loss and RDEM development of concrete with air-entraining agent are presented in Figs. 6 and 7.

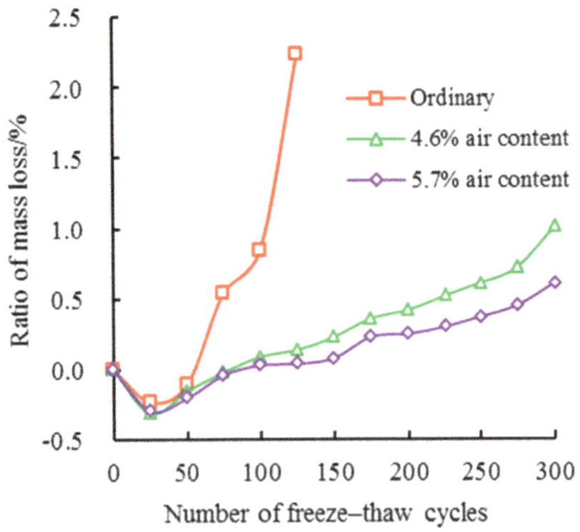

Figure 6. *Mass loss of concrete with different air content rates.*

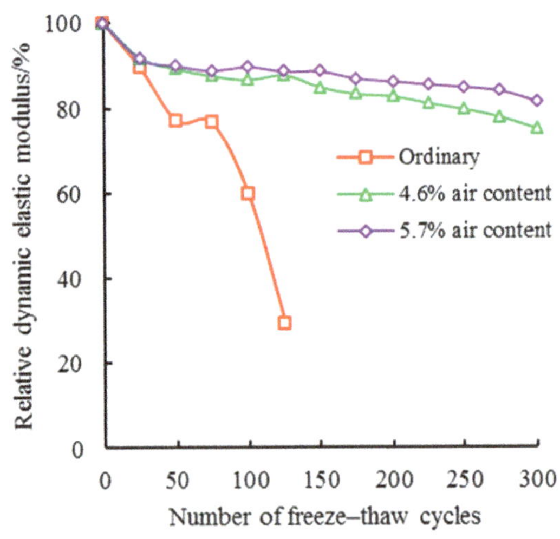

Figure 7. *RDEM Development of concrete with different air content rates.*

The mass loss of concrete along with the Na_2SO_4 solution freeze–thaw cycles decreased after the addition of an air-entraining agent, as shown in Fig. 6. After 125 freeze–thaw cycles, the mass loss ratio of the specimens without the air-entraining agent was 2.24%, whereas the mass loss ratios of the specimens with 4.6% and 5.7% air content were only 0.14% and 0.04%, respectively. The mass loss ratios of these two air-entrained specimens dropped to 94% and 98% compared with that of ordinary concrete, and were still very small even after 300 freeze–thaw cycles. For the loss of RDME, the value dropped to 59.8% after 100 freeze–thaw cycles for the ordinary concrete, which was lower than the specified failure criterion of 60%. At this point, the RDEM losses for specimens with 4.6% and 5.7% air content were 86.6% and 89.6%, which increased by 45% and 50% compared with those of ordinary specimens. Even after 300 freeze–thaw cycles, the RDEM losses for the air-entrained specimens remained above 80%. Therefore, adding certain air-entraining agents can effectively improve the resistance of concrete to Na_2SO_4 solution freeze–thaw cycles.

The air-entraining agent brings a number of uniformly distributed small bubbles into the concrete; these bubbles block the channel that the sulfate solution passes through and optimize the internal pore structure of the concrete [11]. As a result, the PYA and PYB specimens exhibit good resistance to sulfate solution freeze–thaw cycles. This finding is consistent with those reported by other researchers [12] [14] [18].

The compressive strengths of the PY0, PYA, and PYB specimens after 28 days were 53.4, 46.3, and 43.3 MPa. Concrete compressive strength declined after the addition of the air-entraining agent, and higher air content usually increases the concrete strength loss [17]. Compared with non-air-entrained concrete, concrete with 4.6% and 5.7% air content respectively showed a 13.3% and 19% decrease in compressive strength after 28 days. Therefore, to ensure both compressive strength and resistance to salt solution frost, the air-entraining agent should not be added abundantly to the concrete.

3.4. Service life Prediction of Concrete Under Freeze–Thaw Cycles in Na_2SO_4 Solution

According to the conclusion obtained in section 3.1, the RDEM loss of concrete often plays a controlling role in the failure modes of concrete subjected to Na_2SO_4 solution freeze–thaw cycles. Thus, a relationship can be established between RDEM loss D and ultimate number of freeze–thaw cycles NF. This relationship can be used to predict the service life of concrete that is subjected to sulfate frost. With the adverse effects of adding fly ash or slag on concrete performance to resist Na_2SO_4 solution freeze–thaw cycles, the regression equations are made only for normal concrete and air-entrained concrete in Figs. 3 and 7 based on the least squares principle. These regression equations are listed in Table 3.

Table 3. *Regression equations of concrete RDEM loss D and freeze–thaw cycle number N.*

Item	Regression equation	R^2
P0.4	$D=-5.3\times10^{-7}N^3+1.1\times10^{-4}N^2+3.8\times10^{-4}N$	0.99
P0.5	$D=-7.4\times10^{-7}N^3+1.4\times10^{-4}N^2+6.5\times10^{-4}N$	0.99
P0.6	$D=-7.0\times10^{-7}N^3+1.3\times10^{-4}N^2+1.6\times10^{-4}N$	0.99
PYA	$D=-3.0\times10^{-8}N^3+1.4\times10^{-5}N^2+2.5\times10^{-3}N$	0.96
PYB	$D=-3.0\times10^{-8}N^3+1.4\times10^{-5}N^2+2.3\times10^{-3}N$	0.90

As shown in Table 3, the correlation coefficients R^2 of all regression equations are more than 0.90. This value indicates that the regression equations are reliable. The number of rapid freeze–thaw cycles that concrete can bear before failure can be calculated based on the regression equations in Table 3 and the failure criterion D=0.4. Li et al. suggested that 1 rapid freeze–thaw cycle for concrete in the laboratory is nearly equal to 12 freeze–thaw cycles for concrete in the natural environment [20]. Moreover, the relationship between the predicted service life of concrete T and the maximum number of rapid freeze–thaw cycles N_F can be calculated using the following equation:

$$T = \frac{BN_F}{kM} \quad (4)$$

where B is the coefficient of rapid freeze–thaw experiment, namely, the number of 1 rapid freeze–thaw cycle equivalent to the natural freeze–thaw cycles, which is generally set at 12; k is the safety factor for concrete structure design, which is commonly set at 1.5; and M is the annual average number of freezing and thawing cycles of concrete structures in the natural environment. According to [20], the value of M in the northwest region of China can be set at 118.

The expected service life of concrete under sulfate freeze–thaw cycles can then be calculated. The predicted service life of concrete with w/c ratios of 0.4, 0.5, and 0.6 is equivalent to 5.1, 4.3, and 4.0 years, respectively. The average life span is 4.47 years. Decreasing the w/c ratio from 0.6 to 0.4 can extend the expected service life by 27.5%. The average life of concrete with 4.6% and 5.7% air content is approximately 25 years, which is nearly 5.6 times that of the non-air-entrained concrete. Therefore, adding the air-entraining agent can significantly extend the service life of concrete in a sulfate frost environment.

4. Conclusions

The following conclusions are obtained from the present experimental study:

1) The appearance damage and mass loss of concrete subjected to freeze–thaw cycles in Na_2SO_4 solution are usually unnoticeable until failure, and the DEM losses often play the controlling role to determine the failure of the specimens. A sudden fracture failure in the middle occurs easily for specimens with low w/c ratio. In general, decreasing the w/c ratio of concrete can slightly increase its resistance to Na_2SO_4 solution freeze–thaw cycles.

2) Adding fly ash or slag reduces the concrete resistance to Na_2SO_4 solution freeze–thaw cycles, and a higher replacement ratio increases the adverse effects. In particular, the adverse effects of slag powder are slightly less than those of fly ash. Therefore, adding fly ash or slag into concrete is not recommended for the environment of Na_2SO_4 solution freeze–thaw cycles.

3) Adding an air-entraining agent to concrete can significantly improve the resistance to Na_2SO_4 solution freeze–thaw cycles. The concrete with 4.6% and 5.7% air content has a service life that is 5 times longer than that of ordinary concrete under the Na_2SO_4 solution freeze–thaw cycles.

Acknowledgments

This project is supported by the Fund for Key Problems and Protective Measures on Durability Degradation of Concrete Bridges in Xinjiang Region, the Transport Department of Xinjiang Uygur Autonomous Region (No. 2014-11), and the open fund project of Jiangsu Collaborative Innovation Center for Building Energy Saving and Construction Technology (No. SJXTY1509).

References

[1] Wang, P. L., W. Q. Zhang, Y. K. Yang and Y. T. Liang. Investigation and analysis of corrosion failure of cement concrete in the saline lake area of Qinghai. Journal of Qinghai University (Natural Science Edition), 2003, 21(6): 57-59. (in Chinese)

[2] Zhou, G., S. R. Li, Z. J. Wang and C. F. Wang. Investigation and analysis on the corrosion of concrete in saline soil area. Journal of Architectural Science and Engineering 2011, 28(4): 121-126. (in Chinese)

[3] Neville A. The confused world of sulfate attack on concrete, Cem. Concr. Res. 8(2004) 1275-1296.

[4] Amini A., Tehrani S. S. Combined effects of saltwater and water flow on deterioration of concrete under freezing-thawing cycles, J. Cold Reg. Eng. 25(2011) 145-161.

[5] Sotiriadis K., Nikolopoulou E., Ysivilis S. Sulfate resistance of limestone cement concrete exposed to combined chloride and sulfate environment at low temperature, Cem. Concr. Comps. 8(2012) 903-910.

[6] Łaźniewska-Piekarczyk B. The frost resistance versus air voids parameters of high performance self compacting concrete modified by non-air-entrained admixtures, Constr. Build. Mater. 48(2013) 1209-1220.

[7] Mu R., C. W. Miao, J. P. Liu and W. Sun. Effect of NaCl and Na_2SO_4 solution on the frost resistance of concrete and its mechanism. Journal of the Chinese Ceramic Society 2001, 29(6): 523-529. (in Chinese)

[8] Yu H. F., W. Sun, L. H. Yan and B. Yang. Freezing thawing durability of high strength and high performance concrete exposed to salt lakes. Journal of the Chinese Ceramic Society 2004, 32(7): 842-848. (in Chinese)

[9] Yuan L. D., D. T. Niu, L. Jiang, Y. Z. Sun and Q. N. Fei. Study on damage of concrete under the combined action of sulfate attack and freeze–thaw cycle. Bulletin of the Chinese Ceramic Society 2013, 32(6): 1171-1176. (in Chinese)

[10] Zhang Y. Q., H. F. Yu, W. Sun and J. Y. Zhang. Frost resistance of concrete under action of magnesium sulfate attack. Journal of Building Materials 2011, 14(5): 698-702. (in Chinese)

[11] Jiang L., D. T. Niu, L. D. Yuan and Q. N. Fei. Durability of concrete under sulfate attack exposed to freeze–thaw cycles. Cold Regions Science and Technology 2015, 112: 112-117.

[12] Ge Y., W. C. Yang, J. Yuan, B. S. Zhang and A. L. Xiong. Freezing resistance of concrete in sulfate solution. Concrete 2005, 8: 71-73. (in Chinese)

[13] Mu R., C. W. Miao, J. P. Liu W. Sun. Effect of sodium sulphate solution on the frost resistance of concrete. Journal of Building Materials, 2001, 4(4): 311-316. (in Chinese)

[14] Yu H. F., W. Sun, L. H. Yan and Q. Wang. Research on freezing-thawing durability of air-entrained concrete exposed to salt lakes. Journal of Wuhan University of Technology, 2004, 26(3): 15-18. (in Chinese)

[15] Du J. M., Y. Chen, G. X. Yu and Y. S. Ji. Research on the sulfate corrosion resistance of fly ash concrete in adsorption area. Journal of China University of Mining & Technology, 2014, 43(4): 600-605. (in Chinese)

[16] Ghafoori N., M. Najimi, H. Diawara and M. S. Islam. Effects of class F fly ash on sulfate resistance of Type V Portland cement concretes under continuous and interrupted sulfate exposures. Constr. Build. Mater. 2015, 78: 85-91.

[17] Peng G. F., Q. Ma, H. M. Hu, R. Gao, Q. F. Yao and Y. F. Liu. The effects of air entrainment and pozzolans on frost resistance of 50-60 MPa grade concrete. Constr. Build. Mater. 2007, 21: 1034-1039.

[18] Su X. P., Q. Wang, W. H. Wang and H. Y. Sun. Frost resistance and durability mechanism of concrete under saline-alkali condition in seasonal frozen soil area. Journal of Jilin University: Earth Science Edition 2014, 44(4): 1244-1253. (in Chinese)

[19] Hossack A. M. and Thomas M. D. A. The effect of temperature on the rate of sulfate attack of Portland cement blended mortars in Na_2SO_4 solution. Cem. Concr. Res. 2015, 73: 136-142.

[20] Li J.Y., X. P. Peng, Z. G. Deng, et al. Quantitative design on the frost-resistance of concrete. Concrete, 2000, 9: 61-65. (in Chinese)

8

Effects of Road Improvement on Safety: A Case Study of Nairobi-Thika Superhighway

Chelugo Nicholas Chepchieng[1, *], Abiero Gariy[1], Mwatela Josphat K. Z.[2]

[1]Department of Civil, Construction and Environmental Engineering, Jomo Kenyatta University of Agriculture and Technology, Nairobi, Kenya

[2]Department of Geomatic Engineering and Geospatial Information Systems, Jomo Kenyatta University of Agriculture and Technology, Nairobi, Kenya

Email address:

engchelugo@gmail.com (C. N. Chepchieng)

Abstract: Road safety is one of the main transportation problems in developing countries. The major goal of the transportation system is to enhance mobility and road traffic accidents are unwanted by-products which have to be minimized while achieving the primary goal. Road safety engineers are faced with the challenge of addressing safety issues within the three major traffic safety pillars: human, vehicle, and infrastructure.All three aspects must be part of a traffic safety plan and dealt with subject to budget limitations. The project involved redesigns of all intersections and increment in the road lanes to enhance capacity and improve safety. The research study aimed to review the occurrences of crashes on the improved Thika-Nairobi superhighway in three phases; before, during and after construction. This research sought to assess the role of incorporation of different safety measures in road projects in Kenya and is anticipated to improve public awareness on the use of road safety provisions with the aim of reducing road traffic accidents and subsequent casualties. The approach used to realize this study involved gathering data from traffic police stations, conducting interviews to get the opinions of road users on safety and making observations on the turning movements of vehicles and pedestrians. The interactions were explored, analysed and modelled using Chi-squared distribution. These were then used to identify accident black spots and assess the effectiveness of road safety measures installed in the new super highway. This study revealed that the large number of accidents occurred in the years before and at the start of construction, reduced during construction and increased again after completion. The primary type of accidents along the highway is vehicle to pedestrian and vehicle to vehicle. Githurai 45, Safari park and G.S.U were established to be black spots. The most efficient road facility was found to be foot paths, and foot bridges.

Keywords: Black Spots, Nairobi-Thika Superhighway, Road Crash, Road Safety, Traffic Accidents

1. Introduction

Many countries have set ambitious quantified road safety targets [1]. Targets alone cannot guarantee enhanced safety but they can enhance implementation of cost-effective safety measures. To be able to build up a realistic and effective safety programme, the measures must be selected using latest scientific knowledge on their effects. Scientific research has produced lots of good estimates on the effects of safety improving measures. The effects are usually presented as a percentage change in the number of injury accidents. So it is vital to be able to evaluate reliably what would have happened if the road safety measure had not been implemented. In addition experience suggests that the greatest mistakes in evaluating the safety effects of road improvements are done while evaluating the current safety situation. It is a challenge to develop effective measures for improving road safety – especially when resources are scarce and economic means are limited. Nevertheless, a major target for European policy as well for as for national, regional and local decision makers is to improve road safety significantly in our roads [2].

The Nairobi-Thika Highway is one of three major corridors linking downtown Nairobi to the suburbs and satellite towns. Traffic demand on this road is almost twice the existing capacity. Accident data shows that over the past five years more than 700 accidents occurred on the Nairobi

Thika road of these 227 were fatal. The upgrading of the highway, therefore, will provide adequate capacity and considerably decrease the accident rate by minimizing vehicle conflicts with traffic interchanges and by providing separate service roads for local and non-motorized traffic. The completed 45km Nairobi-Thika superhighway design highlights comprises of six to eight-lane divided highway with full access control between Nairobi and Thika Town built services roads along the Highway built with nine traffic interchanges built. The overall objectives of the improvement project were to Contribute to and improve the accessibility, affordability, and reliability of the transport infrastructure system to promote economic growth and socio-economic development in Kenya Contribute to regional integration in the eastern and horn of Africa regions,Improve road transport services along the Nairobi-Thika corridor and enhance urban mobility within the metropolitan area by reducing traffic congestion, Contribute to the development of a sustainable urban public transit system for the Nairobi Metropolitan Area and finally to promote private sector participation in the management, operation, and financing of road infrastructure in Kenya.

Rapid urbanisation and motorisation in Kenya have fuelled urban mobility demand, threatening long term sustainability of urban areas. Nairobi's population has grown significantly from 350,000 in 1963 to 828,000, 1,325,000 and 2,137000 in 1979, 1989 and 1999 respectively.Its night time population by 2007 is estimated at 3.05 million and is growing at a rate higher than even the national population growth (4.8 per cent as compared to 3.4 per cent.The current population of Nairobi Metropolis is projected at 4.73 million [3].Notably, Nairobi's population represents about 21 per cent of Kenya's urban population.A feature not to be ignored also is Nairobi's primacy; Mombasa the second largest city has a population of 660,000 representing approximately 25% of Nairobi's population.Projections indicate that this population is expected to hit the 8 million mark by the year 2030. With the expected population increment it is expected that demand for fast, safe, efficient and secure transport system shall also go up and consequently increase in chances of road traffic accidents and crashes. According to [4] road transport sector has been increasing from 36.9% in 1997 to 65.3% in 2009 as a share of total transportation sector hence the need to have a reliable road crashes data system which is very important for desirable road safety practice. It is, therefore, paramount to examine how road improvement as exemplified by Thika-Nairobi Superhighway affects safety on the roads.

2. Statement of the Problem

Globally, it is estimated that on average, road traffic accidents cost up to 1% of a country's Gross Domestic Production (GPD), and the Sub-Saharan Africa has the highest road accident death rates in terms of fatalities per registered vehicle, as compared to any other region in the world [5]. It is also reported that road safety in the African region is a major social and economic concern in that although the region has about 4% of the world's motor vehicles, 11% of the world's reported fatalities are due to RTAs. Further, the cost of road traffic accidents was crudely estimated to be US $ 3.7 billion in 1997 in this same report [6]. The road accidents impact massively on both the health and insurance sectors whereby the cost implication is very high and slices a huge share of Kenya's GDP. In spite of this, there is poor accident and crash costing which downplays the critical role road safety plays and thereby undermines the urgency and vigour with which safety measures should be undertaken.

In Kenya road accidents account for the loss of more than 3,000 lives annually, a significant proportion of these are young and productive men and women in the 15-45 years age bracket. The economic cost of such road accidents and fatalities has been estimated to be in excess of 4 billion shillings. Kenya has one of the highest rates of accident in the world at present with 510 fatal accidents for every 100,000 registered vehicles. With the high growth of road accidents in Kenya, it is essential that adequate sums of money be spent in dealing with the problem. In the absence of an estimate of accident-related economic issues, it is difficult to identify the sums of money that should be invested each year on road safety countermeasures. Nevertheless, Theestimation of the national explicit costs of road accidents as well as the implicit value of the accidents is rarely highlighted to policy makers. In light of this, the present study is aimed to shed more light on the cost of traffic accidents whereby the findings will help to evaluate the effectiveness of road safety programs as well as in the proper allocation of the country's limited financial resources for the most effective prevention plans.

3. Study Objective

To assess the effects of road improvement on safety along the 45 km Thika Superhighway.

4. Research Hypothesis

H_0: Road rehabilitation improves road safety and reduces road crash rates and frequencies.

5. Literature Review

5.1. Characteristics of Crashes

The Highway Safety Manual (HSM) defines safety as "the crash frequency and/or crash severity and collision type for a specific time period, a given location, and a given set of geometric and operational conditions". Safety is defined in similar terms by the American Association of State Highway and Transportation Officials [7]. Roadway safety is usually defined and evaluated in terms of the number of recorded crashes. Crash severity also plays an important role in understanding roadway safety. For example, one site may experience considerably more crashes than another; however, the second site may have a much larger proportion of severe, particularly fatal, crashes. Therefore, both crash frequency and crash severity are essential in determining the safety of a facility. In order to understand how to reduce crash frequency and crash severity, it becomes important to first understand the factors behind crashes.

5.2. Causes of Crashes

Crashes represent a very small proportion of all of the events that occur on a transportation system. There have been numerous proposed theories that try to explain the causes of crashes. The Handbook of Road Safety Measures [1] provides a more in-depth look at various proposed theories. In this research, it is sufficient to emphasize four conclusions that can be made from these theories. First, it is probable that all proposed crash theories have an element of truth in them. However, while crash cause theories do include portions of the truth, none of the theories provide a complete understanding or explanation of why crashes occur. One of the key reasons it is difficult to understand why crashes occur is because crashes are usually not the result of one factor, but the combination of multiple events, circumstances, and [8]. There are three categories of factors that contribute to crashes [7]. These include:

i. Human – including age, judgment, driver skill, attention, fatigue, experience and sobriety
ii. Vehicle – including design flaws, safety features
iii. Roadway/Environment – including geometric alignment, cross-section, traffic control devices, surface friction, grade, signage, weather and visibility.

The combination of multiple events can severely alter the amount of risk a driver may face. For example, imagine a deer runs in front of a driver on a rural highway. Driver A, driving during the daytime, is at considerably less risk than Driver B, driving at night during a snowstorm. Driver A may have to deal with factors such as reaction time, stopping distance, and brake wear. Driver B would be affected by the same factors in addition to reduced surface friction and visibility. Understanding this point helps to understand that crashes are the outcome of a vastly complex random process [8].

Secondly, there are certain roadway, vehicle, and driver trends that make the occurrence of a crash more likely. The exact impact these trends have on roadway safety has long been the focus of research. Difficulty arises due to the fact that some are known, while others are not. Third, it is also important to understand that even though improvements are continually made to reduce crash frequency, no system is entirely perfect. Drivers are fallible and thus still subject to error in judgment, whether recklessly or not. This provides great difficulty in determining the effectiveness of an improvement [8]. Finally, even if it were possible to account for all crash factors, the ability to predict a crash is not absolute. The reason is that crashes are still to some extent a random event [8].

The key principle is that understanding the nature of crashes is a vastly complex and random process when considering just the known factors. It is important to remember that there are factors that contribute to crashes that are unknown. The premature assumption might then be made that limited understanding of the contributing factors of crashes make it extremely difficult, if not impossible, to determine the proper remedy for crashes. The Handbook of Road Safety Measures, [1] provides valuable insight in understanding the concept of crash causes. The handbook states "the logic of the argument that you need to know the causes of a problem in order to solve it seems irresistible. Yet, as far as crashes are concerned, there is not necessarily a very close connection between the causes of the problem and its solution" [1]. The complexity of known and unknown contributing factors can be overcome through the development and use of proper statistical tools that correctly model crash characteristics and behavior.

5.3. Crash Rate

In order to model crashes, it is necessary to first define what exactly is being measured. Traditional practice has been to use crash rates as a measure of safety [9]. The crash rate is the frequency of crashes adjusted to account for volume or exposure. The general relationship between crash frequency and crash rate is explained in Equation 1.

$$Crashrate = \frac{averagecrashfrequencyinperiod}{duringtheperiodofexposure} \quad (1)$$

Crash rates for road segments are typically reported in crashes per million vehicle miles traveled (MVMT) or per hundred MVMT. Crash rates for intersections are typically reported in crashes per million entering vehicles (MEV). Equation 2 shows the crash rate equation for a section of roadway [9].

$$\text{CRsec} = \frac{\text{N}}{\text{Vsec} \times 365 \times \text{L}} \times 10^6 \quad (2)$$

Where:

CR_{sec}= crash rate for section (in crashes per MVMT),

N = number of crashes per year,

V_{sec} = average annual daily traffic (AADT) of road section, and

L = length of section (in miles)

Equation 3 shows the crash rate equation for intersections.

$$\text{CRint} = \frac{\text{N}}{\text{Vsec} \times 365 \times \text{L}} \times 10^6 \quad (3)$$

Where:

CR_{int} = crash rate for intersection (in crashes per MEV), and

V_{int} = sum of average daily approach volumes of intersection.

When using crash rates, an assumption is often made that the relationship between frequency and exposure is linear. Recent studies have shown that this assumption is not always valid [10]. It has been determined that the use of crash frequency is a more accurate indicator of roadway safety than the use of crash rates. Research shows that the relationship between traffic volume and crash count is more complex and relates to quantities such as the distribution of traffic throughout the day and the types of crashes experienced. Some studies have indicated that there is indeed a relationship between the number of crashes and traffic volume [11]. The exact form, however, is still unknown and likely depends on crash type. Models using aggregate data (not separated by crash type) and exposure as inputs ignore significant variation in highway crashes resulting from hourly volume changes and human behavioral changes throughout the day. A study performed using disaggregate data (crashes broken down by type) revealed how the relationship between crashes and traffic volumes varies from location to location and by crash type [12]. New approaches are also being developed for incorporating traffic volumes in crash rate analysis and forecasting studies [8].

5.4. Crash Frequency

One solution that overcomes the non-linear relationship between crash frequency and exposure associated with crash rates is to use crash frequency as the fundamental basis for safety analysis and measurement of treatment effectiveness [7]. The use of crash frequency as a measure of safety eliminates the inclusion of exposure altogether. A crash frequency is obtained by counting the number of crashes at a certain site of interest, usually a roadway segment or intersection, over a certain period of time. Crash frequency and crash severity are important elements of crash history and form the basis of quantifying and measuring safety. Along with understanding these elements of crash data, it is also important to understand trends exhibited within the data so that an appropriate statistical approach can be used in analysis of crashes.

6. Research Methodology

Research methodology encompasses the step-by-step procedure that is followed to arrive at findings that are in tandem with the study objectives. In the context of the present study, scientific approach was adopted whereby it was thought better to regard the phenomenon of RTAs as a chance process. At any time, the road, the traffic, the user and the vehicle is a set of circumstance have a certain chance or probability of leading to an accident. The following sequence was followed during the study:

i. Field observations were conducted to establish road conditions that pose a threat to safety.Road characteristics that were of interest during the survey included cross-section, drainage, roadside, safety barriers, alignment, junctions, bus and, parking facilities, access to properties, speed management, and speed management, and pedestrian usage characteristics and provision of pedestrian facilities.
ii. Conducting interviews, and
iii. Collection of road accident data and traffic count data for the study section and its interpretation.

6.1. Sampling

An assumption was made that the population is unknown, thus Cochran's equation (Equation 4) was used in the determination of the sample size

$$n_0 = \frac{Z^2 * P * Q}{e^2} \tag{4}$$

Where: n_0 = Sample Size
Z = Z score (area under the normal curve)
P = Estimated proportion of an attribute (Standard Deviation)
e = Desired level of precision

$$Q = 1 - P$$

With a 95% desired confidence level, standard deviation of 5 and a desired margin of error of ± 5%, the sample size for this study was calculated as follows:

$$\frac{1.96^2 * 0.5 * (1 - 0.5)}{0.05^2} = 385 \text{ respondents}$$

6.2. Secondary Data Collection

Road accident data used in this study was obtained from police records.The Kenya Police routinely collect and store data on motor vehicle accidents mainly for prosecution and insurance compensation purposes. The accident data used in this study was collected from Kasarani, Ruiru, and Muthaiga police stations and the traffic police headquarters. Road accident data for the year 2005 to 2013 for Thika Road; between Muthaiga roundabout and Ruiru were used. The data collected covered the pre-construction period, construction period and post-construction period of Thika Superhighway. The expected frequency and severity of crashes caused by each safety issue have been identified and rated according to the categories shown in Table 1.

Table 1. *Severity Rating.*

Typical Crashes Expected	Expected CrashSeverity	Severity Rating
Crashes involving high speeds or heavy vehicles, pedestrians, or bicycles	Probable fatality orincapacitating injury	Extreme
Crashes involving medium to high speed;Head-on, crossing, or run-off-road crashes	Moderate to severe injury	High
crashes involving medium to low speeds;Left-turn and right-turn crashes	Minor to moderate injury	Moderate
crashes involving low to medium speeds;Rear-end or sideswipe crashes	Property damage only orminor injury	Low

7. Analysis and Discussions

7.1. Number of Accidents

The total number of accidents recorded in the span of 8 years was 993 (Table 2). The most number of accidents occurred in the years before and at the start of construction, that is 2006 to 2009. In the following two years of construction, 2010 and 2011, the number of accidents reduced remarkably. This can be explained by the reduced vehicle speeds during construction. At the end of construction, that is year 2012, when most of the construction was completed, there was a 79% increase in the number of accidents over the previous year. However, the linear graph (Figure1) indicates that, in general, the number of accidents reduced over the five year period.

Table 2. *Accident Data Before, During and After Construction.*

ACCIDENT DATA								
YEAR	**Ped to Veh**	**Veh to Road**	**Cyclistto Veh**	**Veh to Veh**	**Slight**	**Fatal**	**Serious**	**Total No of Accidents**
2006	96	35	15	62	32	53	72	208
2007	84	42	21	45	56	62	58	192
2008	53	16	6	34	29	55	26	109
2009	76	18	7	58	28	76	55	159
2010	30	27	7	28	14	36	40	92
2011	32	8	3	14	16	12	27	57
2012	76	9	2	15	7	44	55	102
2013	48	10	3	13	5	31	48	74
Grand Total	495	165	64	269	187	369	381	993

Source: Kenya Police (2013)

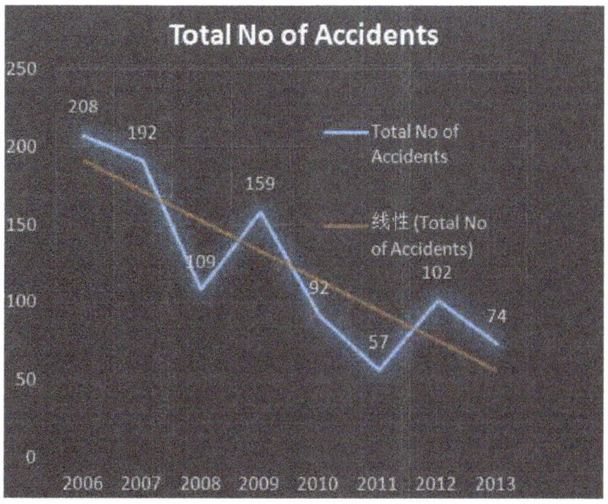

Figure 1. *Total Number of Accidents and their Trend.*

7.2. Traffic Accident Severity

Out of the 993 accidents, 39% were fatal and 41% were serious. The trend of slight accidents generally decreased during the 8 year period.However, that of fatal and serious accidents reduced then later increased after the completion of the project as shown in Figure 2. The fact that fatal accidents are still the most type accidents shows that additional safety measures(policy and physical)still need to be put in place and effected.

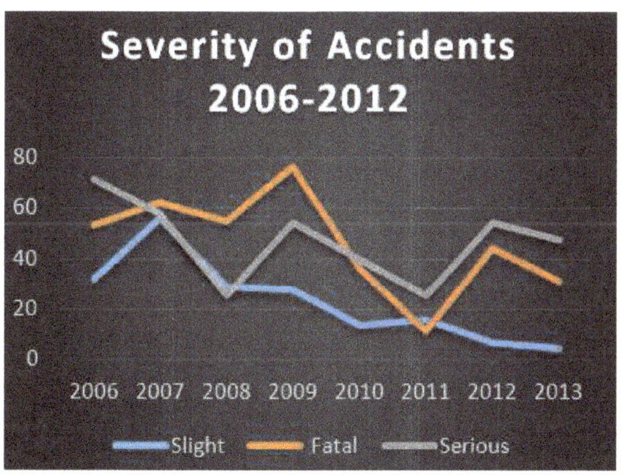

Figure 2. *Severity of Accidents (2006-12).*

7.3. Types of Accidents

The primary type of accident along the highway was found to be Vehicle to Pedestrian, 50%, followed by Vehicle to Vehicle, 27% (Chart 4.4). Pedestrians continue to be the most vulnerable road users despite installation of foot bridges, zebra crossings bumps and pedestrian foot paths along the highway. The number shot after completion of the project because of the tendency of drivers to over speed because when using a new pavement. In addition, at that time some pedestrians had the tendency of not using the facilities (foot bridges) provided for them. Some considered them cumbersome and time consuming. Vehicle to Road accidents have reduced significantly owing to the new features in the highway like additional lanes and grade intersections (Figure 3).

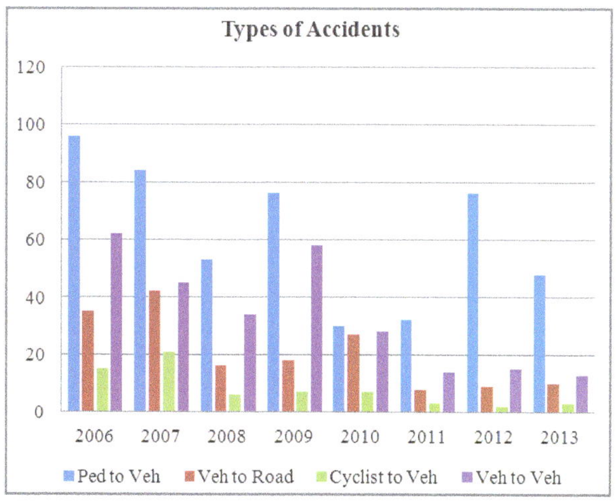

Figure 3. *Types of Accidents.*

7.4. Traffic Accidents by Vehicle Type

Small cars were the leading vehicle type in fatal, severe and slight accidents, with 40% (Table 3). This was partly because small cars and four wheel drives tend to be the dominant vehicle type in the daily traffic volume using the highway. Public service vehicles, that is, matatus and buses accounted for 16% and 10% respectively.

Table 3. Traffic Accidents by Vehicle Type.

Vehicle Type	Fatal	Serious	Slight	Total
Bus	14	17	10	82
Car	70	82	52	355
Lorry	21	21	7	69
M/Cycle	4	9	1	34
Matatu	26	39	31	136
Pick Up	7	17	4	34
Van	5	6	5	26
Minubus	8	7	1	65
Unknown	22	7	8	50
Others	2	1	1	23

Source: Kenya Police (2015)

7.5. Number of Accidents Versus Time of Day

This analysis determined whether or not the number of accidents was related with the time of day. Table 4shows a calculated χ^2 - value of 125.73 for 3 d.f. and a critical χ^2 - value of 19.675 at 0.05 confidence level. Since the calculated χ^2 - value was found to be greater than the critical χ^2 - value, the null hypothesis, that is, accidents are uniformly distributed over the hours of the day is rejected. Thus, it indicated that accidents were not uniformly distributed over the hours of the day before after and during construction.

Table 4. Chi Square Analysis for Goodness of fit (a).

Hrs	Observed Frequency	Expected Frequency	D.F	Critical Chi. Sqr	Calculated Chi. Sqr	
0000-230	15	43.17			18.38	
230-430	14	43.17			19.71	
430-630	30	43.17			4.02	
630-830	37	43.17			0.88	
830-1030	48	43.17			0.54	
1030-1230	57	43.17	11	19.675	4.43	125.73
1230-1430	45	43.17			0.08	
1430-1630	66	43.17			12.07	
1630-1830	62	43.17			8.21	
1830-2030	86	43.17			42.49	
2030-2230	40	43.17			0.23	
2230-0000	18	43.17			14.68	
	518	518				

7.6. Number of Accidents Versus Month of the Year

In this analysis, whether or not the number of accidents was related to the month of year was analyzed. Table 5shows a calculated χ^2 - value of 23.34 for 11 degrees of freedom and a critical χ^2- value of 19.675 at 0.05 confidence level. Since the calculated χ^2 - value was greater than the critical χ^2 - value, the null hypothesis, that is, accidents are uniformly distributed over the year was rejected. Thus, it is indicated that accidents were not uniformly distributed over the year before, after and during construction.

Table 5. Chi-Square Analysis for Goodness of Fit (b).

Month	Observed Frequency	Expected Frequency	D.F	Critical Chi. Sqr	Calculated Chi. Sqr	
January	35	42.25			1.24	
February	33	42.25			2.03	
March	37	42.25			0.65	
April	49	42.25			1.08	
May	40	42.25			0.12	
June	26	42.25	11	19.675	6.25	23.34
July	58	42.25			5.87	
August	57	42.25			5.15	
September	41	42.25			0.04	
October	39	42.25			0.25	
November	46	42.25			0.33	
December	46	42.25			0.33	
	507	507				

7.7. Number of Accidents Versus Location

Table 6 shows a calculated χ^2 - value of 2008 to 2012 respectively, for 10 degrees of freedom and a critical χ^2- value of 10.307 at 0.05 confidence level. Since the calculated χ^2 - value was greater than the critical χ^2 - value, the null hypothesis that stated accidents are uniformly distributed over the whole road section was rejected. Hence, it was concluded with a 95.95% certainty that accidents black spots do exist at specific areas along Thika Road.

Table 6. *Chi-Square Analysis for Goodness of Fit (c).*

	Critical	2006	2007	2008	2009	2010	2011	2012	2013
K.U Engen Kahawa Barracks Githural 45 Clayworks Roysambu Safaripark Roasters G.S.U Crescent P.S Total P.S	10.307	39.88	39.84	22.06	38.56	36.6	23.9	73.54	17.51

7.8. Speed and Traffic Count Data

An important traffic characteristic that affects safety is speed variance. A major influence on speed variance is the difference between the design speed and the posted speed limit. It has been shown that speed variance will be minimised if the posted speed limit is correlated with the design speed. This emphasises the link between posted speed limit and design speed, and supports the recommendation that speed limits should be set by a traffic engineer, based on spot speed studies. The posted speed limit along Thika road varied from 50km/hr to 60Km/hr. Table 7 shows the speed distribution over the different vehicle classes and the number of vehicles recorded over a 24hr period.

Table 7. *Speed and Traffic Data Matrix.*

Speed (Km/hr)	Motorcycles	Cars	Pickups/Jeeps/4WDS	Matatus/Buses	Buses	Light Trucks	Medium Trucks	Heavy Trucks	TOTAL
10-20	299	1712	31	226	68	27	17	31	2411
20-30	415	8498	233	950	194	96	56	132	10574
30-40	618	27636	440	2199	272	228	63	112	31568
40-50	730	32423	326	1576	104	167	63	32	35351
50-60	590	34090	97	1017	73	63	29	9	35968
60-70	951	11011	30	455	18	15	9	1	12490
70-80	363	3887	3	139	4	10	4	1	4411
80-90	325	1044	2	48	3	8	0	0	1430
90-100	62	295	15	14	0	2	0	0	388
100-110	1	37	12	3	0	0	0	0	53
110-120	1	9	5	3	0	0	0	0	18
120-130	0	2	0	4	0	0	0	0	6
130-140	0	2	1	2	0	0	0	0	5
140-150	0	4	1	1	0	0	0	0	6
150-160	0	9	0	3	0	0	0	0	12
TOTAL	4355	120659	1126	6640	736	616	241	318	134691

Source: KenHA (2013)

From Table 7, 43% of vehicles were traveling at an average speed of 55Km/hr whilst 23% of the vehicles were traveling at an average speed of 35Km/hr. The former speed was mainly achieved during the middle hours of the day and the later during the morning and evening hours when vehicle flow and density were at a maximum. Pedestrians have a 90% chance of surviving a car crash at 30km/hr or below but less than 50% chance of surving impacts of 45km/hr or above

Using data in Tables 2 and 7, the current crash rate for Thika Road using was found to be:

$$C_r = \frac{74}{134691*365*50} * 10^6 = 0.03 \text{ Per Million Vehicle Kilometres Travelled}$$

The crash rate for Thika Road before construction was:

$$C_r = \frac{208}{60000*365*46} * 10^6 = 0.2 \text{ Per Million Vehicle Kilometres}$$

Accident rates did not necessarily increase with an increase in average speed, but have been found to increase on all classes of roads with an increase in speed variance.

8. Conclusions

This study examined the effects of road improvement on safety. Therefore, the analysis compared the trends of road accidents, characteristics of accidents and road features that affect safety before, during and after Thika Road construction. The study established that indeed road rehabilitation, in the case of Thika Road, improved safety by minimizing and/or eliminating causes of accidents like: road geometry, lack of pedestrian facilities, bad road surface and inadequate carriageway width and at-grade junctions.Road rehabilitation, however, was found to enhance speed related accidents. This paper has clearly confirmed that safety of road users can be achieved in various ways. Hence with the government's intent to rehabilitate and construct over 10,000Km of road networks, safety of road users is expected to increase enormously. It was concluded that the driving habits ofmost motorists was not up to standards hence this paper recommends the enhancement of skills drivers and to upgrade the requirements, and competence of PSV drivers on the road.

Recommendations

The study recommends that the conditions of other roads across Kenya be improved as one way of addressing road safety issues. The traffic department of the National Police Service (MPS) is advised to erect speed guns along the Nairobi-Thika Superhighway and other refurbished roads across the country. This would enable the police to apprehend the motorists who abuse the speed limits, given that this is the major cause of road accidents along improved highways. The study further recommends that all road users particularly motorists should be sensitized on traffic rules. This would hugely reduce traffic road accidents in Kenya.

References

[1] R. Elvik, and T. Vaa. "TheHandbook ofRoad Safety Measures." Elsevier, Oxford, UK, 2004.

[2] A.T. de Blaeij, M. Koetse, Y. Tseng, P. Rietveld, and E. Verhoef. "*Valuation of safety, time, air pollution, climatechange, and noise*; Methods andestimates for variouscountries." Reportfor the EU project ROSEBUD. Vrije Universiteit, Amsterdam, 2004.

[3] Kenya National Bureau of Statistics. "Kenya: 2009 Population and Housing Census Highlights. Nairobi:" *Kenya National Bureau of Statistics; 2009.*

[4] E.M. Aligula. "A critical Look at management of Transport Sector for efficiency: The challenges and the way forward." *A paper presented to the 13th ICPSK Economic Management Symposium, 27th October 2010 at the Sarova Panafric Hotel, Kenya*, 2010.

[5] World Health Organization. "Global Burden of Disease: *2004 Update. Geneva, Switzerland:*" World Health Organization, 2008.

[6] J.K.Z. Mwatela. "Application of Geographical Information Systems(GIS) To Analyze Causes of Road Traffic Accidents (RTAS)" – "*Case Study of Kenya, Presentation in International Conference on Spatial Information for Sustainable Development Nairobi," Kenya 2–5th October 2001.*

[7] AASHTO. American Association of State and Highway Transportation Officials. "*Highway Safety Manual,*" *Washington, DC*, 2010.

[8] G.S. Grant, J.T. Daniel, N.O. Andrewand C.R. Shane. "Transportation Safety Data and Analysis;" Volume1: "*Analyzing the Effectiveness of Safety Measures using Bayesian Methods." Report No. UT-10.12a*, 2010.

[9] R. P. Roess, E. S. Prassas, and W. W. McShane. "Traffic Engineering," *3rd Ed., Pearson Education, Upper Saddle River, NJ*, 2004.

[10] E. Hauer, D. W. Harwood, F.M. Council, and M. S. Griffith. "Estimating safety by the empirical Bayes method: A tutorial." *Transportation Research Record. 1784, Transportation Research Board, Washington, DC*, 2002, *126-131.*

[11] S. Miaou. "The relationship between truck accidents and geometric design of road sections: Poisson versus negative binomial regressions." *Accident Analysis & Prevention, 26(4), 471-82.* 1994.

[12] X. Qin, J. N. Ivan, and N. Ravishanker "Selecting exposure measures in crash rate prediction for two-lane highway segments." *Accident Analysis and Prevention*, 36(2), 183-191, 2004.

9

Application of Progressive Collapse Analysis in a High-Rise Frame-Core Tube Structure

Hu Kai[1, *], Qi Sanlin[2], Qu Ge[1], Jiang Li[1], Lu Zheng[3], Yang Shengsheng[2]

[1]China Shipbuilding NRDI Engineering Co., Ltd, Shanghai, China
[2]CCCC Fourth Harbor Engineering Co., Ltd., Huangzhou, China
[3]College of Civil Engineering, Tongji University, Shanghai, China

Email address:
whocannet@163.com (Hu Kai), SLQi730@163.com (Qi Sanlin), qu75675@163.com (Qu Ge), youlin_popo@126.com (Jiang Li), luzheng111@tongji.edu.cn (Lu Zheng), yshengsheng@cccc4.com (Yang Shengsheng)
[*]Corresponding author

Abstract: Frame-core tube structure system is commonly adopted in high-rise office buildings and landmarks, which will cause a large range of damage and serious loss of life and property once the progressive collapse occurs. To analysis the progressive collapse resistance of a high-rise frame-core tube practical project, numerical simulation analysis based on alternate path method and explicit dynamics theory was conducted to study the dynamic response of the remain structure after removing a key member. In addition, large-scaled tests for the frame-core tube structure under static and dynamic loads were carried out to study the behavior and damage situation of the structure. The numerical simulation and test results proved that progressive collapse would not occur when local damages occurred in the bottom corner column, and the structure still has a good seismic performance under dynamic actions. And the feasibility and reliability of the selected numerical simulation method of progressive collapse analysis is proved according to the test results.

Keywords: Progressive Collapse Analysis, Frame-Core Tube, RC Structure, Alternate Path Method, Numerical Analysis, Collapse Test

1. Introduction

Security issues are the primary problems of structural design, and many engineering accidents happened because of the progressive collapse caused by incidental loads. In recent years, such kind of accidents emerge in endlessly, such as the collapse in the terrorist attacks of 11 September 2001, the "Loudoadao" event in shanghai, as well as the "Loucuicui" event in Fenghua, Zhejiang Province, which once occurs, would cause a large range of damage and serious loss of life and property, increase the later repair difficulty of the buildings, or even lead to the structure completely discarded.

In order to effectively avoid such collapse accidents, many researches on progressive collapse resistance of RC structures have been carried out. The collapse mechanism of a plane frame structure by pseudo-static collapsed destruction test and numerical simulation analysis with LS-DYNA software [1]. Dynamic responses of frame structures with different members removed respectively were studied based on OpenSees platform [2-3]. Force mechanisms of frame structure with bottom columns removed was studied by test [4-5]. And the collapse process of a high-rise frame-core structure under seismic actions was analyzed use numerical simulation methods with TECS program [6]. According to the current studies of progressive collapse analysis of RC structures, continuous collapse resistance analysis based on numerical simulations and tests is feasible, and the reliability was verified in multiple examples of plane frame structures and space frame structures, which lays the foundation for the analysis of practical engineering projects.

Due to the complexity and diversity of engineering projects and the release of new specifications of China, higher requirements of structural design were put forward, and continuous collapse resistance analysis of the structure has become an effective approach to examine the rationality and stability of the whole structure to provide the basis for structural safety evaluation thus effectively reduce or even avoid such accidents.

In this paper, based on a high-rise frame-core tube practical project, the dynamic response and damage degree of the structure with key member removed were analyzed by numerical simulation method and large-scaled collapse tests to verify that this project has a good progressive collapse resistance under incidental loads.

2. Engineering Situation

This project is located in Haizhu District of Guangzhou, including a commercial podium and a high-rise office building which are separated by seismic joints. The main building covers an area of 95,385 m2, and is 198.9m high (B level height) with 43 floors on the ground and 3 floors underground, as shown in Fig. 1. And the plane size of the main building is 47.5m × 47.5 m.

Based on the specifications and regulations of China, RC frame-core tube was adopted as structure system, and the main features of the structure are as follows:

1) The height of the main building is nearly 200m which exceeds the appropriate maximum height of buildings.

2) The structure belongs to super high-rise buildings with several irregularities, characterized by torsion irregular and floor slab discontinuity.

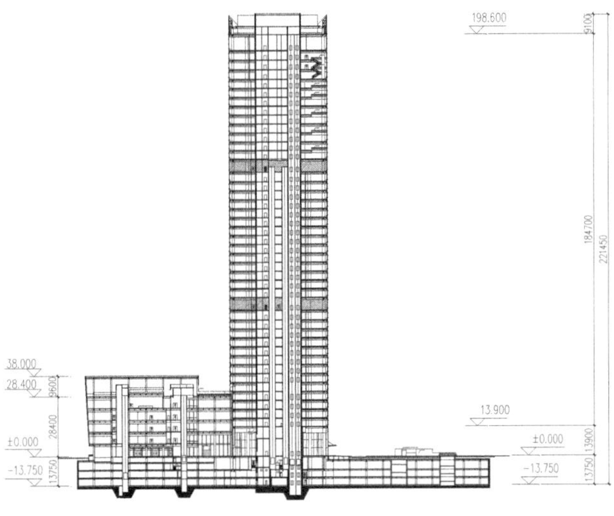

***Figure 1.** Architectural section.*

Due to the height and huge volume of this project, the security problem was especially considered during the structure design. And taking into account that local damages and failures of outer frame components in frame-core tube structures may occur and further lead to the transmission route change of vertical loads in the unexpected events and disasters, progressive collapse analysis of this project was focus on the dynamic response of the structure with the failure of the bottom corner column under the representative value of gravity load.

Based on the studies results of progressive collapse analysis of RC structures, numerical simulation and test methods were adopted respectively to study the continuous collapse resistance of the main structure.

3. Numerical Simulation Analysis and Main Results

3.1. Simulation Principle

As a complex numerical process of gradual translation from the elastic continuum to plastomer or even non-continuum in principle, collapse simulation raises high requirements for the selections of several aspects such as collapse analysis method, numerical simulation method, and numerical simulation algorithm, and the simulation principles used in this paper are as follows.

3.1.1. Collapse Analysis Method

Refer to design processes and suggestions of specifications such as GSA2003, DoD2010 and JGJ3-2010 [7-9], the alternate path method with clear principle and strong operability was chosen, which is to establish the finite element model of the complete structure and then "kill" a key member to analyze the dynamic response of the remain structure. And the key members such as corner columns, middle columns at long side or short side and columns where the geometry of structure plan is obviously changed which are usually identified as the important vertical members in the force transmission system of the structure could be selected as the removed member [10-11].

3.1.2. Numerical Simulation Method

Considering the applicability of multi structure systems and completeness of calculation process files and large amount of applications in practical engineering [10-12], finite element method was chosen as the analysis method.

3.1.3. Numerical Simulation Algorithm

As the structural collapse process is always accompanied by serious nonlinear problems such as large displacement, large strain, large rotation and material nonlinearity, the explicit dynamics algorithm based on central difference method with the advantages of small step size, stability and better convergence performance was chosen to capture the internal stress wave propagation in small time steps.

3.2. Collapse Analysis

Based on these simulation principles, ANSYS/LS-DYNA software based on the explicit dynamics was chosen for the collapse simulation analysis. The model contains 40027 nodes, 24394 beam elements and 36364 shell elements, and the elements were joined by common node connection. In order to simulate the destruction of RC members, strength criterion and deformation criterion were mainly used as failure judgment criterion. The plastic kinematic material model was used with the parameters set as follows:

1) The equivalent yield strength value of beams and

columns were set considering the reinforcement ratio of members which he reinforcement was converted to equivalent concrete.

2) The member was considered as failure when the tensile rate of the reinforcement steel bar was more than 10%, that is, the failure of the members was controlled by effective strain.

To analysis the static and dynamic response of the frame-cure tube structure with member losing under accidental load, corner columns at one side of the main structure from the first floor to the third floor were removed to analysis the remain structure as shown in Fig. 2.

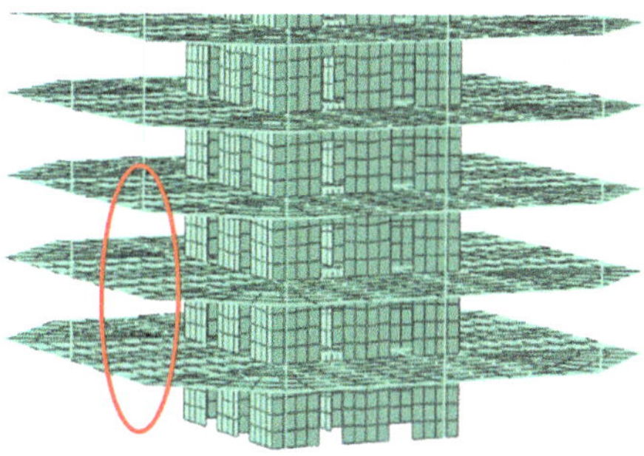

Figure 2. *Numerical model with corner column removed.*

3.3. Analysis Results

The first five natural vibration frequencies and periods of the complete model are shown in Table 1, and the vibration models are obtained as shown in Fig. 3.

Acceleration time-histories of three natural seismic waves as input: El Centro, Taft and Chi-Chi were chosen, and the horizontal peak acceleration of the primary direction was set as 310 cm/s^2 and the secondary direction as 263.5 cm/s^2 in accordance with the requirements of IV class site and 7 degree rare earthquake intensity in the "Code for Design of Concrete Structures" [13], namely the acceleration amplitude ratio of two horizontal directions is 1:0.85.

Table 1. *Structural dynamic characteristics.*

Modal	Frequency/Hz	Period/s
1	0.238	4.2012
2	0.238	4.2012
3	0.6415	1.5588
4	1.0705	0.9341
5	1.0706	0.9341

Under the bi-directional seismic actions of 7 degree rare earthquake intensity, responses of the complete model and the corner column removed model of the main structure were analyzed respectively. The maximum story drift and the acceleration peak of each floor were calculated and compared, and the results are as follows:

1) Under the earthquake actions, the main structure had a good seismic performance as the maximum story drift of the nonlinear model with corner columns removed was nearly 1/200 which was less than the specification limit of 1/100.

2) Although the distribution laws of the story drift varied under different seismic wave input, most of the maximum story drifts appeared in the middle-upper part of the structure.

3) Removing of the corner columns had a little negative influence on the displacement response of the structure as the maximum story drift was increased by only 3%~5% compared to the complete model, while the acceleration peak distribution was affected greatly as the acceleration peak was nearly doubled in some layers.

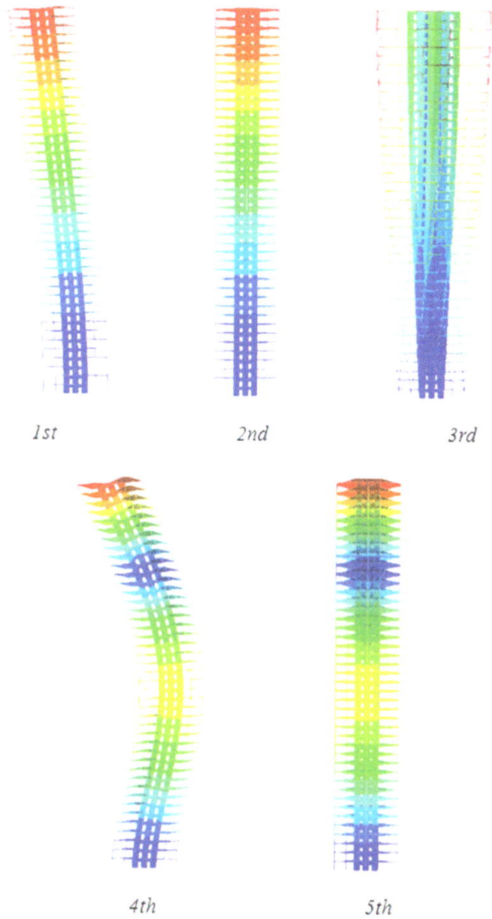

Figure 3. *The first five vibration models.*

In addition, the deformation of the whole structure and plastic strain of the vertical members of the two models were analyzed, and the deformation comparison in the Chi-Chi wave condition is shown in Table 2. The results show that after removing corner columns, the whole structure was inclined to a certain, while the members' removal had a little influence on the horizontal displacement response as a whole. The plastic degree of the concrete shear wall at the column removed side was deepened and the damage scope was enlarged under the dynamic actions, which reflected that failure of the key member affected the damage accumulation process. Progressive collapse did not appear during the loading procedure, which indicates that the main structure would not collapse during the dynamic actions under 7 degree rare earthquake intensity when bottom corner column were removed, and it could be determined that the structure had a good continuous collapse resistance under this situation according to the numerical analysis.

Table 2. *Deformation and internal force comparison under 7 degree rare earthquake intensity of Chi-Chi wave.*

Column 1	Complete model	Corner column removed model
X-deformation	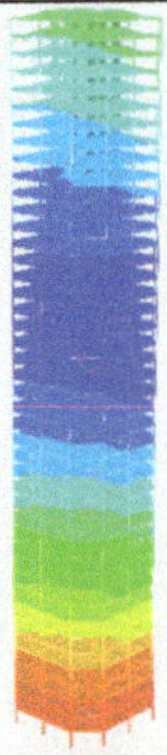	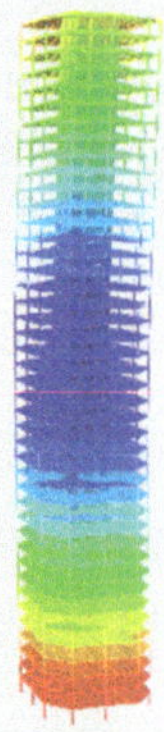
Y-deformation	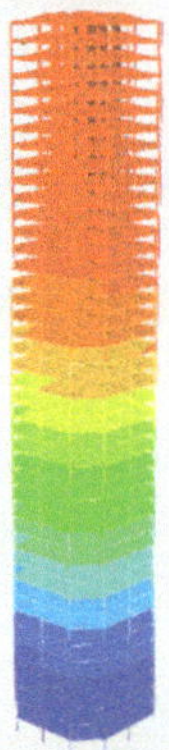	
Z-deformation	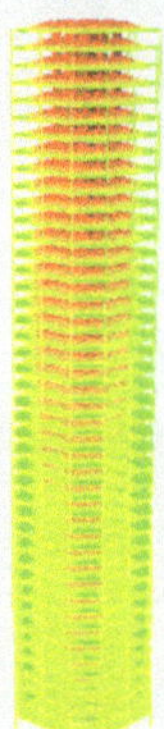	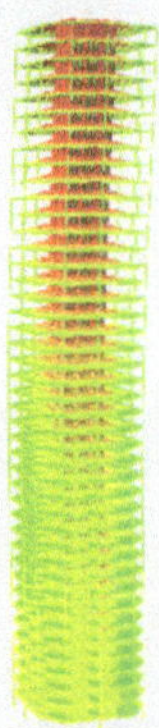
Plastic strain of core tube		

4. Test Study and Results Comparison

For further study of the continuous collapse resistance of frame-cure tube structure, the static test and the dynamic test were conducted respectively [14]. And Taking consideration of the size and load capacity of the shaking table in Tongji University, the simplified model was proposed from the main structure of the practical engineering, and a 1/15 scaled 7-floor test model was established based on the similarity theory. The removed column was replaced by a jack for the simulation of column failure during the static test study, and the test model is shown in Fig. 4.

Figure 4. *Test model.*

During the static test condition, the jack was unloaded slowly to simulate the failure process of the bottom corner column, and the progressive collapse did not happen after the deformation of the structure achieved stability as shown in Fig. 5, which proves that the failure of the corner column would not lead to the structure progressive collapse under gravity loads.

The shaking table test was performed in situ after the static test, 5 seismic waves as EL Centro, Taft, Wenchuan, Chi-Chi and SHW01 were chosen for bi-directional and three-directional input by stage of frequent earthquake, basic earthquake and rare earthquake of 7 degree intensity, as the working conditions are listed in Table 3.

Figure 5. *Static test study.*

Table 3. Structural dynamic characteristics.

Earthquake level	Frequent earthquake	Basic earthquake	Rare earthquake
White noise	1	12	23
El Centro	2,3	13,14	24,25
Tafe	4,5	15,16	26,27
Wenchuan	6,7	17,18	28,29
Chi-Chi	8,9	19,20	30,31
SHW01	10,11	21,22	32,33

After the frequent earthquake stage, no crake was appeared on the model and the natural frequencies of vibration were almost the same to the value before this stage, which proves that the structure with one corner column removed could still remain in flexible working stage under the seismic action of frequent earthquakes.

During the basic earthquake stage, a few vertical crakes mainly appeared at the bottom wall, and inclined crakes appeared at the end of coupling beams. As one corner column was removed, the damage degree of the shear wall adjacent to this position was higher than other shear walls.

Until the 7th input at the rare earthquake stage, crakes on the bottom vertical supporting members developed further as shown in Fig. 6, while the whole structure did not collapse with the story drift below the specification limit of 1/100, which indicates that the structure still has a certain seismic capacity.

Figure 6. Damage of bottom structure under 7th input at rare earthquake stage.

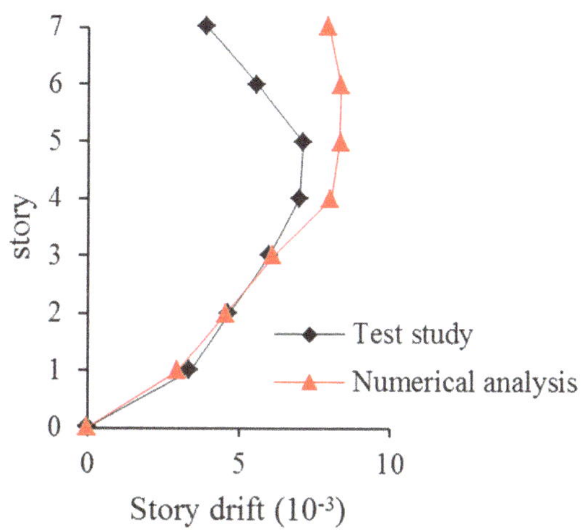

(a) X-direction

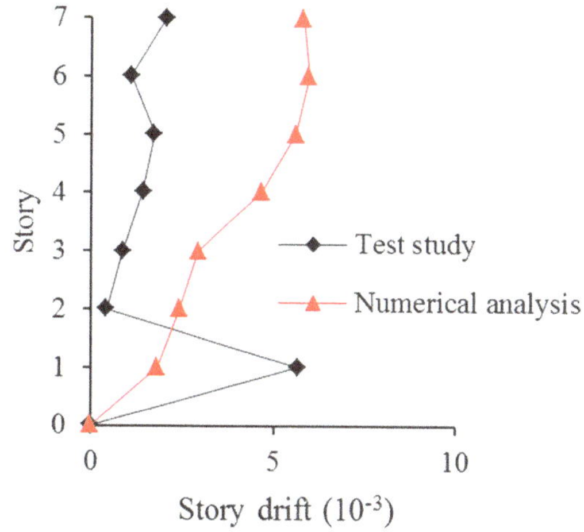

(b) Y-direction

Figure 7. Story drift comparison in 7 degree rare earthquake intensity condition.

In addition, the simulation results fit well with the test result as shown in Fig. 7, which proves that the numerical simulation method could reflect the dynamic response of the structure with local damage under seismic actions. Finally, the structure collapsed with the accumulation of damage and the further damage of the vertical components. And the bottom shear walls and key joints of the main structure were reinforced according to the results.

5. Conclusion

In this paper, progressive collapse resistance of a high-rise frame-core tube project was analyzed by numerical simulation based on the alternate path method and explicit dynamics to study the dynamic response of the whole structure with key members removed. And in order to verify the accuracy of the numerical simulation results, a large scaled test was performed to study the behavior of the frame-core tube structure under static and dynamic loads. The results are as follows:

1) Loss of bottom corner column has a little influence on the displacement response of the structure while the acceleration peak distribution was affected greatly, and plastic degree of the vertical members especially the concrete of the bottom shear-wall was deepened and damage scope was enlarged. With the damage accumulation of the vertical members, overall collapse would happen according to the shaking table test results. And based on the numerical simulation and test study results, the bottom shear walls and key joints of the main structure were reinforce.

2) Under the seismic action of 7 degree rare earthquake intensity, the main structure has a good seismic performance even though the bottom corner column was removed, which checked the security and ability to resist progressive collapse of the structure in this kind of incidental accident and indicated that the structure could meet the seismic fortification target of No Collapsing in the Strong Earthquake.

3) Compared with the test and numerical simulation results, the adopted numerical method could accurately simulate the nonlinear behavior of the structure with local damage under seismic actions, which could provide the basis for continuous collapse resistance analysis of practical engineering. And through the progressive collapse analysis as one of the performance-based analyses, the structure could be more reasonable, reliable and economical.

References

[1] He Qingfeng, Research on Progressive Collapse Behavior of RC Frame Structures. Hunan University: 2009, pp. 19-37.

[2] Ma Gao, Li Hui, OU Jinping, "Dynamic response and progressive collapse resistance of RC frame structures subjected to element removal," Technology for Earthquake Disaster Prevention, vol. 5, pp. 62-72, 2010.

[3] Wang Debin, Li Hongnan, Zhang Jie, "Response analysis of RC frame structures in progressive collapse after the member failure," Chinese Journal of Applied Mechanics, vol. 31, pp. 116-121, 2014.

[4] He Qingfeng, Yi Weijian, "Experimental study of the collapse-resistant behavior of RC beam-column sub-structures considering catenary action," China Civil Engineering Journal, vol. 44, pp. 52-59, 2011.

[5] Xiong Jingang, Wu Zhaoqiang, He Yinong, etc, "Experimental research on progressive collapse performance of RC spatial frame structures," Journal of Nanchang University (Engineering & Technology), vol. 34, pp. 229-238, 2012.

[6] Lu Xinzheng, Lin Xuchuan, Ye Lieping, et al, "Numerical models for earthquake induced progressive collapse of high-rise buildings," Engineering Mechanics, vol. 27, pp. 64-70, 2010.

[7] Liang Yi, Lu Xin Zheng, Li Yi, etc, "Verification and analysis on foreign progressive-collapse-resistance design methods of RC frame structures," Building Structure, vol. 40, pp. 8-12, 2010.

[8] Lu Xinzheng, Li Yi, Ye Lieping, Theory and Design Method for Progressive Collapse Prevention of Concrete Structures. Beijing: China Architecture & Building Press, 2011.

[9] JGJ3-2010 Technical Specification for Concrete Structure of Tall Building. Beijing: China Architecture & Building Press, 2010.

[10] Liu Chengqing, Chen Linya, Su Qiwang. "Research on the progressive collapse resistance performance of Pengshan gymnasium," Steel Construction, vol. 30, pp. 1-5, 2015.

[11] Wei Yong, "Design of multi-story height steel cantilever truss for Handan culture and art center," Progress in Steel Building Structures, vol. 18, pp. 44-54, 2016.

[12] Sun Jian. Progressive Collapse Analysis on the Suspension Structure of Tianjing Large Theatrer Building and the Steel Structure of Tianjing Library. Tianjing University: 2012.

[13] GB50011-2010 Code for Seismic Design of Buildings. Beijing: China Architecture & Building Press, 2010.

[14] Wang Peng, Lu Xilin, Yang Zi, et al, "Shaking Table Test Study of Earthquake Collapse of a Reinforced Concrete Frame-core Tube Structure," Structural Engineers, vol. 30, pp. 122-129, 2014.

10

Effects of Crude Oil Contaminant on the Engineering Properties of Concrete

George Abednego, Onungwe Ishmael, Oba Achemie, Akpan Paul, Sarogoro Samuel

Department of Civil Engineering, Rivers State Polytechnic, Bori-Ogoni, Rivers State, Nigeria, West Africa

Email address:

tonsoky@yahoo.com (G. Abednego), ishmael.onungwe@gmail.com (I. Onungwe), achemie25@gmail.com (O. Achemie), paulynciap06@yahoo.com (A. Paul), ndesarosr@yahoo.com (S. Samuel)

Abstract: The artificial application of crude oil in various percentages in concrete mix constituent to under study the service interaction of concrete in contact with oil was carried out. The research contained herein present the investigation, analysis and comparative results obtained from a total of 96 concrete cubes sample consisting of 12 control cubes, 12 0% contaminated cubes cured in Bodo crude oil water and 72various percentages of crude oil contaminated cubes cured in potable water. The tests conducted were slump and flow rate in terms of freshly mixed concrete and compressive strength in terms of hardened content. The OPC concrete mix which adopted 0.5 water cement ratio was cured for 7, 14, 21 and 28 days in its cubes form and the results obtained indicated that concrete is highly susceptible when in contact with crude oil. Its contact with crude oil causes it to deterioration as well as slow strength development and increased percentage flow rate. The difference between the control mix and the 25% crude oil contamination was 16N/mm^2 at 28 days, which explicitly spells out the consequences of crude oil effect on the properties of concrete structures.

Keywords: Crude Oil, Contaminated Water, Compressive Strength, Slump, Concrete Flow Rate

1. Introduction

The amount of crude oil contamination of soil and the environment has continually increased and presently, it constitutes a significant fraction of waste material in the environment [1]. Contaminants in aggregates or concrete constituents does not only affect the appearance of concrete in terms of color and rotting but also the strength, durability, abrasion, water tightness, volume stability, freeze and thaw resistance and resistance to deicing chemicals. The effect of crude oil on some engineering properties of concrete in a crude oil contaminated area had been considered by some authors and one of such authors [2], on a general note observed that there was high reduction in the compressive strength and about 11% reduction in splitting-tensile strength of concrete when soaked in crude oil. Concrete deterioration and cracking in marine environment is more severe than in any other terrestrial environment, hence it has been observed that deterioration occur as a result of such factors as physical and chemical characteristics of repair compound, initial curing periods, environmental conditions, among other factors [3], in his work on some properties of crude oil, soaked concrete exposed at ambient temperature observed variations in mechanical properties with time. Furthermore, he reiterated that the variation in compressive strength length change and absorption characteristics of typical offshore concrete structures exposed to crude oil for 365 days compromised the durability and serviceability of such structure. However, studies by [4] showed that factors significantly affecting concrete properties include conditions of curing prior to exposure, moisture condition of concrete at the time of exposure, storage temperature of crude oil as well as the type of cement. The occurrence of concrete deterioration and cracking as a result of crude oil spill, has remained a green area in literature. Observation showed that for four decades, Nigeria has continued to experience remarkable increase in operational activities in her oil and gas exploration, exploitation, refining and product marketing which is concentrated in the Niger Delta region and that the region has been mired by various degrees of health and environmental pollution problems. Regular crude oil spillage on surface and subsurface water sources, erosion and drainage problems of the built environs culminating to incessant failure of buildings and other onshore structures have become a regular news item

[5]. [6] inferred that the optimum crude oil contamination for the achievement of normal compressive strength is as low as 0.3%. The research of [7] on the effect of crude oil contamination on the index properties, strength and permeability of lateritic clay showed that the liquid limit, plastic limit and plasticity index of the soil increased as its crude oil content increased. The specific gravity, optimum moisture content, maximum dry unit weight, CBR and permeability of the soil decreased as its crude oil content increased. Crude oil contaminated soil requires stabilization or remediation before using it as a construction material. The studies of [8] also showed that A 18–90% compressive strength was lost due to 2.5–25% crude oil contamination, respectively. These and other physical observatory factors has informed the research contained herein which focuses on artificially contaminating concrete sample constituent in terms of fine aggregate in various percentages and crude oil contaminated water sample and subjecting these samples when constituted as concrete to investigation on the effect of the contaminations on the properties of the concrete both in fresh and hardened state which actually models the situation in Bodo due to oil facility failure, crude oil spillage or vandalism as most constructional materials or concrete constituents are extracted from contaminated river bodies or most contaminated water are used for construction.

2. Materials and Method

2.1. Materials

The cement used for this study is one of the available commercial brands of Ordinary Portland Cement (OPC), however care was taken in the cement material purchase as well as the conducting of the practical to ensure test reliability. Aggregates used were coarse aggregate of 12 mm maximum size conforming to [9] which was obtained from Akamkpa quarry in Cross River State, while the fine aggregate was natural white color river bed sand obtained in Bori Rivers State. The fine aggregate was oven dried having determined the moisture content to achieve dry surface condition in order to ensure the actualization of materials void of saturation so as not to affect the water cement ratio, thereby bringing the aggregate to conform to [10] specification. Other tests and procedure carried out in this research work include grain size analysis, slump test, flow table test, cubes casting, curing and compressive strength test using a digital crushing machine at Rivigo JV Nigeria Limited, Ogoni-Andoni-Opobo Road project site base. The sieve analysis was carried out in accordance with [11] in order to determine particle grading curve of the aggregates from where it was deduced that fine aggregate fell in zone 2 consisting of fine, medium and coarse sand with coefficient of uniformity (Cu) of 1.69 and coefficient of curvature (Cc) 0.84 indicating a well graded fine aggregate. The coarse aggregate fell into zone 3 consisting of fine and medium gravel with (Cu) of 1.96 and (Cc) of 1.01 indicating a well graded coarse aggregate. The crude oil used was obtained from one of the oil fields within the locality.

2.2. Methods

Concrete mix design of 1:2:4 at water cement ratio of 0.5 was adopted and batched in weight in accordance with British Standards. The mixed design covers three sets of sample including samples of 0% contaminated cubes cured in portable water which serves as control, 0% contaminated cubes cured in Bodo crude oil contaminated water and different levels of artificial contaminated cubes of (2.5%, 5%, 10%, 15%, 20% and 25%) cured in fresh water. These three samples were compared in terms of freshly mixed concrete and compressive strength in terms of hardened concrete. The concrete constituents were thoroughly mixed in a clean dry manual tilting concrete mixing drum with a total of 96 cubes produced. The material mix proportion is as shown in table 2.

Tests on Fresh Concrete Sample

Basically the tests conducted on the freshly mixed concrete sample were slump and flow table test. Results obtained from these tests are as shown in table 3 and 4 respectively. The slump test was carried out on both the control and contaminated samples to check workability in accordance with the procedural steps as given in [9]. The flow table test was conducted to observe the concrete sample consistency and cohesiveness as well as the degree of segregation. The flow table test was carried out in accordance with the procedure outlined in [12]. The hardened concrete cube test conducted on the sample was the compressive strength test using a digital crushing machine at Rivigo JV Nigeria Limited. The cubes tested for totaling 96 cubes comprising of 12 0% contaminated cubes as control and 84 contaminated cubes of 150mmx150mmx150mm.The cubes were all removed from its mould within 24 hours of its cast and cured in a water curing tank within a room temperature and later removed from the curing tank according to the duration of crushing and was tested for compressive strength in accordance with [13]. The results obtained are as shown in table 5.

3. Results

Table 1. *Properties of Nigerian Crude oil.*

S/No	Parameters	Rate/Unit	Magnitude
1	Gravity Degree	API	<35.00
2	Specific Gravity	15°C	0.812
3	Sulphur Content	% by weight	0.30
4	Moisture Content	% by volume	0.40
5	Wax Content	% by weight	7.0
6	Carbon Residue	% by weight	2.10
7	Melting point	°C	57
8	Viscosity	21°C	6.81
9	Acidity	Mg/KOH/g	0.05

*Source: [14].

Table 2. Materials Mix Proportion.

S/No	Cement (Kg)	Water (Kg)	Fine (Kg)	Coarse (Kg)	Crude oil * (Kg)	% Contamination
1	135	67.5	270	540	0	0
2	135	67.5	270	540	6.75	2.5
3	135	67.5	270	540	27	10
4	135	67.5	270	540	40.5	15
5	135	67.5	270	540	54	20
6	135	67.5	270	540	67.5	25

* The crude oil contamination value was obtained with respect to fine aggregate.

Table 3. Slump Test Result.

S/No	% Contamination	Slump (mm)	Workability
1	0	20	Very Low
2	2.5	45	Low
3	10	70	Medium
4	15	85	Medium
5	20	110	High
6	25	135	High

Table 4. Flow Table Test Result.

S/No	% Contamination	Initial Concrete Base Diameter (cm)	Average Concrete Diameter (cm)	Flow Rate (m/s)	Flow Remark
1	0	25	29.0	16	Very Low
2	2.5	25	32.5	30	Low
3	10	25	35.5	42	Low
4	15	25	37.0	48	Low
5	20	25	40.0	60	Medium
6	25	25	42.5	70	Medium

Table 5. Average Compressive Strength.

Description	Age (Days)	Strength (N/mm^2)
Control mix cured in portable water	7	30
	14	31
	21	32
	28	35
0% Contaminated mix cured in bodo crude oil water	7	26
	14	32
	21	32
	28	32
2.5% Crude oil mix cured in portable water	7	27
	14	29
	21	30
	28	31
5% Crude oil mix cured in portable water	7	25
	14	28
	21	30
	28	30
10% crude oil mix cured in portable crude oil water	7	23
	14	30
	21	30
	28	30
15% Crude oil mix cured in portable water	7	27
	14	30
	21	26
	28	29
20% Crude oil mix cured in portable water	7	24
	14	25
	21	25
	28	26
25% Crude oil mix cured in portable water	7	19
	14	20
	21	22
	28	20

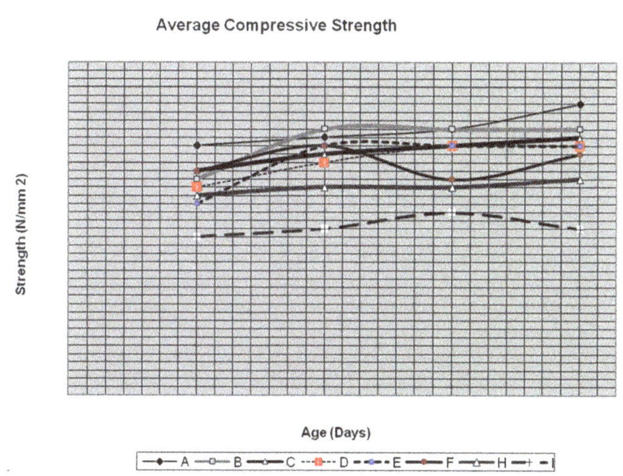

Figure 1. Graph of Average Compressive Strength against Age of Concrete.

Legend
A-Control mix cured in portable water
B-0% Contaminated mix cured in bodo crude oil water
C-2.5% Crude oil mix cured in portable water
D-5% Crude oil mix cured in portable water
E-10% crude oil mix cured in portable crude oil water
F-15% Crude oil mix cured in portable water
H-20% Crude oil mix cured in portable water
I-25% Crude oil mix cured in portable water

4. Discussion

The results interpreted and presented in figure 1, tables 3, 4 and 5 respectively are explicit, but for the purpose of results discussion and interpretation clarity, it will be observed that the control mix for 28 days cured samples indicated the highest compressive strength of an average strength value of 35N/mm^2 as compared to the 0% contaminated cube samples cured in Bode crude oil water with an average strength value of 31N/mm^2 and the results kept on decreasing with increase in various percentage contamination of crude oil in the concrete constituent which finally brought the lowest average compressive strength to a value of 19N/ mm^2 which represent the 7 days 25% crude oil contamination mix. It may be interesting to note that the 28 days of the same 25% crude oil contaminated mix gave an average result value of 20N/ mm^2 which is 1N/ mm^2 difference from the 7 days of same 25% crude oil contaminated mix. This represents the processes which concrete structures under goes when it gain contact with oily substances, specifically in terms of crude oil which forms the basis of this research. In essence, the reduction in strength can be linked to the crude oil content in various percentages in the concrete constituent which has absorbed into the microstructure of the matrix of the concrete leading to delayed process in the gel of the cement as well as weakening the cohesiveness of the binder paste which is in conformity with the works of [15] and [3]. Also, other than the current observation of the reduced or slow compressive strength gain of the samples, the effect of crude oil on concrete structures worsens with time resulting to rotting of the structure. On the averages of the compressive strength results which fall from 32N/mm^2 control sample to 30N/mm^2 Sample cured in Bodo contaminated water down to 29N/mm^2 for 2.5%, 28N/mm^2 for 5%, 28N/mm^2 for 10%, 28N/mm^2 for 15%, 25N/mm^2 for 20% and 20N/mm^2 for 25% respectively for the crude oil contaminated samples. The flow table test is practically suitable for fresh concrete with high workability, in subjecting the sample to this test, it was clearly observed that the presence of crude oil in the concrete aided the workability as the percentage of flow rate of the control sample was very low indicating a value of 16% as compared to the highest level of percentage crude oil contaminated sample with percentage flow rate value of 70% as shown in table 4. On the other hand, the slump values when compared with the highest contaminated sample of 25% have a difference of 120mm.

5. Conclusion

Having conducted the various practical and analysis in order to ascertain the strength of the concrete with respect to the effect of crude oil, the following conclusions were reached: That when concrete gain contact with liquid substances other than portable or fresh water, it may be considered injurious to the cement base material of such concrete, especially when it is OPC, as it has weak resistance to environmental and chemical reaction effect which causes slow strength development and ultimately low compressive strength when hardened as well as leading to concrete early rotting.

Finally, to ensure durability, rapid strength gain, avoidance of cracks, water tightness, abrasion resistance, volume stability, resistance to freeze and thaw and as well as resistance to deicing chemicals of OPC based structures, location of concrete constituents collection, mixing and curing water should be void of crude oil or other lubricants as such undiluted substances enhances high deterioration of concrete structures.

References

[1] Ajagbe, W.O., Omokehinde, S. O., Agbede, O.A. and Alade, G.A. (2011).*Effect of Crude Oil Impacted Sand on Compressive Strength of Concrete*. Construction Building Materials, Vol. 26, Issue 1 pp 9-12.

[2] Abdul-Ahad, R.B. and Muhammed, A.A. (2000).*Compressive and Tensile Strength of Concrete Loaded and Soaked in Crude Oil*, Engineering Journal of the University of Qatar, Vol. 13, pp 123-140.

[3] Onabolu, O. A. (1989). *Effect of Hot Crude Oil on Concrete Offshore Storage Application.* PhD Thesis, University of London. pp: 298.

[4] Matti, A. M. (1976). *Some Properties and Permeability of Concrete in Direct Contact with Crude Oil.* PhD Thesis, Sheffield University, UK pp: 230.

[5] Ukoli, M. K. (2001). *Environmental Factors in Management of the Oil and Gas Industry in Nigeria.* Posted in web. http://www.warmfloor.co.uk/pages/environmentalpdfbrochure.pdf.

[6] *So O. and Nwankwo E.(2015).Effect Of Crude Oil Contamination On The Compressive Strength Of Concrete,Nigerian Journal of Technology*, Vol. 34, No 2 (2015) pp 259-265 *http://dx.doi.org/10.4314/njt.v34i2.7*

[7] Akinwumi I. I., Diwa D and Obianigwe N (2014). *Effects of Crude Oil Contamination on The Index Properties, Strength and Permeability of Lateritic Clay*, International Journal of Applied Sciences and Engineering Research, Vol. 3, Issue 4, 2014, Received on July. 12, 2014; Accepted on August 6, 2014; Published on Aug. 29, 2014 http://DOI:10.6088/ijaser.030400007

[8] Wasiu A., Olusola S., Gabriel A. and Oluwole A. (2012). *Effect of Crude Oil Impacted Sand on Compressive Strength of Concrete*, Journal of Construction and Building Materials, 2012; 26(1):9–12.University of Ibadan, Ibadan, Nigeria http://DOI: 10.1016/j.conbuildmat.2011.06.028

[9] British Standard Institution (1983).*Method for Determination Slump Test.* (BS 1881: Part 102) London: British Standard Institution.

[10] British Standard Institution (1983).*Method for Making Test Cubes from Fresh Concrete.* (BS 1881: Part 1). London: British Standard Institution.

[11] British Standard Institution (1989). *Method for Determination Particle Size Distribution.* (BS 1881: Part 103).London: British Standard Institution.

[12] British Standard Institution (1983).*Method for Determination Flow Table test.* (BS 1881: Part 105) London: British Standard Institution.

[13] British Standard Institution (1983).*Method for Determination Compressive Strength of Concrete Cubes from Fresh Concrete.* (BS 1881: Part 116).London: British Standard Institution.

[14] Ejeh, S. P. and Uche, O. A. (2009).*Effect of Crude Oil Spill on Compressive Strength of Concrete Materials*. J Appl Sci Res: 17, 55-61.

[15] Bangham, D. H. (1936).*The Swelling and Shrinkage of Porous Materials and the Role of Surfaces in Determining Technical Strength*. Symposium, Society of Chemical Industry; London. pp 269-333.

Lean Construction in Civil Engineering and Project Management: Case Study Analysis of UT Arlington College Park

Ashwin Amarshi Maru

L.E.College, Civil engineering department, Saurashtra University, Rajkot, State: Gujarat, India

Email address:
maru.ashvin@gmail.com

Abstract: Lean development method, which has been applied extensively in the manufacturing industry, holds enormous potential for the construction industry as well. Projects in the construction segments are complex with both financial level and work level risks. This research critically analyzes the benefits in using lean methods in construction. The lean construction management called the Last Planner System is applied here. The research uses a case study methodology, the case study of UT Arlington College Park Constructions are selected. The Last Planner system and its techniques are applied to Case Study. The benefits of the Last Planner implementation are proved by the recorded PPC value before and after implementation of the last planner system. The PPC values increased. This research concludes that Last Planner system must be further investigated for benefits in Civil Engineering and Project Management, and also recommends it use in Construction Projects.

Keywords: Lean Construction, Last Planner System, LPS, Production Management, Project Management, PPC

1. Introduction

The Construction Industry is one of the most rapidly evolving sectors in Civil Engineering. The Projects are quite large in terms of the outputs needed to be produced and the budgets and time are very stringent as well. In this context the stress applied in terms of the project managers on completing the contract or project within time and within budget constraints are enormous. This study investigates the application of lean construction methodology-the last planner system on construction projects so as to increase efficiency of these projects and also to increase the benefits on the project management end.

Management of projects in the construction industry is a task of immense magnitude. The projects itself are big, and in addition the task divisions and more that are being assigned are also quite complex in itself. Researchers cite that the construction industry is one that suffers huge financial setbacks based on time element. In this context it becomes extremely necessary that the industry have an efficient way to manage its tasks such that productivity is increased and at the same time the project management process also becomes efficient [1]. This is because one might not exist without the other. Project management starts from project procurement tasks, to efficient collation of material and men, developing collaborations between the various project units, arranging systematic task flow towards project completion, managing interdependencies, managing resources, specifying project durations, buffer, critical mean paths and more. Koskela argues that the construction industry and its wasteful practices are creating many struggles with all the stakeholders involved. This was also the time as indicated in literature, when the construction industry started taking strategic hints and planning systems from the manufacturing industry (Koskela, L. 1992).

Lean Technology is a defined technology that is currently being practiced in the workplace and that which simplifies project management tasks and increased productivity in the workplace. Lean technology professes the need for documentation and usage of visualization tools in order to improve process efficiency. This simple documentation and accountability will enable the processes to become more effective. It was a component of manufacturing industry and is now adapted for the construction industry [4]. It derives its production planning and control processes from the manufacturing segment.

2. Research Background

Lean Construction is a novel way in civil engineering. It is a combination of the most adaptable lean construction practices. By lean construction practices the meaning denoted here is that the end to end design becomes more cost effective and productive. It is a process adaptation from the manufacturing industry when the manufacturing industry went into its decline. The manufacturing industry according to Koskela was argued as being ridden with wasteful practices and struggles. This was creating detrimental effects on the organizations and also the people involved [6]. It was in this context that lean technology was arrived at. Lean technology is more a philosophy and a set of principles that was rooted in the manufacturing sector.

There was much complexity in the adoption of lean construction techniques from its counterpart in manufacturing. In the construction industry the end products are not being transported directly to the final users, here the product or onsite production is regulated in a fixed site. The value of the construction project is increased on site, so it is necessary that the contractor equipments and more are to be installed and coordinated for work, at the site. Secondly in the case of production there are many components that come together into production, still since the manufactory unit is quite large, it is necessary to customize the equipment for some standard component manufactory [11]. On the other hand in the case of the construction industry it is seen that the construction industry consumers will play a big role in the project cycle. Customizations are possible to a great extent and the owner of the project will be able to modify the project based on some individual requirements (Adamu & Hamid, 2012).

Thirdly the complexity that is introduced in the case of the manufacturing and the construction industry is also quite different. In the case of the manufacturing industry the suppliers are usually selected quite early, they are then utilized based on the necessary strategic advantages. Reliable flow of the product that is manufactured is hence assured. On the other hand in the case of construction projects it is seen that the projects are usually quite complex. The systems that are involved are very dynamic and although a plan is in place it become necessary that the plan be adjusted based on some of parameters at site. There are overlapping activities of the constructors which furthermore increases the complexity (Adamu & Hamid, 2012). These differences have to be taken into account when the lean production process is applied to construction. It is this very complexity that warrants the need for a construction technique that will be efficient for the worksite. Lean construction as argued by Koskela et al is hence a system that will be able to work with the complexity in the site and still be able to minimize the wastage. The system to hence implement lean construction is possible only with the collaboration of the Owner, A/E, Constructors, Facility Managers, and End-users for the project. In addition the respective on site extra managers, facility managers and more will be included in the process.

The Last planner is a lean construction technique and as such incorporates the benefits of lean construction methodology. It is a technique which "Last Planner is part of a new *production management system* for one-off project-based production such as that in construction and design. This business strategy allows project managers to significantly improve productivity and client/end-user satisfaction when compared to the equally consistent old way of doing business" (p.6) [8].

3. Literature Review

The complexity in construction is increased because of the need for coordination among participants, the material-process flow adequacy and the architecture production concerns. This complexity easily leads to financial losses. The Lean technology system might help remedy these issues by solving potential problems in construction beforehand (Lichtig 2004). Lean technology has one major principle it is discarding of the waste and enhancing profits in that process. The lean technology is the emerging innovative technology that will address all the redundancy in the processes and enable companies to come up with new innovative solutions for the current times. As Lichtig (2004) presents, the production process design can be implemented using lean production principles that will help manage issues based on the information inputs that are created from diverse human resource ends. It will not only be the higher level management, but also site foreman, managers and more that will be involved and hence a wide range of solutions can be arrived at for these issues.

Management efficiency is a concept that is touched upon by authors dealing with lean construction principles. Melhado (1998) identifies the principles that could serve as guidelines for lean construction in the workplace. These principles as argued by Melhado are built on the exact reasons that traditional construction techniques fail [7]. One of the primary reasons for failure as suggested by Melhado is that, the traditional approach relies heavily on the decision making of the management. The traditional management does not cater to the emerging needs of the end user, monetary concerns changes with each project and the current companies need to face up to the challenges brought on by the competitors. There is an inherent need to bring a new paradigm shift from traditional practices. This also emphasis the need for new innovative practices like the Lean technology these issues leads to management deficiency issues in civil engineering. These same issues are also the major reason for the construction industry being reluctant to take up newer practice such as lean construction [7]. A similar approach to expounding on the benefits of lean construction technique in project management and construction sector has been presented by researcher Suresh et al (2012). Here they state management awareness to be the issue. In the adoption of lean technology such as the last planner system, the management comes into contact with the people that they work with on a daily basis. Daily Huddle style meetings assure this.

By adopting Lean technology the companies are able to add more visualization elements into their project planning and implementation. These visual aids assist in creating of Commitment chart, Mobile signs and project markers. Owing to these visual elements accurate depiction of the elements and increased accountability of the processes is created [2]. By implementing lean technology visualization tools the companies are able to stay within budget and adopt robust innovative techniques in the process

Huddle meetings are proposed in this process. This is very different from conventional official conferences. The foremen all gather around and discuss the issues for a maximum of ten minutes before the start of their construction work. This saves time. Apart from these meetings continual assessment is done at regular intervals. This enables the quality of the deliverables at each stage in the construction process. A preliminary examination, which is done in the beginning, is used to determine the processes that require a large amount of resources and costs [3]. Then innovative ideas to reduce any wastage in the existing process are discussed. This is known as First run Studies process. This is a process in lean technology that reduces waste.

By adopting the lean technology process the construction industry needs to follow five main factors [10]. They are categorization of the materials, straightening of the materials, development of standardization of the material layout; clustering the zone of work. When these factors are followed in every step of the process high quality of deliverables is ensured.

The efficacy of the lean technology process is largely dependent on the people implementing and the people following it. This technology is only a tool and it depends on the developer of the plan to ensure that it is robust and innovative. The plan development using lean technology is dependent on numerable factors to name some they are the geographical factors, the size of the project, the cost involved for the project, foremen, contractors, sub-contractors and all the stakeholders involved in the project and the cultural practices of the place [6] Based on these factors the lean technology plan will differ. It has been observed that there has been an overall improvement in all of the processes involved by adopting lean technology.

Mossman explains that "Last Planner" is an important tool of the Lean contruction technique. Last Planner is also acknowledged as "Reverse Phase Scheduling". It is used to remove the redundancy of the processes and enables timely delivery of the projects and the projections are pragmatic. It is also used as a predictive tool that foresees the issues beforehand. The Variance plan and the Percentage Plan finalized plans can be generated from this tool. Weekly Work Plan(WWP) is also done in this process. As the name suggests it is documentation that is done on a weekly basis. This aided in increasing of the accountability and easy idenitification of the anomalies that existed in the system. Mossman presents some of the key benefits of the last planner that will be useful for both civil engineering projects and project management are:

1. It helps in the decentralization of the planning in an organization. Most often organizations push the process of decision making on the management end. While this is the classical approach, in construction industry this leads to a waste in time and available resource management. However where a decentralization supportive tool like the Last Planner is used, then the trade foremen, site men and design-team leaders are all provided with the authority and space required to make instant decisions based on the work site requirements [8]. This not only improves the human resources but also helps increase productivity and improved crises management.
2. CPM is no longer the best way to manage a program as the Critical Path planning system is more predominantly a push form of system [9]. It pushes work into production. The CPM scheduler uses the pre-determined dates for the start of the project and the completion of the project in order to measure progress for the work. However the push measure fails because of the assumptions made on the work done and the plan [8]. People assume that other people are working according to the plan, however what should be done and what is being done are not always that matched. A ceiling contractor might end up working on the ceilings before the M&E contractor finishes their work above them. The work therefore become guess work, and while some guesses are spot on, others might not be leading to disastrous prodcutivity consequences.
3. Last Planner System LPS has the potential to reduce the stress that is being created on the management staff. Project Management Staff ad the advisory staff are usually under tremendous stress when they have to stay engaged with the deadlines, the last planner system helps the Project management stay spot on with the deadlines and at the same time reduces the risks at the workplace so as to increasing the workplace productivity [8].
4. Last planner tool does not look at issues from a unilateral viewpoint rather it encompasses a number of different issues while generating the results thereby adopting a comprehensive approach while providing solutions.
5. The realistic approach to the solutions that this tool provides is the preeminent feature of this tool.
6. The methodology used here is a case study approach. The case study approach is a predominantly qualitative research, which has been used here so as to investigate the application of lean construction methodology-the last planner system on construction projects and its efficiency in increasing benefits to project management.

The case study approach has been used here as it helps in the investigation of a phenomenon where multiple variables of interest are being studied. The Last planner system application to a case study system, helps understand the benefits that it brrought to the system. In this research the

Last Planner system application to the construction project in UT Arlington College Park located on the University of Texas at Arlington Campus is studied. The framework followed by the research article is presented below.

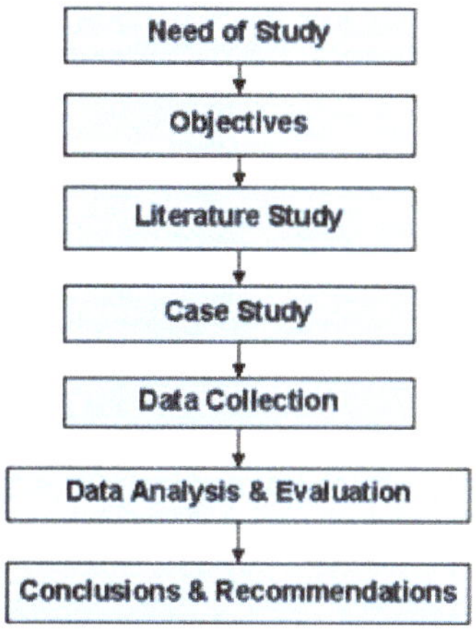

Figure 1. *Research Method*

4. Case Study: Data Analysis and Discussion

The Last Planner system was implemented a project with UT Arlington College Park. Estimated at a value of US $ 65 million, the project involved the completion of three multistory parking garages with car capacity of 1800, residential halls amounting to around 500, central welcome center and sports allotments. To be completed in two phases, Phase 1 was to take a time period of 12 months, while the second was to be completed in 13 months.

Percent Part Complete (PPC) values were calculated as a measure of the workflow reliability.

$$\text{Percent Part Complete (PPC)}\% = \frac{\text{Number of completed assigned activities for the project}}{\textit{Total number of activities defined for the compleation of project}} X\ 100$$

The calculations for the system indicate that the Percent Part Complete (PPC) had improved by more than 60% in the first week, and was at 88% in 4th week and around 83% at the end of the 17 weeks. This gave an average value of 73%, which was higher than the 60 % achieved at Phase 1. The PPC ration for Phase 1 also increased only after the inclusion of the Last Planner system.

Below given table summarizes the project details.

Table 1. *Summary of Project Details*

Project Phase	Value of Contract	Approx. Time of completion	% Time elapsed when the Last Planner System Implemented	Subcontractors involved in the last planner system implementation
UT Arlington College Park located on the University of Texas at Arlington Campus				
Phase – I	US $ 65 Million	12 Months	80%	None
Phase – II		13 Months	0%	Mechanical Architectural Structural Electrical Plumbing

The last planner implementation activity was completed in four stages. Stage 1 was allotted for workshop and observations on current practice and this was also the period where the Project Manager level discussions were conducted. The Second stage involved the Pull planning phase with the look ahead planning (a critical component of Last Planner). Reverse scheduling of the process is also done at this stage. This backward process was then visually depicted using the Gantt chart. Gantt Chart was developed for each building in the projects. Gantt Chart was developed by asking the inputs of the stakeholders in the construction. This enabled the analysts to find out the redundant processes and the processes that require high costs and time. This enabled a reduction of wastage. This step increases accountability of the process and also enables in determining the project in realistic conditions. The Third stage involved the short terms weekly planning. The Third stage is actually the longest stage of the processes. For this project this stage took about 17 weeks. It uses the short-term weekly planning compares it with the PPC. This helps in repetitive evaluation of the processes that are occurring. From this any discrepancy in the process or any issues can be determined in the preliminary phase of the issues and can be rectified. Any process, which was not completed within the preceding week, was carried over to the next and successfully completed in the current week. This enabled easy rectification of the issues. It was observed that by adopting this technique there was better performance in all the relevant processes. This benefited the contractors, companies and all the other stakeholders involved in this process. The final stage was the evaluation of the Last Planner system. For the evaluation the people involved in the actual processes was questioned orally and the people were also asked predefined questions in the form of survey questionnaire. This evaluation technique enabled the companies to identify the vulnerable areas. The problem areas was identified and rectified. The people were also questioned about their inputs. This would aid the future projects greatly. This is an important final stage, which allows the stakeholders to retrospect. All the four stages were given equal importance.

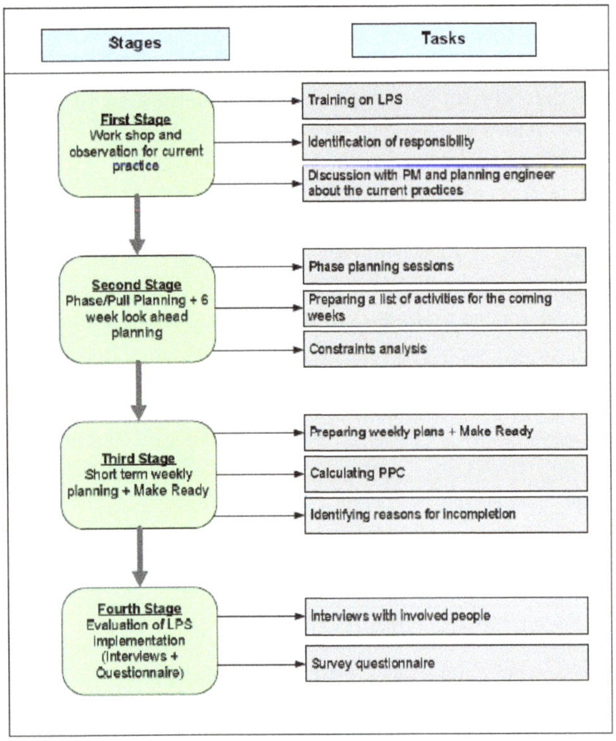

Figure 2. *The Last Planner System Implementation Strategy, UT Arlington College Park located on the University of Texas at Arlington Campus. (Source: Garza, and Leong, 2000) [5]*

5. Conclusion

The improvement of the PPC values indicates that the project performance improved with time. This improvement was noted after the Last Planner system which is a lean construction system was implemented in the Phase 1 and 2 of the projects. The case study hence indicates that the Last Planner system is able to improve efficiency in productivity. In improving project management processes by means of individually planned phases and look ahead weeks it also helps reduce the stress on project managers and hence is vital to construction project management. It should be noted that the last planner system does not focus on the issues from a unilateral angle rather it takes into account a multitude of factors thereby improving the processes. Also the documentation of events and process has greatly improved the accountability.

References

[1] Adamu, S., Hamid, R, A. (May 2012), Lean Construction Techniques Implementation in Nigeria Construction Industry, *Canadian Journal on Environmental, Construction and Civil Engineering*, 3(4), 186-193

[2] Arditi, D., and Gunaydin, H. (1997) Total quality management in the construction process." *Int. J. Proj. Manage.*, 15(4), 235–243.

[3] Ballard, G. (1994). The Last Planner. *Spring Conference of the Northern California Construction Institute, Monterey, CA*, April 22-24.

[4] Ballard, G., & Tommelein, I. (2012). Lean management methods for complex projects. Engineering Project Organization Journal, 2(1-2), 85-96.

[5] Ballard, G. (1997). Look ahead Planning: The Missing Link in Production Control, *Proc. 5th Annual Conf. Int'l. Group for Lean Constr.*, IGLC 5, July, Gold Coast, Australia, 13-26.

[6] Garza, J. M. D. L., and Leong, M. (2000). "Last planner technique: A case study." *Construction*

[7] Koskela, L. (September, 1992), Application of the New Production Philosophy to Construction, *CIFE Technical Report*, 72, 1-75.

[8] Melhado, S, B. (1998), *Designing for Lean Construction,* Proceedings IGLC.

[9] Mossman, A. (2013). *Last Planner: 5 + 1 crucial & collaborative conversations for predictable design & construction delivery.* The Change Business Limited

[10] Rother, M & Shook, J, 2003, *Learning to See*, Lean Enterprise Institute Inc., USA

[11] Salem, O.,Solomon, J.,Genaidys, A. & Minkarah, I. (October, 2006), Lean Construction: From Theory to Implementation., *Journal of Management in Engineering*, 168-175, Birmingham.

[12] Dave, B., Koskela, L., Kiviniemi, A., Tzortzopoulos, P., & Owen, R. L. (2013). *Implementing lean in construction: lean construction and BIM.*

[13] Garza-Reyes, J. A., Parkar, H. S., Oraifige, I., Soriano-Meier, H., & Harmanto, D. (2012). An empirical-exploratory study of the status of lean manufacturing in India. *International Journal of Business Excellence*, *5*(4), 395-412.

[14] Gao, S., & Low, S. P. (2014). The Toyota Way model: an alternative framework for lean construction. *Total Quality Management & Business Excellence*, *25*(5-6), 664-682.

[15] Howell, G. A. (1999, July). What is lean construction-1999. In *Proceedings IGLC* (Vol. 7, p.

[16] Koskela, L. (1997). Lean production in construction. *Lean construction*, 1-9.

[17] Melles, B. E. R. T. (1997). What do we mean by lean production in construction. *Lean Construction*, 24-29

[18] Ohno, T, (1988), *Toyota Production System: Beyond Large Scale Production*, Productivity Press, New York.

[19] Sarhan, S., & Fox, A. (2013). Barriers to implementing lean construction in the UK construction industry. *The Built & Human Environment Review*, *6*.

[20] Tommelein, I. D. (2015). Journey toward Lean Construction: Pursuing a Paradigm Shift in the AEC Industry. *Journal of Construction Engineering and Management.*

Stakeholder Management on Construction Projects: A Key Indicator for Project Success

Joseph Ignatius Teye Buertey[1, *], **Daniel Amofa**[2], **Felix Atsrim**[1]

[1]Department of Built Environment, Pentecost University College, Kaneshie, Accra, Ghana
[2]Department of Project Management, Accra Institute of Technology, Private Mail Bag, Accra, Ghana

Email address:
jbuert@yahoo.co.uk (J. I. T. Buertey)
[*]Corresponding author

Abstract: Participation is now an international agenda for ensuring full representation of people in terms of their ideas, feeling and decision on matters concerning their development. It has been observed that most projects fail after implementation not due to poor execution but rather due poor stakeholder consultation and engagement. The purpose of this study was to determine the barriers to stakeholder involvement in developmental projects at the grassroots level and examine the impact of stakeholder involvement on the success of projects implemented. Data was gathered through structured questionnaires distributed to ordinary citizens, community leaders and local authority staff in selected district assemblies in Ghana. Analysis of structured questionnaires revealed that there was inadequate explanation of the background, technical and material justification for the project to the stakeholders prior to project initiation. Stakeholders held that they had difficulty in participating in technical discussions and there was the perceived unwillingness of project implementers to involve them during decision making, to this end, the impact of stakeholders towards project success was significant. To overcome the challenge of stakeholder involvement and meaningful impact to projects, stakeholders must develop capacities to contribute meaningfully in discussions or delegate their concerns to professional representatives. To this end, projects implementers must acknowledge the value of stakeholders and embark on stakeholder outreach to solicit their involvement for enhanced project success.

Keywords: Stakeholder Involvement, Project Success, Development, Local Government, Communication, Community

1. Introduction

The project stakeholder is sine qua non for the measuring of project success. According to [1], stakeholders are individuals and organisations actively involved in the project, or whose interest may be affected as a result of the project execution or completion. Due to the interest of stakeholders on the project, they may exert influence on the project's objective and outcomes. To ensure a successful project, the project team must identify and engage all stakeholders, determine their requirements and expectation and manage their influence in relation to their requirements. Amponsah [2] postulates that Ghana as any developing country has been recording failure in most of their development projects attributable to poor stakeholder engagement. It is further argued that [3] and [4] holds that project failure is strongly related to stakeholders' perception of a project and their involvement in it.

The purpose of this study was to review the extent to which project actors (executors) intrinsically manages their stakeholders, identifying impediments to their involvement in project execution and expatiate solutions and innovations to guide and positively influence daily practice of stakeholder involvement for development at the local level. The results would enable the leaders at the lower level to be more aware of the extent of stakeholder involvement and to find the appropriate steps in involving them much more. Furthermore, the result of the study is also beneficial to the stakeholders at the district level so that in future projects, they will make themselves available for project planning and implementation. This will ensure that they are made part of the decision of choosing projects for the community and effectively support the project and improve the acceptance criteria for the project. In other words, organizations can anticipate what their stakeholder's reaction to the project is likely to be so that they can build into the plan of action to win the stakeholders involvement.

Stakeholder literature has offered various conceptualizations and descriptions. According to [5], a stakeholder is a person, group or organization that has interest or concern in an organization. Ward and Chapman [6] postulates that *"voluntary stakeholders bear some form of peril as a consequence of having invested some form of capital, human or financial, something of value, in a firm. Involuntary stakeholders are placed at risk as an outcome of a firm's activities. But without the constituent of risk there is no stake"*. Thus stakeholders have an interest in the actions of an organization and have the ability to influence it. Others like [7] expatiate on the need to emphasize the authenticity of stakeholder relationships in the stakeholder definition as: having some legitimate, non-trivial relationship with an organization, such as exchange transactions, action impacts, and moral responsibilities. This description highlights the environment of relationships between stakeholders and the organization.

The relationships between stakeholders and firms have also been distinct either more broadly or more intently. These views approve a strategic perspective and emphasize the fact that companies have only limited resources and imperfect time that they can spend on dealing with their stakeholders. Therefore, it is in the concentration of management to categorize and pay attention to those stakeholders who have significance on organizations economic interests. Stakeholders can also be defined through their casual relationships and moral claims towards the business. These views regarded as the growth and sustainment of moral relationships with stakeholders as the firm's responsibility. Studies have suggested a diversity of stakeholder classification schemes. Altman [7] mentions that stakeholders can be classified as either claimants or influencers and consider the potential of stakeholders to threaten or cooperate with the organization. Stakeholders can also be separated into internal and external stakeholders, or primary or secondary stakeholders. Internal stakeholders are, for example, employees, customers and shareholders, while examples of external stakeholders are community activists, media, advocacy groups and other non-governmental organizations [8]. Secondary stakeholders are not directly connected with the crucial business because they lack a "formal contractual bond with the firm" or "direct legal authority" over the firm. However, primary stakeholders, such as employees and customers, are in an undeviating association with the firm, indulged in transactions with the firm or have direct legal authority over the firm. In turn, secondary stakeholders are not directly engaged with economic activity, but are still able to impact an organization. Moral and legitimate claims are often emphasized in correlation with less important stakeholders.

Differently stated, [8] however held that stakeholders may also been separated into strategic and moral stakeholders. Strategic stakeholders are measured to be able to impose decisions on the firm, thus the management of their interests is essential. Moral stakeholders are those who are rather affected by the firm. Other works however disagrees with earlier stakeholder conceptualizations and categorizations for uncertainty and suggests that a difference should be made between stakeholders, stake watchers and stake keepers. In terms of classification, stakeholders are those who have a tangible and real stake in a company. Stake watchers, in turn, do not actually have a stake themselves but they guard the interests of actual stakeholders. Examples of stake watchers are unions and community pressure groups. Finally, [8] postulates that stake keepers are the autonomous regulators who have no stake in the firm but have influence and control, such as governments, regulatory agencies and certification organizations [9].

2. Empirical Review

2.1. Stakeholder Involvement in Developmental Projects

The role of the stakeholders is to produce a system that best meets their needs, be willing to work with others, particularly those outside your chosen specialties, share all information, including "work in progress", and to actively expand their knowledge and skills. It can be upheld that the role of project stakeholders may include but not limited to: provide resources (time, money, etc.) to the project team, educate project implementers about their community, spend time to provide and clarify requirements, be specific and precise about requirements, make timely decisions with respect to project implementers assessment of cost and feasibility, review and provide timely feedback regarding relevant work artifacts of project implementers and promptly communicate changes to requirements [5]. These roles effectively define the relationship between project team and its stakeholders, a relationship that must be honored for a project to be successful.

The quantity of alternatives to top-down governance is myriad; however, participation of stakeholders in all stages of projects (agenda setting, policy formulation, decision making, policy implementation, policy evaluation) appears to have gained extensive endorsement [9]. Stakeholder involvement is a procedure which captures both specialist and lay knowledge in project management. Although many shades of participation exist, it is noteworthy that all forms embrace one kind of stakeholder involvement or another. The critical question is which variety of stakeholder involvement is appropriate for a particular setting and what level of intensity is deemed sustainable [10]. Numerous researchers argue for instance that stakeholder involvement can improve project management process [11]. Another school of thought questions this assertion and suggests that stakeholder involvement is not necessarily an improvement on the original top-down governance approach. It cannot be overemphasized that stakeholder involvement requires consensus-building, negotiation, conflict resolution, trade-offs and holistic thinking and these issues are frequently time consuming and expensive, irrespective of the scale [10]. The opposing views in the on-going debate imply that there is the need to look at the issues at stake from the arena of complexity since either way, a mere reversal of governance approach seems to be insufficient to tackle multi-dimensional problems.

In the case of stakeholder involvement in projects, this is instrumental in the sense that it is a means to an end. The end or goal is behavioural change on the part of the stakeholders. It is however simplistic to assume that information access through stakeholder involvement will necessarily translate into behavioural change. Regardless of one's position on the matter, it seems evident that a certain gauge for evaluation is necessary to inform the position an individual takes on the success or failure of the participatory process. According to [12], any kind of evaluation has to capture either the process goals or the outcome goals. An evaluation which focuses on process goals may consider the participatory exercise triumphant if the socio-economic dissimilarity gap is bridged. At the same point in time, an assessment which focuses on result goals may see the exercise as ineffective because the project was a disappointment. In either case, the conclusion is subjective because it hinges on the values of the individual conducting the assessment. Owing to the malfunction of development projects in the 1950s and 1960s, social workers and field activists began to call for the inclusion of populations concerned with development in project design and implementation [13]. The perception then was that such projects were ineffective because local populations were left out of the decision making process; a state of affairs that tended to perpetuate social inequality. Hence stakeholder involvement was proposed as an apparatus to uphold equality through empowerment inclusion. Remarkably, even though stakeholder interest was projected to address inequality, the literature is stuffed with cases where it instead perpetuated inequality [14].

Central to the course of participation is the issue of mediation, which involves a discursive and impulsive connection between the law and democracy [15]. According to [16], the concept of discursive democracy within which stakeholder involvement resides, is becoming the normal practice in the management of projects. The conjecture of discursive democracy transcends forthcoming action as it encapsulates a theory of law and democratic institutionalization. The theory considers *'democracy as not being ingrained in civic society or popular autonomy, but in the structures of communication, for which Habermas takes for granted the prospect of consensus and argumentative discourse'* [15]. In discursive democracy, equal access to the discursive platform is a fundamental notion [16]. However, [17] argues that this assumption in reality may not always be the case. For instance, in Ghana, the history of stakeholder involvement shows that this is not the case [8]; therefore, effective monitoring of the involvement process is indispensable to ensure equivalent entree to the discursive territory.

Delanty [16] argues that Habermas perceives popular or representative democracy as inadequate in grasping the complications in modern plural societies. In the same vein, it is also argued that popular democracy fails to 'take account of the actuality of multi-cultural pluralism, which challenges both the concept of unity of the civic community and the appeal to legitimacy on the basis of popular sovereignty' [15]. These shortcomings of popular democracy underpin Habermas' proposal of the theory of discursive democracy of participation. The theory is noteworthy for a number of motives. First, it recognizes the authenticity of multi-cultural value systems. Next, it takes into account the problem of complication in modern societies. Finally, it identifies the question of law and institutionalization [17]. In essence therefore, the theory of discursive democracy of stakeholder participation takes into account both agency (of individuals to make decisions) and structure (the establishment of new institutions through participatory practice) [15].

2.2. Stakeholder Involvement in Different Stages of Projects

Development projects often start through multiple stages in which different stakeholders may be involved [18]. Active stakeholders have a significant influence in the community and on the project, make choices that directly affect others, and are necessary for recruiting other stakeholders. In contrast, passive stakeholders are influenced by the choices of active stakeholders. Primary stakeholders are active decision makers that have great influence in the community and also on the project, and must be involved in all the project stages. Secondary stakeholders are mostly passive, undertaking decisions made by primary stakeholders and have little influence during the implementation phase.

The planning and design stages of projects which is supposed to be the most important phase, paradoxically involve only few stakeholders and are most often the project gatekeepers. This phase of the project interestingly takes the shortest duration, mostly neglecting the primary stakeholders. The implementation stage mostly involves almost all the key stakeholders including the primary stakeholders. The monitoring and evaluation stage of developmental projects, which is an evolving stage of the implementation phase include all stakeholders involved in the prior stages in an attempt to gauge the long-term success of the project, generally ensuring that most stakeholders are considered and encourages their active involvement in the project [19]. Though stakeholder involvement must not be fixed or rigid, identifying where and how their input is needed, is necessary particularly in multi-stakeholder projects as it is not possible or practicable to involve all stakeholders passionately at every stage of the project. While it is not all stakeholders that have the same power and influence, it is believed that it is crucial to identify at the early stages of the project, the primary stakeholders who exert most influence and are most significant to the efficient project completion [20].

2.3. Factors That Influence Stakeholder Involvement in Development Projects

Many dissimilar forces are at work, proceeding to, and throughout, a public partaking programme, with intrinsic features moving following features have moved the participatory deliberation forward. People generally resist change, especially when they do not comprehend or approve to the goals, the devices, the sponsor or the timing of the anticipated change. So, keeping the public in the dark is often a recipe for disaster [21]. The citizens are also progressively unenthusiastic to defer to 'expert' government agency opinions, and are unwilling to act as sounding boards for bodies that have already made decisions, particularly when they affect their local communities [22]. These tendencies lead to a lack of public buoyancy and trust, except the relevant authorities take note of them. A far-reaching programme ensued, based on meaningful involvement and fair management for all races, cultures and incomes in project decision-making. The transmittance of 'experience', rather than knowledge, is the critical neglected dimension of decision-making, particularly as it relates to project decision-making [23]. The further the project implementation is from stakeholders, the more pertinent becomes the public participation in same, and in this context, face-to-face meetings are critical to producing real learning and trust.

Jergeas et al., [24] found two factors that influence stakeholder's involvement in development projects; communication and setting common goals, objectives and project priorities. Effective communication is very vital for getting the support and commitment of stakeholders [25]. Regular communication with members of the project community is very important for effective project performance [26]. According to [27], the performance of development projects and its ability to satisfy stakeholders is dependent on decisions that are made and the care taken by policy-makers in stakeholder communication. In the view of [28] the key issue in project stakeholder involvement is about managing and promoting the friendship and relationship between the project and its stakeholders. It can be held that successful relationships between stakeholders and the project are important for successful implementation of projects. Trust and commitment between project implementers and stakeholders can be developed and sustained through an effective relationships management [19].

2.4. Benefits of Stakeholder Involvement in Project Implementation

The actualization of project goals and the improvement of projects benefits to the community is the focal aim of stakeholder participation in projects. It is intended to advance the quality of both the procedure and the end decision. The participation process is also a learning curve for all the stakeholders involved, particularly if there is an unrestricted flow of information between the parties. For the public it can often be the first experience of taking a vigorous part of the project implementation process. Projects benefit from the direct and immediate knowledge held by citizens and business, concerning project implementation environmental in their communities. Encouraging the public and other stakeholders to share their knowledge, with the regulatory authorities, fosters better-informed decisions and decreases the likelihood of project failure. Out of the 25 overseas projects sponsored and evaluated by the World Bank, 13 failed mainly as a result of lack of local input [29]. It should be realized that wisdom is not limited to scientific specialists and government officials, and that rational analysis carried on in ignorance of political reality, may well end up so divorced from social reality, hence may be of little use to anyone [23].

Stakeholder involvement gives broader perspectives on a particular process, and early involvement gives added time to study issues and develop the process, enhances credibility of the decision making process, fosters early identification of the diverse perspectives on the issues of concern and the generation of solution options [29]. Better-designed projects, which avoid costly delays in appraisal and implementation, can also result from early and planned consultations and stakeholder involvement [30]. For the proponent, early stakeholder involvement can have the added benefit of diffusing opposition to a project. If a broad based consensus is built, it can also lead to a public sense of 'ownership' [31]. The public's enthusiasm is potentially a powerful motivating force for project implementation. Stakeholder involvement can also supplement scarce government monitoring, inspection, enforcement and resources [30]. Stakeholder involvement in projects is a learning experience which gives the public insight into the governance process. If the experience is good, it powers the way for future co-operation. The converse can also be true.

2.5. Problems Encountered in Stakeholder Involvement in Project Implementation

There is limited information on the relative costs of using the different methods of stakeholder involvement practice, although some attempt is made to quantify it. Actual costing for specific processes are given in a number of studies [32]. Clearly deliberative processes which are used to engage relevant stakeholders in debate, discussion and deliberation, could only be done as part of another research. The intensity of the stakeholder process, including commitments of time, energy and money, and often, uncertain results, has led to a burnout phenomenon amongst many participants, and from all stakeholder groups [33].

Excessive raised expectations about Stakeholder involvement can lead to frustration and anger with a breakdown in trust. It is essential then to be clear with all participants as to the limits of influence of the process. Where an infrastructure project crosses planning jurisdiction boundaries or where more than one official is involved, the costs to the public, state and advocates are multiplied. The amplified time scale and disbursement involved in stakeholder involvement is of particular worth in projects with a small capital funding. The foremost areas of struggle affecting stakeholder involvement are attitudinal, lack of capability to deliver programmes, lack of clarity about what outcomes are possible and the lack of a legislative framework. The challenge of accessing the necessary resources for a worthwhile stakeholder public participation process, whether by the advocate or the regulator, by the commitment of personnel and finances, is essential. The repercussions for project proponents, of not having a properly planned and funded participatory project process, may prove much costlier than dealing with the backlash. In many circumstances, the choice increasingly is not whether to involve the public, but how to get the best value from the chosen process [32]. To avoid hearing from only the activist or the powerful elite, and in order to get the widest sweep of opinion and information, authorities must reach out into the community [34]. In this regard some activities are best avoided, for example public meetings. In the absence of trained facilitators these can often come down to 'he who shouts loudest wins!'

Some institutions resist change, and communication blocks within stakeholders can prevent them from making head way. The established routines and organizational systems of many bodies seek to promote the 'status quo'. Organizational structures which support sustainability, and therefore public participation, are a threat to 'command and control' style management systems. Managers, who have risen to power in this style of structure, frequently resist the transition to alternative structures, which embrace public consultation [35]. Many opportunities for stakeholder participation are laid down in the environmental permissible framework, where legislation does not specify early stakeholder involvement, then it is at the discretion of the relevant authority, and dependent on the opinion of the relevant public servant. For example, in the growth planning process, the Manager's reporting of the consultative process clearly includes the names and nature of all submissions together with his response to them.

3. Methodology

The study population involved the people of the Ada East and Gomoa East District Assemblies specifically the assembly members, community members and the District Planning Coordinating Unit. A sample size of ninety (90) was determined using statistical of Kumar [36] methods. A structured questionnaire using both open-ended and closed-ended questions were developed and administered by hand delivery to respondents, yielding a response rate of 69%. In cases where the responses to a question were unlimited and could not be adequately answered with few words, or where the responses concerned qualitative and opinion related issues, open ended questions were used to capture the response. This facilitated large collection of data within the shortest possible time. It also provided the opportunity for respondents to express their opinions and views. The questionnaire was designed to cover complete assessment and understanding of the phenomenon under investigation with the following main objectives:

- Understand the factors/barriers that influence/affect the stakeholder involvement process in development projects?
- Examine the impact of stakeholder involvement on the performance (success or failure) of projects implemented?
- Recommend ways of improving stakeholder participation in the decentralization process in Ghana?

4. Analysis and Discussion of Data

The quantitative data was analysed based on the five-point scale ratings provided by the respondents on separate categories. These ratings were combined to deduce the relative importance indices of the factors based on respondents ratings, after which further analysis were made to compute the overall weightings and interpretations based on respondent's view on barriers to stakeholders involvement, and how to improve stakeholders involvement based on the total sample size. The significance testing was used to decide whether to accept or reject the null Hypothesis, Ho. An evaluation of the test statistics (X) was done and the probability (P-Value) of observing a value of the test statistics was also determined. The P value was taken as the smallest value at which the significance level (α) could be present and still have small (lesser than 5%) significance level.

The sample mean for the data in respect of each factor and the effect of variation are shown in the table 1. The 5-point rating (1, 2, 3, 4 and 5) have mean of 18 and varying standard was used based on varying dispersion of respondents' view. The p-value for the test was determined to find out if there was much difference between the null value of μ=18, and the sample means in table 2 to cause the rejection of the Ho. The profitability of observing the sample mean or larger $\mu = 18$ and σ was computed. The test statistics (X) was the central limit theorem, where x is approximately normally distributed with mean μ and Standard Deviation, $\sigma/\sqrt{n}$ where n = number of responses for that factor. The p-value was obtained using the relation below.

$$P = (x \geq \mu) = p\left(z \geq \frac{x-\mu}{\frac{\sigma}{\sqrt{n}}}\right) = 1 - P\left(z \leq x - \frac{\mu}{\frac{\sigma}{\sqrt{n}}}\right)$$

From the cumulative distribution standard normal table, where Fz (z) = P [Z ≤ z], the value of z ranges form -3.9 to 3.9. Any value of z less than -3.9 has a Fz (z) of zero (0) whereas vales of more than 3.9 has Fz (z) of unity (1).

The P value is the smallest level of significance for which the observed data would call rejection of Ho in favor of H_1. The P value gives additional insight into the strength of the decision taken. Thus a relatively small p-value of 0.001 indicates that there is little likelihood that Ho is true. On the other hand a high p- value such as 0.2033 means that Ho is not rejected and there is little likelihood that it is false.

The P-value is often referred to as the observed level of significance for a given level of significance, α

Reject Ho if p-value ≤ α

Do not reject Ho if p-value > α

Thus the larger population has a distribution of P~X [μ = 18, σ =x]

For example the P- value for 'inadequate explanation of background and technical material is calculated as follows

$$P[X \geq 4.696] = p\frac{[307 - 18]}{\left(\frac{7.2}{\sqrt{90}}\right)}$$

$$P\,[Z \geq 1.859] = 0.0955$$

Since the null hypothesis Ho= 0.05< 0.09550, we accept Ho, thus the factor 'inadequate explanation of background and technical material' is an important factor affecting stakeholder involvement in project.

Table 1. *Statistical analysis of Barriers to stakeholder involvement In community Projects.*

No	Descriptors	1	2	3	4	5	Rel Imp	Weight	Rank	Mean	SD	VAR (σ)	STAND (P1)	p-value
A	Respondents view on barriers to stakeholders involvement													
A1	Inadequate explanations of background and technical material?	12	15	12	27	24	307	0.682	4	18	51.84	7.2	1.85	0.0955
A2	Poor Accountability on the part of implementers	6	28	15	29	12	282	0.626	7	18	99.63	9.94	1.62	0.9475
A3	Inadequate stakeholder consultative meetings	18	23	27	11	11	243	0.540	9	18	53.46	7.31	2.05	0.0202
A4	Perceived inability to influence issues	12	11	18	26	23	307	0.684	3	18	44.418	6.9	2.55	0.1100
A5	Lack of time to participate?	11	38	22	12	8	239	0.530	10	18	148.63	12.1	1.22	0.0131
A6	Failure to incorporate stakeholders view during scope definition	25	22	18	14	11	234	0.520	12	18	32.5	5.7	2.58	0.0561
A7	Poor communication between implementers and users (challenges in accessing information)	11	22	21	15	21	283	0.629	6	18	23	4.8	3.39	0.8641
A8	Lack of Technical capacity and support on the part of stakeholders	16	13	15	24	22	293	0.651	5	18	22.5	4.74	3.50	0.9970
A9	Inability of project implanters to identify all stakeholder	39	21	15	5	10	196	0.436	13	18	173	13.2	1.01	0.0846
A10	Low enthusiasm on the part of the local stakeholders (apathy)	22	22	15	14	17	252	0.560	8	18	14.5	3.81	4.02	0.0537
A11	Feeling of powerlessness	10	11	19	25	25	314	0.698	2	18	53	7.28	2.36	0.9960
A12	challenge of accessing resources to enhance stakeholder efficiency	27	15	16	21	11	244	0.542	11	18	38	6.16	2.44	0.0997
A13	Unwillingness of project implementers to involve stakeholders?	7	11	25	14	33	325	0.722	1	18	115	8.3	1.63	0.2500
B	How to improve stakeholder involvement in projects													
B1	Stakeholders must develop their capacities to contribute meaningfully to project discussion?	2	5	2	26	55	397	0.882	3	18	527.715	23	0.85	0.8023
B2	Project implementers must adopt the use of technology to improve stakeholder involvement?	2	7	5	40	36	371	0.824	4	18	332.1	18.2	1.03	0.8707
B3	Project implementers must embark on stakeholder outreach to solicit for their involvement?	2	4	1	34	49	394	0.879	2	18	496.4	22.5	0.87	0.1920
B4	Project implementers must acknowledge the value of stakeholders towards the success of projects?	1	2	2	23	63	415	0.922	1	18	715.63	26.8	0.74	0.2200
C	Respondents view on projects implemented in the community													
C1	Are projects implemented needed by the community?	5	7	22	20	36	344	0.764	1	18	153.9	12.4	1.45	0.3500
C2	How regular are projects implemented in the community used?	13	16	14	21	26	302	0.670	2	18	29.565	5.44	3.10	0.9900
C3	Will you have preferred other projects to the current ones?	11	16	20	14	29	304	0.676	3	18	46.98	6.85	2.47	0.1100

Although participation of stakeholders is seen as critical for the project success, there exist variety of obstacles. To understand the barriers that exist at the district level, a series of questions were developed and administered to respondents. Based on statistical analysis from table 1, it was deduced that the stakeholder opined that the main barrier to their involvement in projects is the unwillingness of project implementers to involve them. With a p-value of 0.0025, the factor can be deployed as a very significant factor. Using relative importance indices, it was weighted 0.722 and ranked first. It can thus be deduced that most project are initiated without the involvement of the end users. These projects are planned from above by the sponsor or financiers and rolled at the lowest level without seeking their involvement and inputs. The research by [37] [38] and [3] affirms the above that the strongest challenge to stakeholder involvement is recognition, this has been postulated to be the attributor to project failure at the lowest level.

With a p-value of 0.0996, analysis of field studies revealed that the end usser stakeholders' feeling of powerlessness is a reason for their non-involvement in project executed. This factor was analysed to be second most significant factor affecting stakeholder involvement in projects with a relative importance of 0.698. The feelings of powerlessness engendered by these hurdles, add to public perceptions of a lack of influence at both lower and higher levels of local development. Consultations of interested parties.... can only ever supplement, and never replace, procedures and decisions of legislative bodies which possess democratic legitimacy. This sense of powerlessness may be part of the reason why, even in cases where considerable energy and resources are expended, to identify individuals and groups, only a small proportion of the public ever attend participation programmes [32]. Stakeholder involvement may actually result in an increased level of conflict. Becoming involved in a consultative process where the decision has already been made, or where the possible outcomes are not made clear at the outset, can lead to a great deal of frustration. Anger may follow when the input of the public is ignored, especially following prolonged constructive engagement. Conflict can also occur between professionals. One source of such disagreements could evolve from the different emphases of 'scientific peer review' and 'social peer review'. The scientific peer review process is well established, and is essential to assess the technical information provided to policy makers. The corresponding social peer review, designed to obtain societal acceptance and legitimacy for the decisions rendered, by contrast, does not have a set of acknowledged professional standards [29]. It should also be remembered that whilst science may strive for the 'truth', the opinions of scientists are colored by their values and beliefs. The resulting deadlock exasperates the capability of planners to accomplish consensus on preliminary issues and contributes to anti-participation attitudes.

Therefore, rather than attitudes as an irrational response to glitches ordinary citizens cannot grasp, the solution is to inspire more participation not less [38]. The misconception that stakeholders are overly emotional, ill-informed and solely encouraged by selfish interests in their impediment of facilities for the common good leads to an information-based strategy, the postulation being that, 'if only stakeholder knew the truth!' This assumes the existence of an objective truth to know [32]. The conflict that arises from having dissimilar bodies making decisions on different parts of the project execution can often confuse the public understanding and increase frustrations.

Closely linked with the second important factor is the perceived inability to influence situation. Analysis of field data revealed that 67% of respondents indicating their opinions and views shared on how to improve the post implementation usage of the project at stakeholder consultative meetings were neither factored any nor reasons given for declining to adopt them. Thus stakeholders perceive that they have very low ability to influence situations since mostly at the execution stage where they are involved, the scope definition for the project has been sealed. According to [38], public 'hearings' often do not comprise 'listening'. In a research by [39], citizen's efforts and ideas are not included in offers and no reason given. Similarly, public hearings are criticized for: being apprehended at times inopportune for the public, establishing an atmosphere that inhibits negotiation, and conducting proceedings that intimidate the public. The lack of an obviously definite purpose can also mean that the role of the public is every now and then uncertain not only to the public but to those delivering the process, with the significance that the participation process chosen is often unsuitable for the specific project [22].

At 5% level of significant, the hypothesis 'inadequate explanation and background of technical materials' was tested to be true, with a p-value of 0.0995. This is closely linked with the fifth most significant factor, 'lack of technical capacity and support on the part of stakeholders. Some significant majority of respondents about (57%) agreed to the assertion that the barrier to stakeholder involvement was perceived to be linked to their low technical capacity. According to a report by [29], there is a universal lack of experience of partaking processes, and very few stakeholders have experienced planned participation processes. There may be suspicion, cynicism, or lack of enthusiasm, but there is unlikely to be previous training. Suspicion about participation could be from politicians, who may feel their power is being diluted, or from NGOs that have very explicit and stationary viewpoints. Cynicism could be from prior public participation processes which were ill managed, leaving participants unenthusiastic to try again. Stakeholders are exasperated when they are treated as antagonists, rather than comfy participants in the project. Lack of technical support for stakeholders, and difficulties in getting access to information can diminish the ability of the public to play an eloquent part in project processes. Stakeholder access to sovereign expertise in a particular technical/scientific field can prove difficult, particularly where the proponent of a project is a monopoly employer. Stakeholders with no technical/scientific experience find it difficult to access, comprehend or evaluate data and information, especially under these short time constraints.

One factor highlighted as significantly important is the poor communication between project implementers and stakeholders. With a p-value of 0.56, the hypothesis: poor communication is a barrier to stakeholder involvement was tested to be true. Interestingly, over 65% of respondents indicated that during project execution, communication between the various stakeholders is very poor. This is an indication that there is the need to undertake proper briefing of stakeholders if their involvement is to be felt and not necessarily the provision of adequate minutes. Also it provides practitioners with an idea about the need for circulation of message to the different stakeholder. That is to say the document for an educated person cannot be the same for an uneducated person. Secondary data presupposes that stakeholders have all the time available but implementers just don't involve them or communicate effectively with them. The challenge of poor communication may be seen as a recipe for the poor accountability the stakeholders highlighted. When people are left in the dark, they make all sort of assumption.

On the question of the need, regularity of use and preference to other projects, an overwhelming majority of respondents (89%) indicated that projects developed in the district were projects which meets their needs. An approximately 82% of the respondents to questionnaire administered further held that the projects developed in the community were effectively in use by the community. Interestingly over 58% of respondents indicated that would have preferred other projects to the current ones.

The above interesting assertion can be explained to mean that though projects executed were important to the local communities, these projects could have been deferred to a later date making way for their actual pressing needs. This interesting scenarios is eminent when governments and people in authority plan for the need of the people at the grassroots without duly engaging them.

On the issue of how to improve stakeholder involvement, solicited and analysed views from stakeholders are skewed towards the need for implementers to acknowledge the value of stakeholders towards project success. At 5% level of significance, this was tested with a p-value of 0.022. closely linked to the above is the request for project implementers to embark on stakeholder outreach to solicit views prior to project commencement.

5. Recommendations and Conclusion

This study provided evidence to support or dispute earlier studies made into the field of stakeholder involvement in the decentralization process in relation to project planning and implementation. Previous studies have examined stakeholder involvement in project planning and implementation [9], [11] and [13], factors influencing their involvement by [20] and [25] benefits of stakeholder involvement [20] and [31]. Other studies have also looked into the problems encountered in stakeholder involvement [32] and [38]. The result from this study suggests that stakeholder involvement in project planning and implementation significantly affect the success or otherwise of the project. This confirms earlier research made indicating that community participation in the design and implementation of a project greatly enhances the likelihood of project success. Further studies also indicate that generally, ensuring that stakeholders are considered encourages their active involvement in the project [19]. However, the study indicate that respondents put emphasis on building trust, agreeing on rules of engagement and adhering to advice from stakeholders as critical to the success or otherwise of the project. Thus, even if the involvement would be limited, the little engagement made should be considered important by implementers and the advice from stakeholders incorporated into the process.

The results of the survey conducted indicate that there are varied factors that inhibit the involvement of stakeholders in the development of projects at the district. According to findings, the key barriers that influence stakeholder's involvement in projects were; inadequate explanations of background and technical material and unwillingness of project implementers to involve stakeholders and not the lack of effective communication or regular communication as suggested in the study undertaken by [26] and [27] respectively. The survey data and analysis indicated that although implementers may have regular meetings with stakeholders, there is little attention paid to the advice provided by stakeholders to the project. Another major finding of the study was that it was not project stakeholders who were unwilling to contribute to the process of project implementation but rather due to the unwillingness of project implementers to involve the stakeholder, the perceived inability of stakeholders to influence issues, unwillingness of project implementers to involve stakeholders amongst others. The study also shows that although stakeholder may be involved in planning and implementation, their involvement is limited, since their views to improve the performance of the project was not acknowledged by implementers. A significant proportion of respondents indicated that though projects executed in the community are regularly used by the citizenry, they would have preferred other projects to the current ones.

There is debate in the literature over whether the performance of development projects and its ability to satisfy stakeholders is dependent on decisions that is made and the care taken by policy-makers in stakeholder communication [28] as against building trust amongst stakeholders and implementers. However, it is suggested that all approaches are important given that the data received from the survey indicated that majority of respondents believed that trust building and communication was an issue between stakeholders and implementers. It is often noted in the research that the feelings of powerlessness engendered by these hurdles, add to public perceptions of a lack of influence at both lower and higher levels of local development.

According to [32], this sense of powerlessness may be part of the reason why, even in cases where considerable energy and resources are expended, to identify individuals and groups, only a small proportion of the public ever attend participation programmes. However the study undertaken proved that not to be the case. A deliberate question was asked targeted at investigating this assertion, however the findings indicated that majority of respondents disagreed with that assertion but rather placed emphasis on the willingness of project implementers.

The results from the analysis of capacity gaps amongst stakeholders confirmed varied studies undertaken which indicate that there is universal lack of experience of partaking processes, with very few stakeholders ever experiencing the planned participation processes. On the other hand, in terms of the ability of stakeholders to comprehend technical documents, the study found out that more than half of respondents interviewed were unable to understand background and technical material of projects and as such found it difficult to evaluate data and information, especially under short time constraints. This constraint was seen as critical in that it was one of the most ranked issue needed for improving stakeholder involvement, with approximately 90% of respondents stating that it was needed.

To improve stakeholder participation and impact on projects implemented, there is the need for:

- Implementers to consciously identify all stakeholders to the project.
- Project implementers to acknowledge, value and engage these stakeholders
- Enhance communication between the stakeholders
- The application of new technologies to entice and improve stakeholder involvement in project development and implementation.
- Improve the technical capacities and support for the project.

With the improvement of stakeholder participation in projects implemented, the study found out that taking advice from stakeholders was more likely to result in projects been used in economic proportions in the district or community for which the project is implemented.

References

[1] Project Management Institute (2013). *A guide to the project management body of knowledge* (5th ed.). Newtown Square, Pennsylvania: PMI Publishing Division.

[2] Amponsah Richard (2012). *The Real Project Failure Factors and the Effect of Culture on Project Management in Ghana.* ICBE-RF Research Report No. 45/12.

[3] Suer, C. (1993). *Why information systems fail: A case study approach.* Henley-on Themes, UK: Alfred Waller.

[4] Lemon, W. F., Browitz, J., Burn, J. & Hackney, R. (2002). *Information systems failure. A comparative study of two countries.* Journal of Global Information Management, 10 (2), 28-40.

[5] Achterkamp, M. C., Vos J. F. J., 2008. *Investigating the use of the stakeholder notion in project management literature, a meta-analysis.* International Journal of Project Management, 26 (7), 749-757.

[6] Ward, B Chapman, A (2008). *A stakeholder approach to strategic management.* In: Freeman, R. E., Harrison, J. S., (Eds.), the Blackwell Handbook of Strategic Management, Blackwell, Oxford, 189-207.

[7] Aaltonen, K.; Jaakko, K.; Tuomas, O. (2008). *Stakeholder salience in global projects*, International Journal of Project Management 26: 509–516.

[8] Botchway, K. (2001). *Paradox of Empowerment*: *Reflections on a Case Study from Northern Ghana.* World Development, 29 (1), 135-153.

[9] Giordano, R., Passarella, G., Uricchio, G. F., & Vurro, M. (2007). *Integrating conflict analysis and consensus reaching in a decision support system for water resource management.* Journal of Environmental Management, 84, 213-228.

[10] Fraser, et al., (2006). *Bottom up and top down: Analysis of participatory processes for sustainability indicator identification as a pathway to community empowerment and sustainable environmental management.* Journal of Environmental Management, 78, 114-127.

[11] Prager, K., & Nagel, U. J. (2008). *Participatory decision making on agri-environmental programmes: case study from Sachsen-Anhalt.* Land Use Policy, 25, 106-115. Germany.

[12] Murdock, B. S., Wiessner, C., & Sexton, K. (2005). *Stakeholder Participation involuntary Environmental Agreements: Analysis of 10 Project XL Case Studies.* Science, Technology and Human Values, 30 (2), 223-250.

[13] Rahnema, S. (1992). *Work Councils in Iran: The Illusion of Worker Control.* Economic and Industrial Democracy, 13, 69-92.

[14] Lizarralde, G., & Massyn, M. (2008). *Unexpected negative outcomes of community participation in low-cost housing projects in South Africa.* Habitat International, 32, 1-14.

[15] Lotz-Sisitka, H., & Burt, J. (2006). *A critical review of participatory practice in integrated water resource management.* Johannesburg: South Africa Water Resources Commission.

[16] Delanty, G. (1997). *Habermas and Occidental Rationalism: The Politics of Identity, Social Learning, and the Cultural Limits of Moral Universalism.* Sociological Theory, 15 (1), 30-59.

[17] Palerm, J. (2000). *An Empirical-Theoretical Analysis Framework for Public Participation in Environmental Impact Assessment.* Journal of Environmental Planning and Management, 43 (5), 581-600.

[18] USAID (2001). *Policy Implementation: What USAID has learned*, at http://www.usaid.gov/our. Accessed 8/10/2014.

[19] Stiglitz, J. (1998). *Towards a New Paradigm for Development: Strategies, Policies, and Processes.* Proceedings from the Prebisch Lecture at UNCTAD at https://ceaemgmt.colorado.edu accessed 08/10/2014.

[20] Bourne, L.; Walker, D. H. T. (2005). *Visualizing stakeholder influence* – two Australian examples. Project Management Journal 37 (1): 5–22.

[21] Connor, D. M (2003). *Preventing and Resolving Public Controversy*. Commission of the European Communities. Brussels.

[22] Breggin, L. and Hallman, H. (1999). *Building Capacity to Participate in Environmental Protection Agency Activities*. Environmental Law Institute.

[23] Barkenbus, J. (1998). *Expertise and the Policy Cycle*. Energy, Environment and Resources Center. The University of Tennessee.

[24] Jergeas, G. F.; Williamson, E.; Skulmoski, G. J.; Thomas, J. L. (2000). *Stakeholder management on Stakeholder management on construction projects*. AACE International Transactions 12: 1–5.

[25] Briner, W.; Hastings, C.; Geddes, M. (1996). *Project Leadership Aldershot*. Gower.

[26] Cleland, D. I (1995). *Project management strategic design and implementation*. McGraw-Hill, New York.

[27] Landin, A. (2000). *Impact of Quality Management in the Swedish Construction process*. PhD Thesis. Department of Construction Management. Lund University.

[28] Aaltonen, K.; Sivonen, R. 2009. *Response strategies to stake-holder pressures in global projects*. International Journal of Project Management 27: 131–141.

[29] E. L. I. Research Report. (1997). *Transparency and Responsiveness: Building a Participatory Process for Activities Implemented Jointly Under the Climate Change Convention*. Environmental Law Institute.

[30] Bisset, R. (2000). *Methods of Consultation and Public Participation*. EIA in Developing and Transitional Countries. John Wiley and Sons.

[31] Acland A. (2002). *Guidelines for Stakeholder Dialogue*. The Environment Council London.

[32] Petts, J. and Leach, B. (2000). *Evaluating Methods for Public Participation: Literature Review*. R & D Technical Report E135. Environment Agency. UK.

[33] Yosie, T. F. and Herbst, T. D. (1998). *Using Stakeholder Processes in Environmental Decision making*. ICF Incorporated.

[34] Canter, L. W. (1996). *Environmental Impact Assessment*. McGraw Hill.

[35] Griffiths, A. (2000) *New Organisational Architectures: creating and retrofitting for sustainability*. Sustainability. Ed. Dunphy D. 226. Allen Unwin N. S. W. Australia.

[36] Kumar, R., (1999) Research methodology: A step by step guide for beginners, London, SAGE Publications.

[37] Jiang, J., Klein, G. (1999). *Risks to different aspects of system success*. Information and Management, 36 (5), 263-271.

[38] Lein, K. L. (2003). *Integrated Environmental Planning*. Blackwell Publication.

[39] C. E. Q. (1997). *The National Environmental Policy Act*. A Study of its Effectiveness after Twenty-Five Years. Council on Environmental Quality, http://ceq.eh.doe.gov.

13

Comparative Analysis of Non-Destructive Strength of Water Cured and Air Cured Concrete

Onungwe Ishmael, Sarogoro Samuel, George Abednego, Akpan Paul

Department of Civil Engineering, Rivers State Polytechnic, Bori-Ogoni, Rivers State, Nigeria, West Africa

Email address:

Ishmael.onungwe@gmail.com (O. Ishmael), ndesarosr@yahoo.com (S. Samuel), tonsoky@yahoo.com (G. Abednego), paulynciap06@yahoo.com (A. Paul)

Abstract: Non-destructive strength comparative analysis was carried out on water cured and air cured concrete cube samples. The investigation was for a period of 7, 14, 21 and 28 days. The concrete cube samples were designed using a mix ratio of 1:2:4 and water cement ratio of 0.55 with batching done in weight. The results reveal a substantial strength gain in the water cured samples to that of the air cured, having taking an average results of four cubes for each group of the samples tested for, using a Schmidt rebound hammer. The 28 day age concrete cube gave a strength difference of 5.1 N/mm^2 between the water cured and air cured samples in favor of strength gain of the water cured cubes, thereby pointing to the significant of curing in terms of concrete strength development.

Keywords: Concrete, Curing, Hydration, Non-Destructive, Strength

1. Introduction

Curing of concrete plays a major role in developing the microstructure and pore structure of concrete. A good curing practice involves keeping the concrete damp until the concrete is strong enough to do its desired job or withstand its proposed imposed load. However, good curing practices are not always followed in most of the cases involving site practices which in turn leads to weak concrete structures. Curing of concrete simply means maintaining moisture inside the body of the concrete during the early ages and beyond in order to develop the desired properties in terms of strength and durability. Curing of concrete is a prerequisite for the hydration of the cement. For a given concrete, the amount and rate of hydration and furthermore the physical make-up of the hydration products are dependent on the time-moisture-temperature history. Generally speaking, the longer the period during which concrete is kept in water, the greater its final strength. It is normally accepted that concrete made with OPC and kept in normal curing conditions will develop about 75 percent of its final strength in the first 28 days [1].The process of curing begins immediately after placement and finishing so that the concrete may develop the intended strength and durability [2]. Curing can be achieved by ponding or immersion, spraying or fogging and wet covering [3]. Hydration of cement is said to be the combination of all chemical and physical processes that takes place after contact of the anhydrous solid with water [4]. For the process of hydration to be effective, it is important to saturate calcium silicate hydrate gels with water [5]. In a similar work done by [6], he pointed out his observations that cement hydration does not just improve when cured at relative humidity below 80% of the concrete vapor pressure. Hence, the importance of saturating the concrete by continuous wetting with water to keep the pores saturated enhances cement hydration. Proper curing enhances the reduction of the rate of moisture loss and as well provides a continuous source of water required for the hydration process which in turn reduces porosity by providing a firm pore size distribution in concrete [7]. The result of [8] revealed that the samples that were cured showed greater curling deflection and these effects increased with the length of curing for the drying conditions investigated. These findings could impact more on construction techniques for volume-change-sensitive structures such as slabs-on-ground in drying environments. [9] recommended that testing of concrete, mortar samples and other research related materials be cured in lime saturated water. A recent concrete strength research by [10] reports a compressive strength loss of 10-20% of concrete cubes that were ambient air cured compared to cubes wet cured. Compressive strength losses were recorded in that research at ages 28, 90 and 180 days for cubes specimens that were air cured compared to those wet cured. Results show that those wet cured were more effective in improving compressive strength at later ages for higher water cement ratio than lower water cement ratio specimen. In like manner, [11] presents a report showing compressive strength increases of concrete cubes cured in water as compared to air cured cubes at 7 and 28 days using Portland cement at cement content of 400kg/m^3. ACI 308 gives standard procedure for curing concrete however, experiment shows that the average strength of concrete can be considerably gained within 28 days. This does not negate the fact that concrete gains strength with age but it is worthy of note that once curing stops, the concrete dries out and the strength gain stops. [12] spells out that if concrete is not cured and is allowed to dry in air, it will gain only about 50% of the strength of continuously cured concrete, if cured for only three days it will reach about 60%, if cured for seven days, it will reach 80% of its strength of continuously cured concrete and if curing stops for some time and then resumes again the strength gain will also stop and reactivate. This however is encouraged within a temperature of 20°C – 21°C as temperature below 10°C (50 ^{O}F) is unfavorable for the hydration hence proper curing enhances effective hydration which in turn increases strength and other desirable properties of concrete structures such as durability, water tightness, abrasion resistance, volume stability, resistance to freeze and thaw and resistance to deicing chemicals [3]. Research showed that increase in the size of the concrete result in decrease in the effect of temperature on the compressive strength of concrete [13]. This paper is aimed at presenting the significant of curing concrete which can be of valuable assistance in adopting good construction practices at site.

2. Materials and Method

The cement used in this study is one of the available commercial brands of Ordinary Portland Cement (OPC). Portland cement type I (normal Portland cement) conforming to the requirement of [14] and clean water from a nearby stream were used. [15]. Also used were coarse aggregate of crushed granite of 12mm size with density 1600kg/m^3 and fine aggregate of natural white color river bed sand with density 1460 kg/m^3. Concrete cube mold, trowel, head pan, weighing balance, water, curing tank, thermometer, bull nose rod (rammer), Schmidt rebound hammer, slump cone and its accessories are also useful materials used to carry out the required practical.

The mixture was batched by weight using a mix ratio of 1:2:4 and it was mix in a clean dry manual tilting concrete mixing drum. Batching by weight was adopted by measuring the constituent materials such as cement, sand, granite stone and water. Batching by mass was achieved by using a weighing balance. This was done for all mix proportions. Water for mixing was also weighed out as a function of the weight of cement used for each mix proportion.

Preparation and filling of moulds, hand compaction of concrete, surface levelling and curing were all done according to the requirement of [16]. The mixed concrete was scooped into a metal mould of 150mm x 150mm x 150mm and then temped in three layers in accordance with [17], having conducted the slump test for the various water cement ratio. The slump test was conducted in accordance with [18] and with reference to the work of [19]. The workability of the concrete mix was satisfactory with a water cement ratio of 0.55. The compacting factor test conducted gave same value of water cement ratio obtained from the slump test. The concrete was properly mixed, transported and placed manually. The concrete cubes produced were properly cured in a pond at room temperature. The strength of cubes produced were determined using a non-destructive device (Schmidt rebound hammer). The strength was determined for day 7, 14, 21and 28 days.

3. Presentation of Results

Table 1. *Composition of Portland Cement.*

Compound	Chemical Formula	Common Formula*	Usual Range by Weight (%)
Tricalcium silicate	3 $CaO.SiO_2$	C_3S	45-60
Dicalcium silicate	2 $CaO.SiO_2$	C_2S	15-30
Tricalcium Aluminate	3 CaO. Al_2O_3	C_3A	6-12
Tetracalcium Alumiinaferrite	4 CaO. $Al_2O_3.F_e$ O_3	C_4AF	6-8

*The cement industry commonly uses shorthand notation for chemical formulas: C = Calcium oxide, S = Silicon dioxide, A = Aluminum oxide and F= iron oxide. Source: [3].

Table 2. *Concrete Mix Proportion.*

Cement (Kg)	Fine Aggregate (Kg)	Coarse Aggregate (Kg)	0.55 WCR
54	105	216	0.55 (29.7kg)

Table 3. *Slump Test Results.*

W/C Ratio	Slump (mm)	Slump Interpretation	Workability
0.45	0mm	No slump	Very Low
0.55	25mm	True slump	Low
0.65	150mm	Shear slump	High
0.75	180mm	Collapse slump	High

Table 4. *Average Strength of Day 7 Cubes.*

Identity	Cube1	Cube2	Cube3	Cube4	Average strength (N/mm^2)
Water cured	23.2	22.9	23.0	23.5	23.15
Air cured	16.1	15.5	16.0	15.7	15.8

Table 5. *Average Strength of Day 14 Cubes.*

Identity	Cube1	Cube2	Cube3	Cube4	Average strength (N/mm^2)
Water cured	26.1	25.9	25.3	25.9	25.8
Air cured	18.4	18.7	18.9	18.3	18.6

Table 6. Average Strength of Day 21 Cubes.

Identity	Cube1	Cube2	Cube3	Cube4	Average strength (N/mm^2)
Water cured	27.4	28.1	28.3	27.7	27.87
Air cured	19.8	20.2	20.6	20.1	20.2

Table 7. Average Strength of Day 28 Cubes.

Identity	Cube1	Cube2	Cube3	Cube4	Average strength (N/mm^2)
Water cured	30.7	32.1	33.0	30.9	31.67
Air cured	21.9	22.1	21.8	22.4	22.1

Table 8. Percentage of Strength Gained by Both Cubes Cured in Water and Air.

Ages	Day 7	Day 14	Day 21	Day 28
Water cured	21.4%	23.8	25.7%	29.2
Air cured	20.6%	24.3%	26.3%	28.8%
% Difference	0.8%	-0.5%	-0.6%	0.4%

4. Discussions

Table 1 showed the proportions of Tricalcium Silicate, Dicalcium Silicate, Tricalcium Aluminate, Tetracalcium Alumina Ferrite that is required for a sound Ordinary Portland Cement. The Tricalcium Aluminate accounts for the initial setting time of the concrete, The Tricalcium silicate is responsible for the strength of the concrete from the first day to the twenty eighth day after which the remaining strength that is expected to develop for the life time of the concrete is accounted for by Dicalcium Silicate. The Tetracalcium Alumina Ferrite accounts for strength and color. The results obtained from the experiments conducted agree with these properties of the cement.

Table 2 showed the results of the weight of the proportion of the concrete mix used to conduct the experiments. The batching was done by weight and the value of the cement, fine aggregate, coarse aggregates and water cement ratios are 54,105,216, and 0.55 respectively.

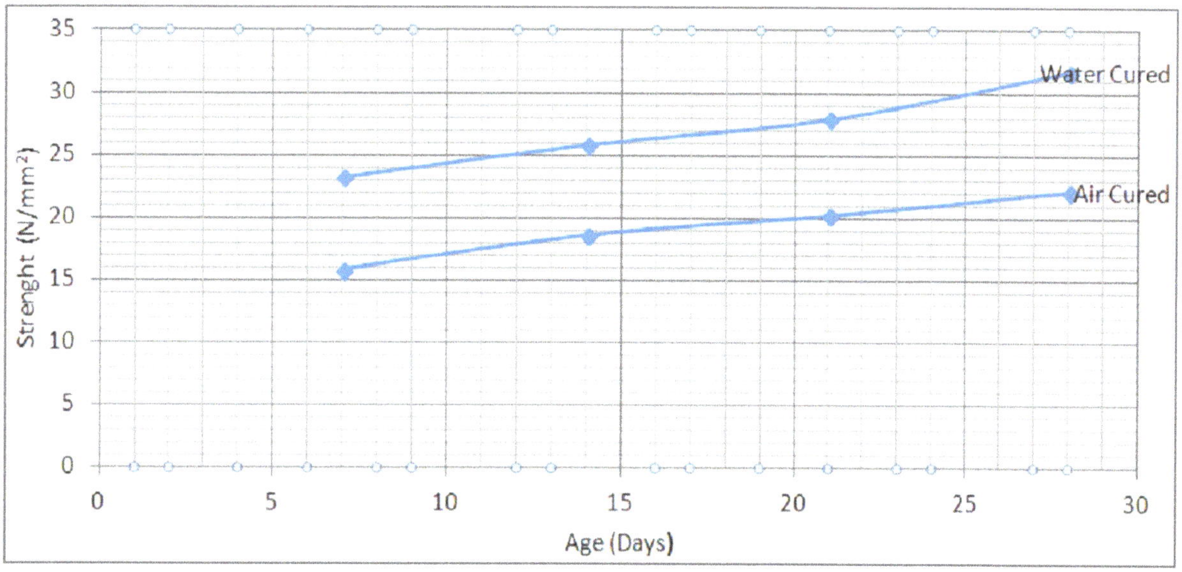

Figure 1. Strength of Cubes versus Age of both Water and Air Cured Concrete.

Table 3 showed the various water cement ratios and their corresponding slump value, slump interpretation and workability. The result of the slump test conducted gave a true slump at 25mm. The workability was low, this result is satisfactory, in view of the fact that compaction of the concrete was done by manual method as explained in the methodology.

From table 4-7, the results obtained from the water curing effect condition on the non-destructive strength test analysis of concrete cubes specimen having water cement ratio of 0.55 at ages 7, 14, 21 and 28 days respectively are compared with concrete cubes specimen of the same water cement ratio tested for, within the same duration or ages but air cured. In like manner, percentage of strength by both cubes cured in water and air are presented in table 8, as well as the variation and consequently graphs of strength of cubes versus age of both water and air cured concrete are as well shown figure 1.

Results obtained from the non-destructive strength analysis of water cured and air cured concrete cubes samples are as shown above in tables 4-7.

From the results obtained from the 7 day age samples as shown in table 4, the average variation between the water cured and air cured concrete samples is 7.4 N/mm^2, hence the average result gotten for the water cured is 23.15 N/mm^2 and the air cured as 15.8N/mm^2 and the percentage strength variation is 21.4% and 21.3% for water cured and air cured cubes samples respectively, thereby resulting to a difference of 0.8%.

The average results differences obtained might seem insignificant here at present but this is as a result of age which represents the duration of testing and the content of water cement ratio at which true slump was achieved as aimed at in this research work, this however will obviously be seen in the behavior of the concrete cubes with respect to performance and durability. Even if the results variation does not differ much, the air cured concrete will be prone to rotting while the water cured will continue in strength gaining by age due to the initial subjection to hydration and the concrete cube microstructure water entrapment as mixing, placing and temperature of concrete has roles to play in concrete strength and durability, as well as design mixture.

Table 5, presents the results as obtained from the 14 days samples from where differences between the water cured and air cured was 7.2 N/mm^2, while the percentage strength variation gave a value of-0.5% which by interpretation indicates that for the adopted water cement ratio the strength for the air cured cube samples however supersedes the water cured in terms of percentage strength variation but not in terms of the actual strength value obtained. In essence, this negative percentage result has no effect on the concrete strength and duration as this is just a commutation of strength result values.

Table 6 and 7 which shows the results obtained for 21 and 28 days cubes samples, presents the following results for the differences between water cured and air cured concrete cubes respectively; 7.7 N/mm^2 and 9.6 N/mm^2. The percentage strength variations are also given as -0.8% and 1.0%. Strictly speaking by interpretation on the values obtained from the 28 day test samples precisely in terms of the strength result, it will be observed that concrete rapidly gains strength with age as the strength variation calls for attention which is in consonance with the work of Stark, (2011). The graphs shown above explicitly provide better understanding to these facts. In essence the strength of the water cured cubes increases from 23.5 to 31.7N/mm^2 from day 7 to day 28 on the first graph series, while the air cured increases from 15.8 to 22.1N/mm^2 for 7 and 28 days indicating the significance of water curing against air curing by interpretation.

5. Conclusion

The results presented in this research work shows that cured concrete structures possesses more strength than uncured or air cured concrete structures as seen from the results, the strength gain of the cubes on average strength scale for 7 day increased from 15.8N/mm^2 to 23.15N/mm^{2}for air cured and water cubes concrete cubes respectively. Similarly, that of day 14 increased from 18.6N/mm^2 to 25.8N/mm^2, day 21 increased from 20.2N/mm^2 to 27.87N/mm^2 and day 28 increased from 22.1N/mm^2 to 31.67N/mm^2 for air cured and water cubes concrete cubes respectively. The strength of cubes versus age of water cured and air cured concrete curve as presented in figure 1, apparently shows the insignificant slope increase from 15.8N/mm^2 to 22.1N/mm^2 between day 7 to 28 on the lower curve which represents the air cured sample and a significant slope increment on the upper curve which represents the water cured cube sample ranging from 23.15N/mm^2 to 31.67N/mm^2 between day 7 to 28. Hence, in order to obtain the desired strength of concrete, it should be thoroughly cured using any of the methods of curing which include ponging or immersion, sprinkling, spraying or fogging, wet coverings using moisture-retaining fabric or membrane such as burlap, cotton mats or rugs and as well as allowing the concrete in its formwork which could be plastic sheeting, ply boards woods or steel sheets. Amidst any of the methods to be adopted, it should be carried out immediately after 24 hours of concrete placement and within a temperature of 20°C – 21°C to enhance effective hydration, rapid increase in strength, avoidance of cracks, durability, water tightness, abrasion resistance in terms of rigid pavement to be more precise, volume stability, resistance to freeze and thaw and as well as resistance to deicing chemicals.

References

[1] Jackson, N. (1976). *Civil Engineering Materials.* London: Macmillan Press Limited.

[2] Kulkarni, S.B. & Clinton, P. (2006).*Significance of Curing of Concrete for Durability of Structures.* Mumbai: Ultra-Tech Cement Limited.

[3] Mamlouk, M. S., & Zaniewski, J. P. (2011).*Materials for Civil and Construction Engineers.* New Jersey: Pearson Books.

[4] Stark, J. (2011). *Recent advances in the field of cement hydration and microstructure analysis.* Cement and Concrete Research. Vol, 41, pp 666-678

[5] Taylor, G. D. (2000). *Materials in Construction, an Introduction.* United Kingdom: Pearson Education Limited.

[6] Spears, R. E. (1983).*The 80 Percent Solution to Inadequate Curing Problems.* Concrete International, Vol. 5, pp 15-18.

[7] Alamri, A. M. (1988). *Influence of Curing on the Properties of Concrete and Mortars in Hot Climates.* PhD Thesis; Leeds University, United Kingdom.

[8] Amir H. and Tyler L. M. (2015). *Impact of Wet and Sealed Curing on Curling in Cement Paste Beams from Drying Shrinkage*, Material Journal of America Concrete Institute, Vol. 112, Issue 1, Pp 79-84 https://www.concrete.org/publications/internationalconcreteabstractsportal.aspx?m=detailsHYPERLINK "https://www.concrete.org/publications/internationalconcreteabstractsportal.aspx?m=details&ID=51686836"&HYPERLINK "https://www.concrete.org/publications/internationalconcreteabstractsportal.aspx?m=details&ID=51686836"ID=5168

[9] Mark B., Kevern J. T., and Eric O. A.(2015). *Effect of Curing Environment on the Strength Properties of Cement and Cement Extenders*, Journal of Materials Sciences and Applications, 2015, 6, 33-39, Scientific Research Publishing Inc., Received 21 October 2014; revised 18 November 2014; accepted 6 December 2014, Published Online January 2015 in SciRes. http://www.scirp.org/journal/msa, http://dx.doi.org/10.4236/msa.2015.61005

[10] Guneyisi, E., Ozturan T., &Gesoglu, M. (2005). *A study on Reinforced Corrosion and Related Properties of Plain and Blended Cement Concrete under Different Curing Conditions.* Cement and Concrete Composites, Vol. 27, pp 440-461.

[11] Alizadeh, R., Ghods P., Chini M., Ghalibafian, M., &Shekarchi, M. (2008). *Effect of Curing Conditions on the Service Life Design of Structures in Persian Gulf Region.* Journal of Materials in Civil Engineering, Vol.1.2.

[12] Kosmatka et al, (2002). *Design and Control of Concrete Mixtures*, 15th Edition, EB001, PCA Engineering Bulletin EB 001, Portland Cement Association , Skokie.

[13] Ali H. H. (2009). *The effect of curing condition on compressive strength in high strength concrete*, Journal of Engineering Sciences, Vol. 02, No. 01, June 2009 35 ISSN 1999 – 8716, Civil Department ,College Engineering , Diyala University Received: 25/10/2008 Accepted: 16/2/2009

[14] BS EN 197-1. Cement-Composition, Specifications and Conformity Criteria for Common Cement, British Standards Institution, 2000.

[15] Gideon O. B., Anthony N. E., Chioma E., Joshua J., Oluwaleke O. and Tajudeen O. (2015). *Assessment of Compressive Strength of Concrete Produced from Different Brands of Portland Cement*, Journal of Civil and Environmental Research, ISSN 2224-5790 (Paper) ISSN 2225-0514 (Online) Vol.7, No.8, 2015 http://www.iiste.org

[16] BS EN 12390-, Testing hardened concrete –Part 2: Making and curing specimens for strength tests, BSI, London, 2000.

[17] British Standard Institution. (1983). *Methods for Determination of Slump*; London. (BS 1881: Part 102).

[18] British Standard Institution (1983). *Method for Making Test Cubes from Fresh Concrete*, (BS 1881: Part 1). London: British Standard Institution.

[19] Emesiobi, F. C. (2000). *Testing and Quality Control of Materials in Civil and Highway Engineering.* Port Harcourt: Goodnews Press.

Exploring Practical Optimal Topology for Reinforced Concrete Moment Resisting Frame Structures

Mehdi Babaei

Department of Civil Engineering, Faculty of Engineering, University of Zanjan, Zanjan, Iran

Email address:
mbabaei@znu.ac.ir

Abstract: Topology optimization of reinforced concrete frames is one of a challenging issue in structural engineering. Frame structures, unlike trusses, must provide useable spaces. Architectural requirements affect greatly the layout of the structure and optimal spans, which are obtained theoretically. Sometimes optimal topologies are not possible for real projects because the calculated optimal spans do not provide architectural requirements. In this article, however, exploring for optimal spans for reinforced concrete frames is investigated considering the architectural limitations. For this purpose, three groups of spans, including 5.6m, 7.5m and 11.2m, are considered based on the parking limitations and then optimal practical sizing for the structures are obtained numerically for 5-storey and 10-storey structures. All models are estimated and considerable differences are shown by diagrams. The optimal topology is proposed for 5-storey to 10-storey buildings with different useable spans.

Keywords: Reinforced Concrete, Frame Structures, Optimization, Span, Topology

1. Introduction

Each year, millions square meters of buildings are constructed in the world. In 2012, it is been reported that the total buildings of about 260 million square meters have been built, which worth more than $63 billion, while the oil income was about $25 billion. Therefore, building construction is a big business and optimal design of buildings affects greatly on the global economy. Since the structural cost is about %20-%30 of total cost of the building, if structural optimization reduces only %20 of the costs, it will be about $3 billion, which is about %10 of Iranian oil income.

Designers should define three aspects of the structure: topology, shape and size. Architects, usually, define topology and shape, then structural engineers calculate size of the elements based on the existing elements and dimensions. However, it is always a big problem to define optimal configuration satisfying all architectural requirements.

One of the vertical and lateral structural systems is moment resisting concrete frame (MRCF), which is most commonly used for mid and low-rise buildings in many countries. Different designs have been proposed and built based on the architectural requirements and structural designer's intuition and experience. Design of reinforced concrete (RC) frames with different ductility levels studied and cost-effective ductility levels proposed for RC buildings by Babaei [1].

One of the challenging issues in structural engineering is to find optimum structures. Researchers developed new methods for optimization and applied to frame and continuous structures [2, 3]. Evaluation of steel structures for optimal design investigated and optimum location and number for trusses obtained using evaluation of many different topologies [4]. A novel hybrid algorithm have been proposed in the literature and applied to topology optimization of steel braced frames [5]. Multiobjective optimization of steel moment resisting frames and braced frames using a hybrid evolutionary algorithm is studied and efficiency of the proposed algorithm reported [6].

Ketkukah et al. presented optimum design sensitivity of a two-bay one-storey RC plane frame [7]. Guerra and Kiousis presented a novel formulation aiming to achieve optimal design of RC structures [8]. Sharafi et al. worked on the optimization of RC structures and developed a heuristic approach for optimum cost and layout design of 3D reinforced concrete frames [9, 10]. Thiruvengadam et al. studied on cost modeling of RC buildings for seismic effects [11].

This article aims to evaluate and compare different practical topologies for MRCF buildings and to obtain optimal structures, which not only provide cost-effective layout but also satisfy architectural requirements.

2. Method and Numerical Models

2.1. Methodology

In this study, to obtain practical optimal topology, which can be used in real projects, some models are considered based on the architectural requirements. The most important requirement for architects is providing parking area, so that the distance between columns has to provide this possibility.

Usually, two, three or four cars could be located between columns. For more than four cars, it needs very large spans for beams, which would not be cost effective. So, in this study only three types of spans for beams are considered as two, three and four cars could be parked next to each other. Floor area for all types is considered the same so that three types of spanning could be possible. Therefore, 23 meter dimension provides the requirements as detailed in the next section. To evaluate the effect of optimal span to the height of the structure, models of five and ten storey buildings are investigated. Therefore, in total, six models are designed and investigated for optimal topology exploration as detailed in the following.

2.2. The Models

Figure 1 displays 3D geometry of the ten-storey models. Table 1 illustrates the details of the geometry for five and ten storey models. The height of all stories is 3.5 meter and the structure's elements are assumed intermediate moment resisting reinforced concrete frames (IMRCF). For models with 11.2m span, a joist is located in each direction at the middle of the span to provide stiffness for the roofs. Three types of spanning are considered as A, B, and C which define 5.6, 7.6, and 11.2 meter span.

Figure 1. *3D view for the ten-storey models.*

Loading is calculated using Iranian National Building Code for loadings (INBC) [12] and a uniform dead of $DL = 700\,kg/m^2$ and live load of $LL = 200\,kg/m^2$ are applied to the models. Earthquake loads are calculated according to the Iranian Standard No. 2800-05 (ICPSRDB) [13], and INBC. Structural elements, however, are designed according to the ACI 318-02 [14].

Table 1. *Geometry specifications of the models.*

Models	Storey Number	Span	Plan	Total Area
10SA	10	5.6	23*23	5290
10SB	10	7.5	23*23	5290
10SC	10	11.2	23*23	5290
5SA	5	5.6	23*23	2645
5SB	5	7.5	23*23	2645
5SC	5	11.2	23*23	2645

3. Results and Discussion

The first group of models contains five-storey IMRCF buildings and the second group includes ten-storey IMRCF buildings. Final cross sections for elements are obtained by trial and error procedure and structural elements are grouped considering requirements for easy construction.

Figures 2-3 display total required concrete for five-storey and ten-storey buildings per unit floor area. As shown in these figures, the difference between type C and other types are remarkable. In the other word, required concrete volume decreases rapidly from type C to B, and then decreases slightly for type A.

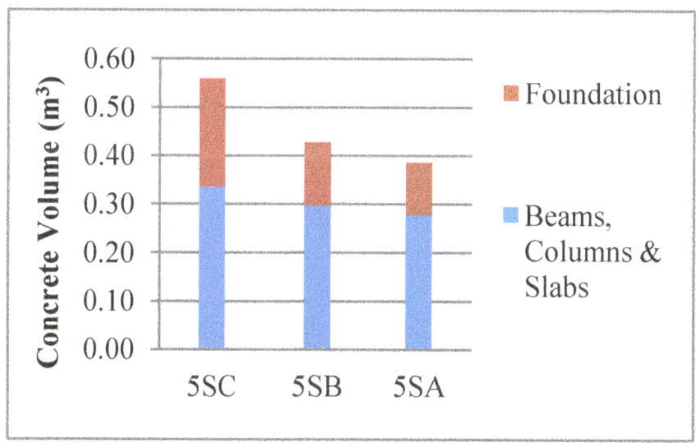

Figure 2. *Total required concrete for five-storey models*

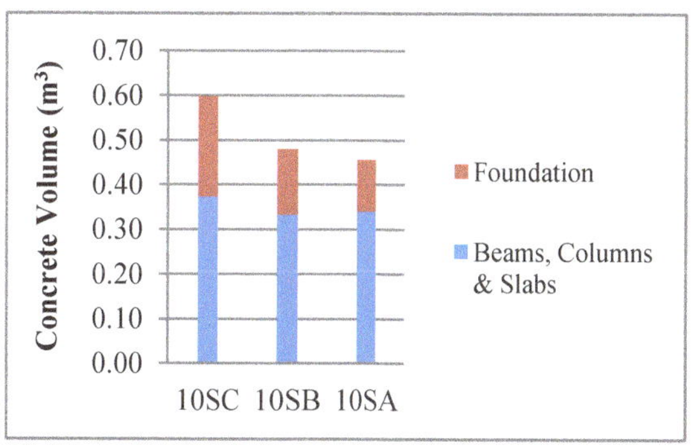

Figure 3. *Total required concrete for ten-storey models.*

Figures 4 and 5 display required reinforcement for five and ten-storey models, while figures 6 to 8 illustrates required concrete volume for structural elements in detail for different storey numbers. Required reinforcements for elements vs. storey number are illustrated in figures 9 to 13.

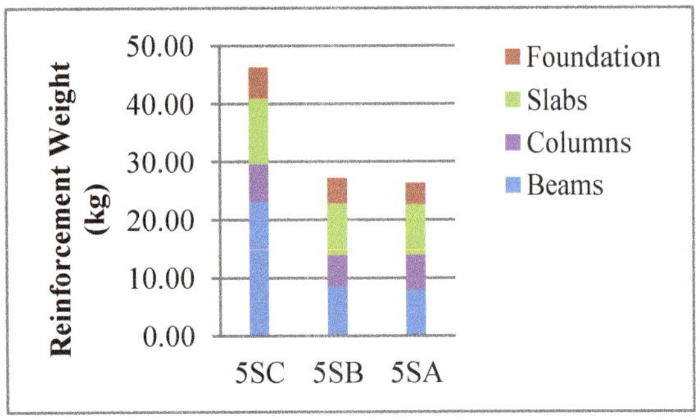

Figure 4. *Total required reinforcement for five-storey models.*

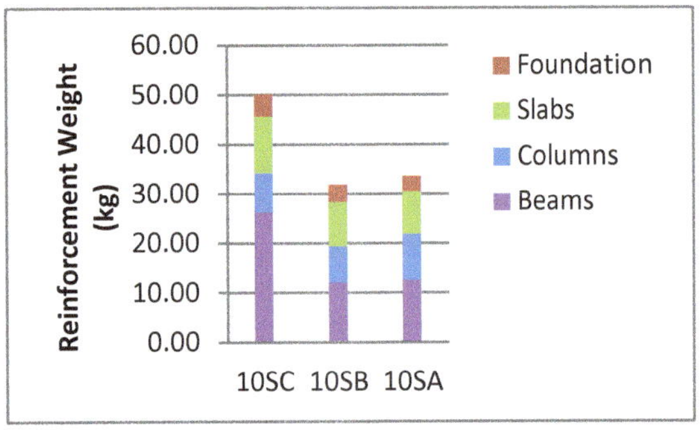

Figure 5. *Total required reinforcement for ten-storey models.*

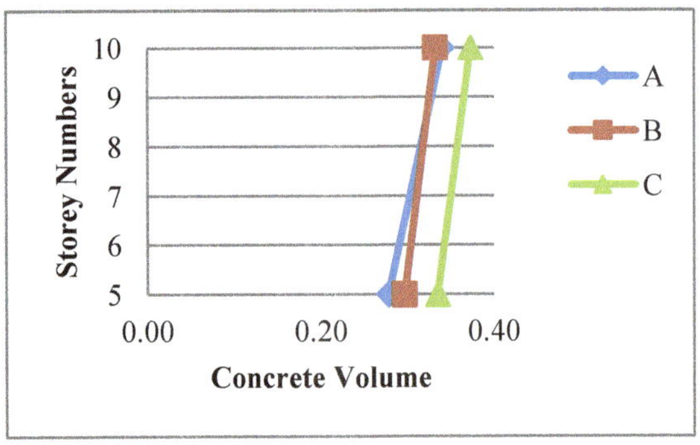

Figure 6. *Required concrete for beams, columns and roofs vs storey number.*

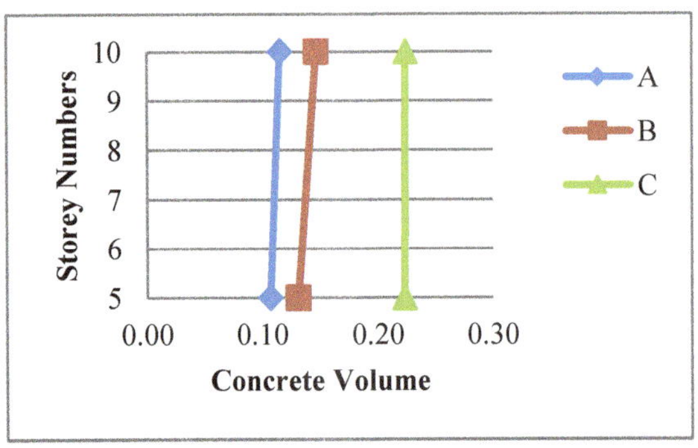

Figure 7. *Required concrete for foundations vs. storey number.*

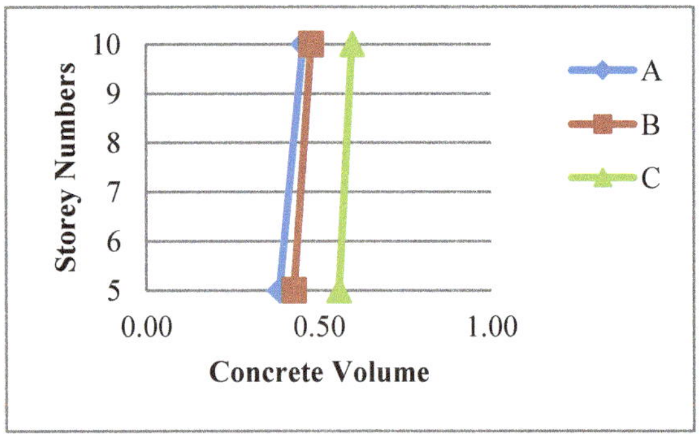

Figure 8. *Total required concrete vs. storey number.*

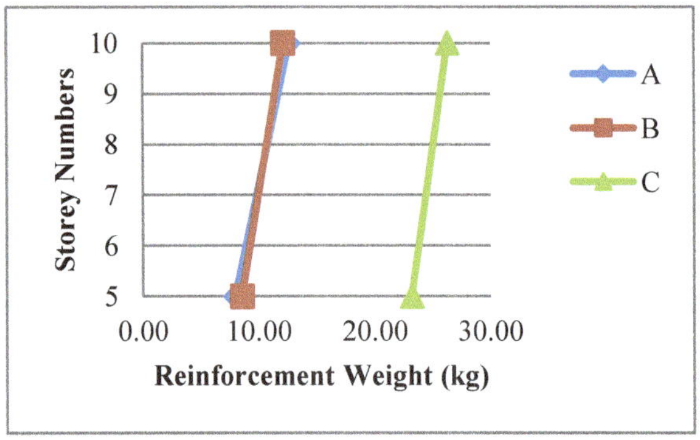

Figure 9. *Required reinforcement for beams vs. storey number.*

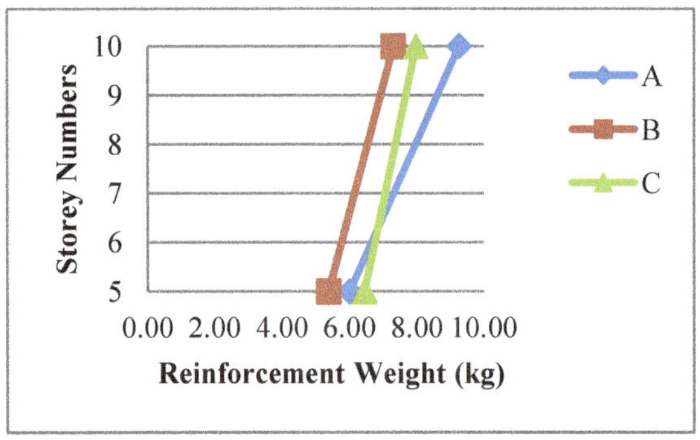

Figure 10. *Required reinforcement for columns vs. storey number.*

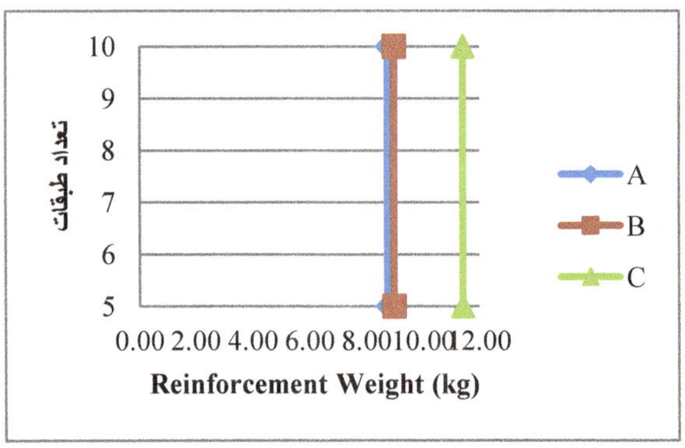

Figure 11. *Required reinforcement for roof slabs vs. storey number.*

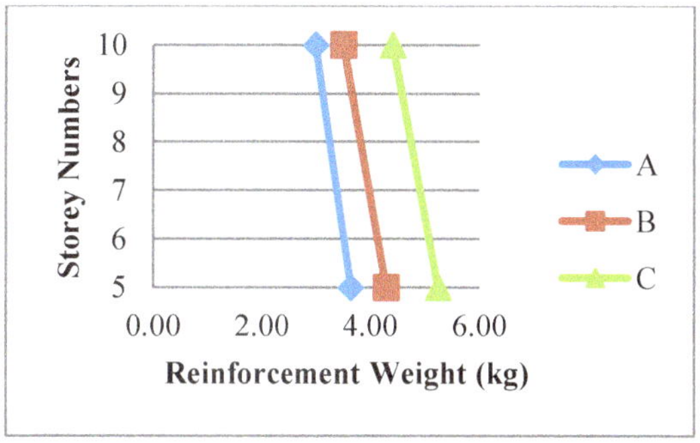

Figure 12. Required reinforcement for foundations vs. storey number.

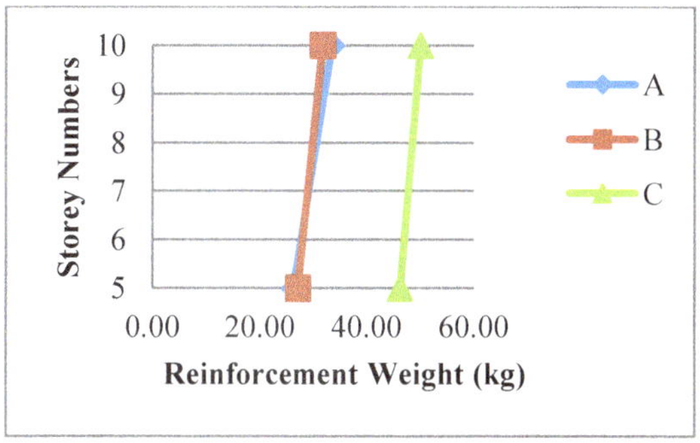

Figure 13. Total required reinforcement vs. storey number.

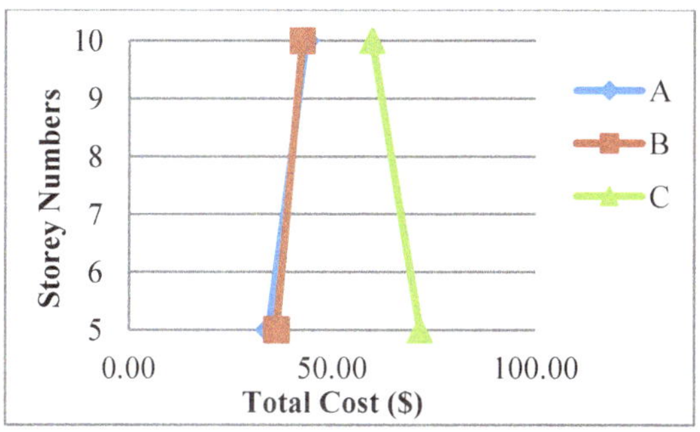

Figure 14. Total cost vs. storey number.

The total cost of models vs. storey numbers is illustrated in figure 14. As shown in this figure, type A is the optimal topology for five-storey buildings, while type B is optimal topology for ten-storey buildings. Optimal topologies for eight to nine-storey buildings are type A and B, and there are two optimal solutions in this case.

Figure 14 also indicates that the total cost for type C decreases when the number of stories increases, and for tall buildings this type, C, could be optimal. In the other words, optimal span increases when the number of stories increases. Moreover, the aspect ratio is an important parameter in this study and there is a relationship between this parameter and span length, which needs more study to obtain specific results.

Total cost, required reinforcement and concrete volume for 5SB model compared to the optimal model, 5SA, only shows only increases of 6%, 3%, and 11% respectively. For model 5SC, however, there is a rapid increase of about 64%, 76%, and 45%, compared to optimal model.

In ten-storey models, type B provides optimal topology, which includes buildings with 7.5m spans. In type A compared to this type, total cost and required reinforcement increases about 2% and 5%, while required concrete volume decreases 5%. On the other hand, in type C total cost, required reinforcement and concrete volume increases 43%, 57%, and 25%, respectively.

In buildings with 11.2m spans (type C) large spans and low stiffness for frames require big beam sections to control displacements. Flexural moments of the beams are large and huge reinforcement is required, as illustrated in figure 9. There are small differences between types A and B for total cost and required materials, which specify 5.6m and 7.5 m spanning.

4. Conclusion

In this article, satisfying architectural requirements practical optimal topology for buildings from five to ten-stories is investigated in the same floor area. Three type of spanning in assumed so that two, three and four cars could be located. Required materials and structural construction cost are evaluated to compare and obtain optimal buildings.

Although for five-storey buildings type A (5.6m spanning) and for ten-storey buildings type B (7.5m spanning) have the optimum total cost, however, there are only 3-5%, 4-10%, and 2-6% difference (increase/decrease) between these types for the required concrete reinforcement, required concrete volume, and the total construction cost of various storey numbers. Type B, however, has a better architectural efficiency compared to type A, which can provide 9 parking locations in each direction, so, in general, buildings with 7.5m spans provide optimum topology.

Acknowledgment

This article is part of a research, which was supported financially by the Ministry of Housing and Urban Development, grant no. 60477.

References

[1] Babaei M., "The economical effect of ductility levels on reinforced concrete frames design", American Journal of Civil and Structural Engineering, 2(1), (2015), 1-6.

[2] Sanaei E. and Babaei M., 2011, "Cellular automata in topology optimization of continuum structures", International Journal of Engineering, Science and Technology, 3 (4), (2011), 27-41.

[3] Sanaei E. and Babaei M., "Topology optimization of structures using cellular automata with constant strain triangles." International Journal of Civil Engineering, 10 (3), (2012), 179-188.

[4] Babaei M., "Multiobjective optimal number and location for steel outrigger-belt truss system", unpublished (under review).

[5] Babaei M. and Sanaei E., "Evaluation of HAG method in optimum topology design of steel braced frames", 1st Regional Congress on Civil Engineering, Malayer University, Iran, (2012).

[6] Babaei M. and Sanaei E., "Multi-objective optimal design of braced frames using hybrid genetic and ant colony optimization algorithm", unpublished (under review).

[7] Ketkukah T. S., Abubakar I. and Ejeh S. P., "Optimum desing sensitivity of reinforced concrete frames", International Journal of Advanced Engineering Research and Technology, 2(5), (2014), 144-158.

[8] Guerra A. and Kiousis P. D., "Design optimization of reinforced concrete structures", Computers and Concrete, Vol. 3, No. 5 (2006) 313-334.

[9] Sharafi P., Hadi M.N.S. and Teh L.H., "Heuristic approach for optimum cost and layout design of 3D reinforced concrete frames", Journal of Structural Engineering, 138(7), (2012), 853-863.

[10] Sharafi, P., Hadi, M. N. and Teh, L. H., "A methodology for cost optimization of the layout design of multi-span reinforced concrete beams", Proceedings of the Fourteenth International Conference on Civil, Structural and Environmental Engineering Computing (p. 124). United Kingdom: Civil-Comp Press, (2013).

[11] Thiruvengadam V., Wason J.C. and Gayathri L., "Cost modeling of reinforced concrete buildings designed for seismic effects", 13th World Conference on Earthquake Engineering, Canada, 2004.

[12] Iranian National Building Code (INBC), Part 6: Loadings, (2013).

[13] Standard No. 2800-05, Iranian Code of Practice for Seismic Resistant Design of Buildings, (2005).

[14] American Concrete Institute (ACI), Committee 318 Building Code Requirements for Structural Concrete (ACI 318-02) and Commentary (ACI 318R-02), Detroit, (2002).

15

Research on Evaluation System of Vibration Induced by Shield Construction

Tao Lianjin[1], Guo Fei[1], Wang Guofeng[2], Huang Jun[3], An Junhai[1], Liu Chunxiao[1], Bian Jin[1]

[1]Key Laboratory of Urban Security and Disaster Engineering of Ministry of Education, Beijing University of Technology, Beijing, China
[2]Beijing Railway Construction and Management Co.Ltd., Beijing, China
[3]CCTEB Civil Defense and Underground Space Institute Wuhan, China

Email address:
ljtao@bjut.edu.cn (Tao Lianjin), guofei2007@163.com (Guo Fei), 728989294@qq.com (An Junhai), 1070672142@qq.com (Liu Chunxiao), bianjin66@163.com (Bian Jin), 18810610118@163.com (Wang Guofeng), 37195274@qq.com (Huang Jun)

Abstract: The research of vibration induced by rail transit should not only consider operation stages, but also extend to the construction phase. Based on the calculation results of site test, the evaluation index system was proposed which is suitable for vibration induced by shield construction. The evaluation indexes were built for the first time and the weight coefficient method was used to determine the weight of each evaluation index. According to actual application of evaluation project, the vibration level was described by Fuzzy language, then the evaluation index system of vibration induced by shield construction was established, which is verified by a project example. The research result shows that: the vibration induced by shield construction can be directly present by the evaluation indexes assessment system which was based on site test, so the leading factors can be effectively traced. Afterwards, a reasonable level of vibration and corresponding risk control measures can be gotten, then the satisfactory assessment result was obtained at last. The evaluation index system of vibration induced by shield construction provides some reference datum for the vibration sensitive area by shield construction.

Keywords: Subway, Environmental Vibration, Evaluation Index, Shield Construction

1. Introduction

In recent years, shield construction has been widely used in China, owing to its safety, construction speed, little disturbance to the surrounding environment and many other advantages. The environmental shield construction induced vibration problems in Beijing western typical sand and gravel stratum resulted in disturbing event, people repeatedly complained petitions and resulted in work stoppage for several months [1]. Vibration rail should not only consider the operational phase, but should extend to the construction phase.

For the vibration induced by subway construction, only have few studies, which focus on the vibration problem in the rock layer caused by TBM tunneling. On the basis of field measurement and numerical modeling, scholars (Tao Linjin [2], Wang Xin [3]) has paid very close attention to environmental vibration problem caused by shield tunneling to the adjacency of railway lines to some sensitive residential areas such as hospitals, research centers, laboratories, or buildings with highly sensitive instruments, relying on specific projects.

In order to study the influence of ground vibration induced by shield tunneling in gravel formation to high precision factory, test was carried out in Taiwan [4]. The scholars put forward the knife dish speed reduction and illustrate the Type of article (Spoke Type) panel, increase rate of the knife dish openings (about 60%, average 45%) measures to weaken vibration.

Rapid metro light rail transit (LRRT) in Buffalo, New York City in the United States, TBM tunneling induced vibration in hard rock. Scholars [5] has carried on the detailed research, and put forward the formula vibration attenuation rate.

In this regard, various aspects of the problems have been investigated in the following three major fields.

Vibration generation: Vibration of the railway track system under a moving load;

Vibration transmission: Transmission of vibrations along the rail to the surrounding environment;

Vibration reception: Response of structures neighboring the tracks to the vibrations.

Existing research shows: people disturbed during construction lots in severe vibration environment during the operation of the inevitable outstanding. Based on current research findings and ideas evoked shield construction civil engineering risk assessment, proposed a comprehensive on-site evaluation of measurement results and risk indicators assessment methods, in order to be able to shield construction induced vibration is given to environmentally sensitive areas quantitative reasonable evaluation, and the corresponding damping action, and to provide reference and guidance for the study of train operations generated vibration.

2. Overall Evaluation System of Vibration Induced by Shield Construction

For shield construction induced environmental vibration assessment, we recommended target system, the law proposed by Richards [6] in 1999 and gradually attracted people's attention.

The most important factor in the evaluation of risk assessment system is to assess the merits of the selection of indicators suggest prominent characteristics of the object, and the reaction of both its nature and general, the assessment should be a comprehensive and representative indicators, this paper using two target system, the choice of degree-level indicators most direct response structural safety factors, two indicators to quantify and subdivision level on the basis of the indicators in order to divide the risk impacts [7].

Overall evaluation system vibration-induced environmental shield construction paper selected and the corresponding values of the standard shown in Table 1.

According to the mechanism of vibration transmission, primary index consists of three main parts: (1) vibration sources; (2) transmission—vibration with the surrounding rock mass propagation; (3) by the vibration body —Sensitivity Buildings.

Shield of penetration [8]: shield advance velocity v and the ratio of the cutter head speed n, expressed with P_{rev}, its physical meaning is the knife dish after every rotation knives of excavation depth, namely the $P_{rev} = v/n$.

Depth: according to the code for metro design (GB50157-2003), shield method construction of tunnel overburden soil thickness not less than outside the tunnel contour diameter, and the Beijing metro tunnel at the top of the overburden soil thickness is generally not more than 20 m.

Soil dynamic weighting modulus of elasticity, excavation face involves the formation of dynamic modulus of elasticity according to the weighted average of the total area, reflects the degree of hardness and dense soil excavation surface·

Horizontal distance: considered rail transit environment vibration attenuation with distance, in a certain distance attenuation will be slow, does not decay, and even distance increases slightly magnified "local amplification effect", score 10 ~ 30 m within the scope of the strengthen [9].

Vibration body: according to the design of buildings, residential buildings and similar protection value of ancient buildings or on vibration sense except for hospitals and research institutes.

Table 1. *The evaluation index system of vibration induced by shield construction.*

Level indicators	Secondary indicators	Classification	Basic score (Rij) Score range	Value	Weight scores (λi)	Assessment score (Xi)
Vibration sources	Maximum total thrust (kN)	$25000<F$	21~30		λ_1	$X_1= R_1\times\lambda_1$
		$18000\le F\le 25000$	11~20	R_{11}		
		$8000\le F<18000$	1~10			
	The maximum cutter torque (kN.m)	$3500<T$	21~30			
		$2500\le T\le 3500$	11~20	R_{12}		
		$1500\le T<2500$	1~10			
	Penetration (mm/rpm)	$50<P_{rev}$	21~30			
		$30\le P_{rev}\le 50$	11~20	R_{13}		
		$6\le P_{rev}<30$	1~10			
	Weighted boring layer elastic modulus (MPa)	$700<E$	21~30			
		$500\le E\le 700$	11~20	R_{14}		
		$E<500$	1~10			
Transmission	Depth (m)	$20<H$	1~10		λ_2	$X_2= R_2\times\lambda_2$
		$10\le H\le 20$	11~20	R_{21}		
		$H<10$	21~30			
	Overburden weighted dynamic elastic modulus (MPa)	$400<E$	21~30			
		$300\le E\le 400$	11~20	R_{22}		
		$E<300$	1~10			
	Vs (m/s^2)	$800< V_S$	21~30			
		$600\le V_S\le 800$	11~20	R_{23}		
		$V_S <600$	1~10			
The vibration body	Horizontal distance (m)	$100<L$	1~10		λ_3	$X_3= R_3\times\lambda_3$
		$10\le L\le 50$	21~30	R_{31}		
		$L<10$	21~30			
	Structure type	Lightweight construction or shallow foundation	21~30	R_{32}		
		Heavy structural or foundation darker	11~20			

3. Vibration Classification

For a specific project, the proportion of the environmental impact assessment index different vibrations, recommended weight coefficient method to distinguish their importance, which is calculated as follows:

$$\lambda = \frac{2n-2m+1}{n^2} \tag{1}$$

Where: λ- weighting coefficient: n- evaluation index number of items; m- number in order of importance, m≤n.

Thus, the shield construction induced environmental vibration assessment value X calculated to determine:

$$X = \sum R_j \lambda_j \tag{2}$$

$$R_j = \sum R_{ij} \tag{3}$$

Where: $i = 1, 2, 3, 4; j = 1, 2, \cdots\cdots, n$.

n is the number of key indicators corresponding to the i level indicators included.

After the calculated X, Chart 2 to determine the level of environmental vibration induced by shield construction.

Table 2. *The classification standard of vibration induced by shield construction.*

Risk level	Assessment score
Level IV (high grade)	X>50
Level III (height)	40<X≤60
Level II (moderate)	20<X≤40
Level I (low)	0<X≤20

Note: The grading standards in accordance with the central limit theorem.

4. Control of Vibration Induced by Shield Construction

For shield construction induced vibration studies in order to control the effects of vibration on the surrounding environment.

Measures ambient vibration damping and noise reduction is generally based on their propagation mechanisms: reduce vibration from the vibration source; the barrier vibration propagation path; enhance vibration body vibration capability by itself.

Based on the results of the assessment of environmental vibration, made with the corresponding vibration and noise reduction countermeasures. As shown in Table 3.

Table 3. *The countermeasure of vibration induced by shield construction.*

Risk level	Acceptance criteria	Control Measures
Level I (low)	Negligible	No action required
Level II (moderate)	Acceptable	By changing the parameters of shield to active vibration reduction
Level III (height)	Do not expect	We must take appropriate damping measures and organizational experts
Level IV(high grade)	Unacceptable	Change the line location or take special measures to reduce vibration

5. Example

5.1. Project Background

The test location was selected in some right line tunnel of Beijing subway, which tunnel structure Outer diameter is 6.0 m and lining thickness is 0.3 m. The depth of shield was about 12~15.5m, covering three main layers, respectively miscellaneous fill soil, silty clay and floated pebble stone. Shield crossing section for gravel layer ④, the circular, continuous gradation, roundness medium, the particle size commonly 2~10cm, maximum particle size of about 19 cm, coarse sand filling in 20%~25%, local with boulders, average particle size of 21~26cm, the biggest size is about 56~70cm, boulders content is about 20%~6%. Layer parameters of court areas as shown table 4. This section used the earth pressure balance shield machine construction, panel cutter head, cutter opening rate was 38%. Shield cutter disc configuration as shown in figure 1.

Fig. 1. *Shield machine.*

Table 4. *Layer parameters of court areas.*

Formation Name	Thickness (m)	P (kg/m³)	*Vs* (m/s)	Dynamic elastic modulus (MPa)
Plain fill	4	1600	168	143
Silty sand	3	1950	173	266
Pebble, Round ravel	7	2100	313	582
pebble	-	2300	345	674

Based on the vibration transmission main controlling factor is the depth of research and formation of consensus, level indicators of the present evaluation system to assess the extent of the order of importance: Pathways> by vibration body> vibration source, based on the weight coefficient calculation formula to obtain the corresponding coefficients were 5/9, 3/9, 1/9.

Table 5. *The evaluation index system of vibration induced by shield construction.*

Level indicators	Secondary indicators	Classification	Value	Weight scores (λi)	Assessment score (Xi)
Vibration sources	Maximum total thrust	F=34000kN	R_{11}=30	$\lambda_1=\frac{5}{9}$	X_1= 43
	The maximum cutter torque	T=5000kN · m	R_{12}=30		
	Penetration	P_{rev}=42mm/rpm	R_{13}=17		
	Weighted boring layer elastic modulus	E=674 MPa	R_{14}=12		
Transmission	Depth	H=14m	R_{21}=14	$\lambda_2=\frac{3}{9}$	X_2=12
	Overburden weighted dynamic elastic modulus	E=389 MPa	R_{22}=18		
	Vs	V_S=242 m/s^2	R_{23}=3		
Vibration body	Horizontal distance	L=1000m	R_{31}=10	$\lambda_3=\frac{1}{9}$	X_3= 3
	Structure type	Frame structure	R_{32}=15		

Therefore, in accordance with the evaluation system of this paper, the environmentally sensitive region overall assessment score X = 43+12 + 3 = 58, belonging to grade III (height). So it may surpass the daytime limit requirements of cultural and educational areas 70dB in *Urban regional environmental vibration standards,* and we must take specific damping measures, organizational experts, then conducted shield construction.

5.2. Field Test

The instruments used in the test were WS-5921 Data Acquisition & Signal Processing System, and 941B ultra-low frequency acceleration sensors and amplifier. All instruments and sensors were calibrated before the field test. Sampling frequency of this test was 200Hz. *Urban regional environmental vibration testing method* (GB10071-88) [10] was strictly executed in the process of testing.

Vibration measurements have been performed on the bedroom, hall and kitchen, both at the basement, the second floor and eighth floor of a 9 floors box shear structure residential building.

5.3. Test Results

5.3.1. A Third Octave Spectrum Analysis

A third octave spectrum, a frequency domain analysis method, has the characteristics of spectral line less bandwidth, which is often used in acoustics, in the human body vibration, mechanical vibration test and analysis. According to the regulations of the current standard, the center frequency of 1Hz, 1.25Hz, 1.6Hz, 2.0Hz, 2.5Hz, 3.15Hz, 4Hz, 5Hz, 6.3Hz, 8Hz, 10Hz,.... A third octave spectrum, lower limit of the relationship between the frequency and the center frequency is as follows.

$$\frac{f_u}{f_l}=2^{1/3},\frac{f_c}{f_l}=2^{1/6},\frac{f_u}{f_c}=2^{1/6} \tag{4}$$

A third octave spectrum bandwidth is $\Delta f = f_u - f_l$: which is f_l lower limit frequency, f_c as the center frequency, f_u for on-line frequency. First FFT transform was carried out on the test data, calculate the amplitude spectrum, then use amplitude spectrum data calculated the average of the data in each of the center frequency bandwidth, and a third octave spectrum value is obtained.

5.3.2. Environmental Assessment Vibrations in GB

According to *Standard of vibration in urban area environment* (GB10070-88) [11] regulation, the vertical vibration to Z level calculated in accordance with the type:

$$VL_z = 20\times\lg(a'_{rms}/a_0) \tag{5}$$

Type, a_0 benchmark for acceleration, average value of 10^{-6}m/s^2; a'_{rms} for vibration acceleration RMS, generally considered $a'_{rms}=\sqrt{\sum a^2_{frms}\times10^{0.1c_f}}$, c_f to vibration acceleration level feeling revised, a_{frms} as the center frequency acceleration RMS.

Table 6. 1/3 octave spectrum calculation of measuring point (dB).

1/3oct/(Hz)	Limit	Basement		Second floor		Eighth floor	
		Vertical	Horizontal	Vertical	Horizontal	Vertical	Horizontal
1	76	43.9	43.7	48.3	49.8	43.4	43.7
1.25	75	45.0	42.4	48.9	49.8	46.3	46.5
1.6	74	46.8	45.9	50.3	52.6	47.4	47.8
2	73	45.6	43.0	45.8	49.0	51.3	50.8
2.5	72	48.9	48.0	48.5	51.1	44.9	45.2
3.15	71	47.7	48.4	48.2	49.3	46.1	46.2
4	70	49.6	49.4	47.0	47.5	48.5	46.5
5	70	54.5	52.9	51.1	51.8	51.4	48.7
6.3	70	57.1	57.0	54.9	55.0	63.1	62.2
8	70	50.8	51.0	66.3	66.7	50.9	52.6
10	72	57.3	56.8	59.8	59.6	60.3	59.6
12.5	74	58.0	59.7	58.8	58.7	60.1	59.4
16	76	58.0	56.6	76.8	76.1	76.6	75.6
20	78	58.1	58.1	66.5	67.6	61.1	61.3
25	80	60.1	60.0	62.2	61.9	57.3	56.2
31.5	82	68.4	67.5	67.2	66.5	63.1	61.7
40	84	63.3	64.0	64.2	64.8	66.1	63.7
50	86	65.5	65.1	67.9	67.7	69.9	63.6
63	88	77.2	76.6	65.0	64.8	68.7	62.4
80	90	65.9	64.0	68.4	68.3	66.7	65.5
VLz		65.5	56.2	72.8	62.5	72.0	60.6

Table 6 for maximum point 1/3 octave calculation, the floors basement ground vibration is given priority to with 31.5Hz and 63Hz center frequency, the second floor is 8Hz and 16Hz center frequency is given priority to, and vertical to the eighth floor with 8Hz~20Hz center frequency is given priority to, among them the second and the eighth floor vibration level at the center of the 16Hz frequency calculation is beyond specification requirements, and the revised point on the 8th floor on the second floor of the most famous level beyond city regional environmental vibration standard day 70dB limit stipulated by the cultural and educational area.

6. Conclusions

(1) This article first puts forward the shield through the vibration of the environmentally sensitive area evaluation index and classification standard, and gives its vibration accept criterion, then set up the evaluation system. This article is the work of environmental vibration assessment before a new way of thinking.

(2) Relying on concrete practical engineering, this paper conduct site test and data analysis according to the national standard. Through comparison with test results, verify the application of this system in shield tunnel crossing the effectiveness of environmental sensitive area vibration prediction.

(3) Vibration induced by shield construction environment evaluation system is a kind of quantitative evaluation methods. The system to a certain extent, to predict the influence of the vibration. The establishment of the system through the environmental sensitive area to provide some reference for shield construction.

References

[1] Guo Fei, Huang Jun, Su Yi, et al. Shield construction process induced vibration source characteristics [J]. Journal of Beijing university of technology, 2014, 40(12): 1820-1827. (in Chinese).

[2] Tao Lian-jin, Guo Fei, Huang Jun, et al. Field tests for environment vibration induced by shield tunneling in sand gravel layer [J]. Journal of Vibration And Shock, 2015, 34(16): 213-218. (in Chinese).

[3] Wang Xin, Han Xuan, Zhou Honglei. Research on environment vibration response induced by metro shield tunneling in central urban area [J]. Civil Engineering Journal, 2015, S2: 309-314. (in Chinese).

[4] Liao Chung-Rong, Wu Shi-Hong, Chen Shue-Cheng. Ground Vibration Induced by Shield Tunneling in Gravel Formation [J]. The 11th strait rain water tunnel and underground engineering and technology seminar 2012, 41-46. (in Chinese).

[5] Nelson, P N, O'Rourke, T D, Flanagan, R F, Kulhawy, F H and Ingraffea, A R. Tunnel boring machine perfor -mance study. Report No. UMTA-06-0100-84-1,448 pp, US Department of Transportation, Washington, DC, 1984.

[6] Richard W. H. Estimation of fault propagation distance from fold shape: Implications for earthquake hazard assessment [J]. Computers and Geotechnics, 2005, 32: 223-244.

[7] An Junhai, Tao Lianjin, An linxuan, et al. Research on seismic risk assessment system of urban large and complex underground structures [J]. China civil engineering journal, 2015, 48(S2): 118-123.

[8] Zhuang Haiyang, Cheng Shaoge, Chen Guo xing. Numerical simulation and analysis of earthquake damages of Dakai metro station caused by Kobe earthquake [J]. Rock and Soil Mechanics, 2008. 29 (1): 245-250. (in Chinese).

[9] Zheng Xin, Tao Xiaxin, Wang Futong, et al. Mechanism of local amplification in attenuation of ground vibration induced by rail traffic [J]. Journal of Vibration And Shock, 2014, 33 (3): 35-40. (in Chinese).

[10] The People's Republic of China state environmental protection agency. GB10071-88 Urban regional environmental vibration testing method [S] Beijing: China standard publishing house, 1988. (in Chinese).

[11] The People's Republic of China state environmental protection agency. GB10070-88 Urban regional environmental vibration standard [S] Beijing: China standard publishing house, 1988. (in Chinese).

16

Composite Concrete Slabs with Profiled Steel Decking: Comparison Between Experimental and Simulation Study

Namdeo Hedaoo[1,*], Namdeo Raut[2], Laxmikant Gupta[3]

[1]Department of Civil Engineering, Govt. College of Engineering, Pune, Maharashtra, India
[2]Department of Applied Mechanics, Govt. Polytechnics, Yavatmal, Maharashtra, India
[3]Department of Applied Mechanics, Visvesvaraya National Institute of Technology, Nagpur, Maharashtra, India

Email address:
nahedaoo@yahoo.com (N. Hedaoo), nv_raut@yahoo.com (N. Raut), lmgupta_vrce@yahoo.co.in (L. Gupta)

Abstract: The research work is to study the behavior of the composite concrete slabs with CRIL DECKSPAN type profiled steel decking by experimental and simulation study. The slab is created by composite interaction between concrete and steel deck with rolled embossments to improve their shear-bond characteristics. However, it fails under longitudinal shear-bond due to the complex phenomenon of shear behavior. Therefore, an experimental full-size tests has been carried out to investigate the shear-bond strength under flexural test in accordance with Eurocode 4: Part 1.1. Eighteen specimens are split into six sets of three specimens each in which all sets are tested for different shear span lengths under static and cyclic loading on simply supported slabs. The full-size finite element (FE) modeling and analysis of the composite slabs is presented, in which the shear-bond interaction between the concreteand steel deck is simulated by the use of interface contact elements. The FE analysis is verified and validated by comparing the experimental results. Comparisons of the experimental and simulation results indicate that the FE analysis agrees well with the test results, and is capable of predicting the behavior and the load carrying capacity of composite slabs.

Keywords: Composite Slab, Full-Size Test, m-k Method, Interface Contact Element, Finite Element Analysis

1. Introduction

Over the years, a composite slab has proved to be one of the simpler, faster, lighter and economical constructions in steel framed building systems. The system is well accepted by the construction industry due to the many advantages over other types of floor systems [1-2]. In the last decade, the construction industry is looking beyond conventional methods and exploring for the better to win over today's challenges and therefore composite slab construction is one of the viable options.

The casting of concrete is carried out on profiled steel decking as permanent formwork which is supported by floor beams. The force transfer mechanism of composite slabs can be characterized by three failure types: (i) flexural failure, (ii) longitudinal shear failure and (iii) vertical shear failure. The longitudinal shear failure is the most common failure type which is affected by the interlock on the concrete-steel deck interface. The strength of the longitudinal shear bond also depends on many factors, among which include the shape of the steel deck profile, type and frequency of embossment, thickness of steel decking, arrangement of load, length of shear span, slenderness of the slab and type of end anchorage. Because of these numerous influencing factors, it is not possible to provide representative design values that can be applied to all slab conditions. This is why the design of composite slabs requires full-size tests to determine the shear transferring mechanism and the longitudinal shear-bond strength. Two design methods for the verification of composite slabs are given in Eurocode 4 Part 1-1: the *m-k* method and the partial shear connection method.

In the experimental study, eighteen full-size composite slab specimens consisting of cold-formed trapezoidal profiled steel decking with rectangular dishing type embossments are built and tested in accordance with the Eurocode 4:Part1.1 to determine: (1) the structural behavior; (2) the load carrying capacity; and (3) to provide the necessary information to validate the simulation procedures. Also to study the load-deflection curves, load-end slip curves and failure modes are subject to imposed loads. The steel decks (CRIL DECKSPAN) are manufactured and supplied by Colour Roof India Limited, Mumbai, INDIA. A one-way, single-span, simply-supported slab specimens are tested using M20 grade of concrete subjected to two equal line loads placed symmetrically at six different shear span lengths.

The full-size FE modeling and analysis is carried out to investigate the behavior of composite slabs considering the intermediate stiffeners on top and bottom flange of trapezoidal profiled steel decking using the software ANSYS 11. Effect of embossments is not considered in modeling. Nonlinear three dimensional models are proposed, in which all main structural parameters (concrete slab, profiled steel sheeting and its interface) and associated nonlinearities are included. The FE analysis requires data from experimental full-size tests as an input to the model. The data, which is the interaction property between the concrete and the steel deck in the form of longitudinal shear bond strengthis the most important factor that affects the accuracy of the results.

The purpose of the current research is to study the FE modeling and analysis results of composite slabs which can predict the load-deflection behavior, load-end slip behavior and failure modes are presented and compared with those results of the experimental full-size tests. A parametric study is carried out and factors like different shear span lengths that influence the behavior and load carrying capacity of the composite slabs are discussed.

2. Review of Literature

Most of the full-size tests done in the past by researchers have shown that the load carrying capacity of composite slabs is governed by the longitudinal shear bond failure. Porter and Ekberg [3-4] have carried out a large number of experimental studies, in which the design equations for the shear bond capacity are derived from the data collected from a series of full-scale performance tests on one-way single span slabs and developed by establishing the linear regression relationship. The method is further verified and supported by Wright et al. [5]. A new simplified method is developed by Crisinel et al. [6] and Mohan Ganesh [7] which provides the design of composite slabs by its critical cross-section and partial interaction method respectively. To simplify full-scale tests and small-scale tests are also proposed by push-out/pull-out tests and slip block test respectively. Marimuthu et al. [8] carried out experimental investigations on full-scale slab specimens to determine the shear bond characteristics for composite slabs.

For composite slabs, various simulation models have been proposed using different finite element software packages. This study has been driven by the significance of providing low cost tools with higher accuracy. Baharom and Xiao [9] performed finite element analysis using software ANSYS to study the effect of longitudinal slip model in ribbed decking composite slab. Full and partial interaction model is used to study the influence of slip in the model and to investigate the difference in composite slab behavior under both modeling techniques. For slip model, the interface element is developed between concrete and steel decking which include link element and spring element. For full-interaction model, concrete and steel decking used the same nodes.

The FE modeling and analysis of the ultimate behavior of one and two-way composite slabs are carried out by Eldib et al. [10] using software COSMOS/M 2.6 and validated by full-size tests. The effects of cold steel straps at the bottom steel deck flange perpendicular to corrugation direction for two-way slabs, embossments flattening, slab aspect ratio and slab slenderness ratio are investigated. The concrete slab is modeled using solid element and steel deck is modeled using shell element. The steel reinforcement mesh, slip interaction and perpendicular interaction at interface between steel deck and concrete is modeled by truss element.

The shear bond interaction between the concrete and the steel deck in composite slabs is a highly nonlinear contact problem where sticking, sliding and frictional phenomena are present at the interface. The shear bond interactions of the composite slabs are treated as a unilateral contact problem and simplified as a two dimensional contact model. The contact pair is constructed by using area to area contact finite element by Tsalkatidis and Avdelas [11]. More recently, a universal FE approach of composite slabs is presented, in which the shear bond interaction between the concrete and the steel deck is treated as a contact problem considering adhesion and friction. The FE analysis using the contact model is further carried out in study of the load behavior of composite slabs in full-size flexural bending by Chen and Shi [12].

The review of literature based on experimental study, shows that the strength of longitudinal shear bond achieved depends on many factors, among which include the shape of steel deck profile, type and frequency of embossments, thickness of steel decking, length of shear span and slenderness of the slab. Currently an accurate determination of load carrying capacity for a new profiled steel decking type is possible only by full-size testing in accordance with Eurocode 4. The purpose of the tests is to provide data for the design of composite slabs using shear-bond method. In the case of simulation study, the FE modeling and analysis is carried out on full-size models to investigate the longitudinal shear-bond behavior of composite slabs. In every case the layout of the models is the following: the concrete is modeled by solid elements, the steel decking is modeled by shell elements and the shear connection between the concrete and the profiled steel decking is modeled by different non-linear interface contact elements.

3. Experimental Study

3.1. General

The experimental study used for the composite slabs with profiled steel decking consists of static and cyclic tests on six sets of full-size slab specimens subjected to six different shear span lengths [13]. For each set of three specimens, one specimen is tested to failure under static loading and other two specimens are tested for cyclic loading [14]. Subsequent sets of tests are conducted in similar manner with remaining shear spans. The aim of the experiments presented here, is to provide the fundamental information on the behavior and load carrying capacity of composite slabs with realistic geometric and material characteristics. As has already been mentioned, the whole experimental procedure is performed to the specifications of Eurocode 4: Part 1.1.

3.2. Preparation of Slab Specimens

All composite slab specimens are cast with full supports on the plain surface concrete flooring in the Composite Testing Laboratory. Figure 1 illustrates the geometric shape of the trapezoidal profiled steel decking sheet. A total of eighteen full-size composite slab specimens are constructed with 102 mm nominal depth, 830 mm width and 3000 mm span. The thickness of the concrete above the flange is 50 mm while depth of the profiled steel deck is 52 mm as shown in Figure 2. All slabs are constructed utilizing M20 grade of concrete obtained from a hand mixing method. The 70 mm depth of slabs are cast first, over which mild steel mesh reinforcement of four steel bars, 6 mm in diameter, are placed at a center to center distance of 250 mm in the longitudinal direction and twelve at a spacing of 250 mm in transverse direction to complete cross sectional dimension of the slab and tied with binding wires [15]. Mild steel mesh reinforcement is used as shrinkage and temperature control reinforcements as specified in the ASCE specification [16]. Remaining 32 mm depth of the slab is cast and the top surface is finished by proper compaction of concrete. The test parameters used in this study are summarized in Table 1.

Table 1. *Test parameters.*

Test no.	Slab designation	Type of loading	Effective span, L (mm)	Shear span, L_s (mm)	Cover width, b (mm)	Nominal depth, h_t (mm)	Conc. comp. strength, f_{ck} (N/mm^2)	Steel deck thickness, t (mm)
1	EX-01ST300	static	2700	300	830	102	20.8	0.8
2	EX-02CT300	cyclic	2700	300	830	102	20.8	0.8
3	EX-03CT300	cyclic	2700	300	830	102	20.8	0.8
4	EX-04ST375	static	2700	375	830	102	20.8	0.8
5	EX-05CT375	cyclic	2700	375	830	102	20.8	0.8
6	EX-06CT375	cyclic	2700	375	830	102	20.8	0.8
7	EX-07ST450	static	2700	450	830	102	20.8	0.8
8	EX-08CT450	cyclic	2700	450	830	102	20.8	0.8
9	EX-09CT450	cyclic	2700	450	830	102	20.8	0.8
10	EX-10ST525	static	2700	525	830	102	20.8	0.8
11	EX-11CT525	cyclic	2700	525	830	102	20.8	0.8
12	EX-12CT525	cyclic	2700	525	830	102	20.8	0.8
13	EX-13ST600	static	2700	600	830	102	20.8	0.8
14	EX-14CT600	cyclic	2700	600	830	102	20.8	0.8
15	EX-15CT600	cyclic	2700	600	830	102	20.8	0.8
16	EX-16ST675	static	2700	675	830	102	20.8	0.8
17	EX-17CT675	cyclic	2700	675	830	102	20.8	0.8
18	EX-18CT675	cyclic	2700	675	830	102	20.8	0.8

Note: Slab designation in the form of "i-j-k-l" where i,j,k and l are variables indicating the experimental study, serial number of test specimen , static or cyclic test and shear span length (mm) respectively. Hence, 'EX-01ST300' refers to the specimen using experimental study, 1st test specimen, static loading and 300 mm shear span.

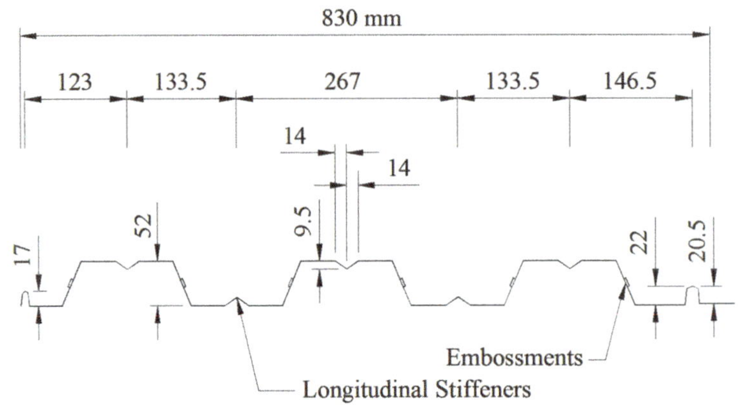

Figure 1. Deck cross-section and dimensions.

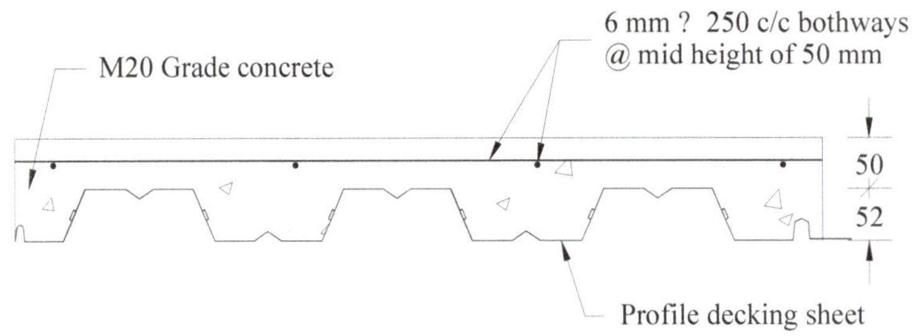

Figure 2. Cross-section of test specimen.

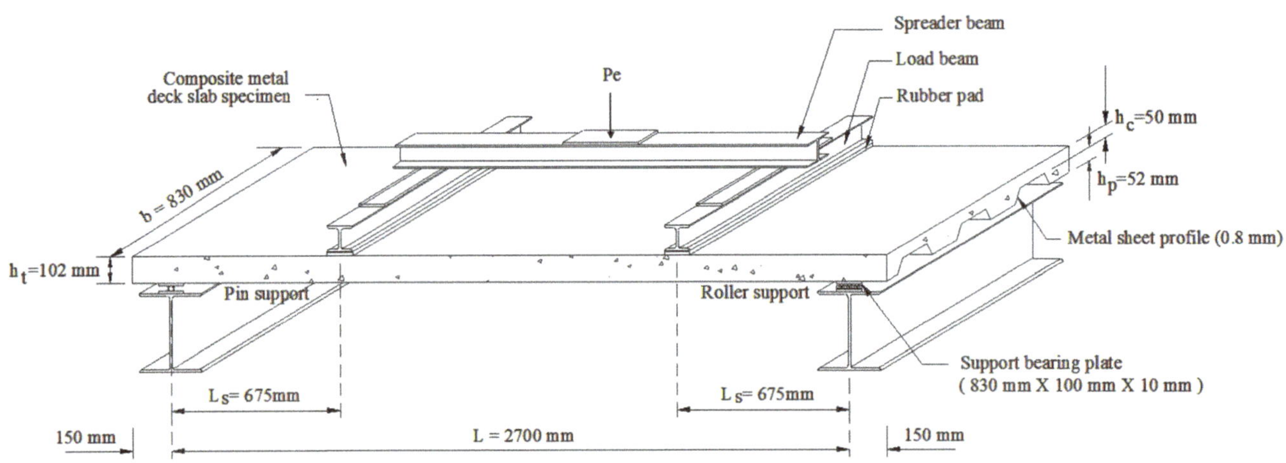

Figure 3. Schematic view of the experimental test set-up.

Figure 4. Experimental test set-up.

The curing period of all eighteen slabs was 28 days. Despite all required preventive measures during transport phase, specimen EX-12CT525 (Figure 16.d) presented premature slippage, probably due to riding procedure. Hence there are no results recorded for the specimen and invalidating test.

3.3. Arrangement

The specimens are placed carefully and symmetrically over the experimental set-up using a crane and belts. The schematic view of arrangement for the simply supported composite slab configuration with an effective span of 2.7 m subjected to two symmetrically located uniformly distributed line loads is shown in Figure 3. In order to simulate simple support conditions, roller and hinge supports are specially fabricated for study. These supports are extended along the whole width of the specimen. Loading is applied by a single hydraulic jack system mounted on structural spreader beam section (ISMB 150), beneath the structural load beams (2 ISMC 100, placed back to back) and load is measured with the help of cell at the point of application. Uniform loading is applied by inflating a 15 mm thick by 100 mm wide hard rubber pad, which is confined by the top surface of the test slab. A steel plate with 10 mm thickness by 100 mm width is placed on the top of the pad. Fig. 4 shows the complete experimental set-up.

3.4. Loading Procedure

One slab of each group is subjected to static loading until failure. The specimens are placed over supporting roller-hinge supports and loading points are marked on shear span. Load is applied incrementally by single hydraulic jack system. Rate of loading is adjusted in such a way that failure did not occur in less than one hour. Rate of loading adopted for static test is 0.1 mm/s. Tests are determined as per the calculated design value, or discontinued when the deflections reaches $L/50$.

Cyclic loading is required to be implemented in the tests prior to the static loading. Hence, two specimens under each shear span are subjected to preliminary cyclic loading. This cycling loading ensures that any kind of chemical bond formed between the concrete and the steel is removed and the static load that is later applied would provide the true indication of the mechanical bond formed by the embossment. Slab is subjected to 3 cycles of loading applied in a time span of three hours according to BS 5950: Part 4:1994 [14]. The next loading sequence is of a static nature.

The vertical mid-span deflection is measured using micro-level equipment. For end-slip measurements, two dial gauges are attached to one end of the composite slab in order to measure the relative slip between the concrete and the steel deck. Average value of the total load at failure (average of one statically loaded and two cyclically loaded) is calculated for each set of specimen.

3.5. Experimental Results

Figure 5 shows the load-deflection curves for shear span (L_s) = 450 mm. For the shear spans, 300 mm, 375 mm and 450 mm, first initial shear cracks formed near the loading point and then flexural cracks formed near center of span at the bottom of the concrete.

Figure 6 shows the load-deflection curves for shear span (L_s) = 600 mm. For the shear spans, 525 mm, 600 mm and 675 mm, first initial flexural cracks formed at the bottom of the concrete near center of span and then shear cracks formed near loading points. Also flexural cracks are formed in between loading points. In Figures 5 and 6, both the static and the cyclic sequence of tests are plotted.

Figure 7 shows the observed typical visible crack formation for L_s = 600 mm.

Figure 8 shows the differential movement of the concrete slab and the steel deck for L_s =600 mm. Slip is observed from both side of profile towards center of slab.

In cyclic test, the behavior and capacity are slightly lesser than obtained in case of the static loading.

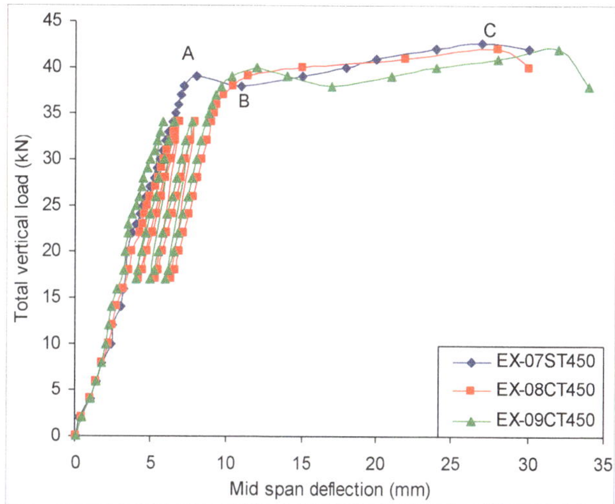

Figure 5. *Load-deflection curves for L_s=450 mm.*

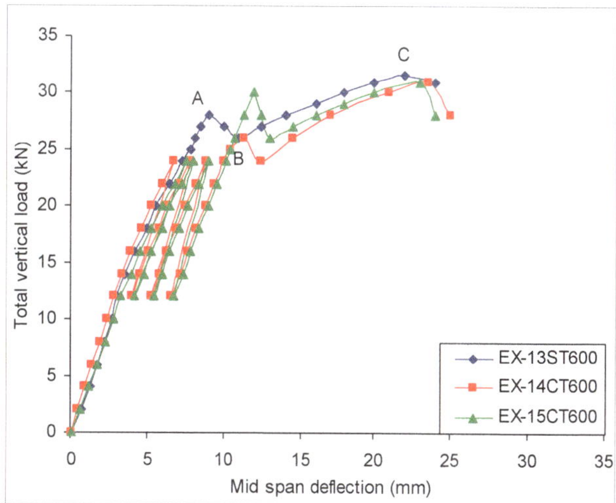

Figure 6. *Load-deflection curves for L_s=600mm.*

Figure 7. *Crack formation for L_s= 600 mm at the ultimate stage.*

Figure 8. *Typical end slips for L_s= 600 mm from left & right side of specimen.*

4. Simulation Study

4.1. Development of the Simulation (FE) Model

An assembly of finite elements represents the concrete slab and profiled steel decking, connected together by interface contact elements and friction phenomena of the surfaces, has formed the basis for modeling of composite slabs. Three-dimensional nonlinear FE modeling and analysis of composite slabs are carried out using ANSYS 11 [17]. Input via a command text-based format is chosen for analysis.

In the preliminary development of the FE model, non-linear material properties of concrete and steel decking are taken from the laboratory test results and values are given in Tables 2 and 3. Contact problems are highly non-linear and require significant computer resources to solve, since an automatic control of time stepping is provided. The behavior of several finite element types is studied under different contact conditions to choose values more appropriate for a realistic simulation.

Selection of proper 3-D point to point contact element and combination element between the concrete and steel decking are tried to determine a suitable combination that produced acceptable results. Once correct material properties and element size are obtained, models are expanded to study the behavior and strength of composite slabs.

4.2. Structural Model

Owing to symmetry in geometry, loading and boundary conditions, only a half of the slabs are modeled as shown in Figure 9. Symmetry exists around the middle of the span and slab width [9, 11]. Final dimensions of the composite slab are length*width*height = 1350*400.5*102 mm and these coincide with those of the experimental set-up.

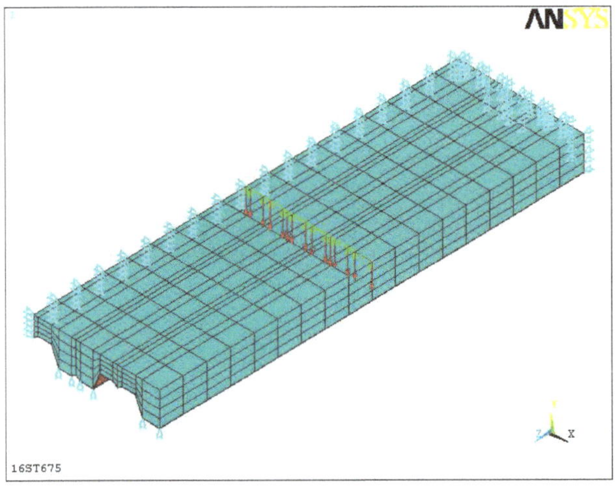

Figure 9. *FE model grid, loading and boundary conditions.*

Concrete slab is modeled with 8-node solid element (SOLID 65) while the profiled steel decking is modeled with 4-node finite strain shell element (SHELL 181) [11, 18]. Steel decking elements located underneath concrete slab elements are offset from the bottom of concrete elements at a distance (0.3746 mm) half of decking thickness [19].

Concrete slab-steel decking interface is modeled with three - dimensional point to point contact element (CONTAC 52) and combination element (COMBIN 40) in order to achieve composite action. If only CONTAC element is used, analysis would fail at an unrealistically low load. The CONTAC element is used for its abilities to prevent overlapping. However, a COMBIN element is used to increase shear bond strength. The contact stiffness values in terms of shear stress are estimated from the results of full - size tests (Figure 4) with the gap initially closed and not in sliding [19 - 22]. To model a composite slab, it is necessary to allow adjacent steel and concrete to slip relative to each other when bond strength or shear bond capacity has been exceeded. Relative movement of steel decking and concrete (when facing each other) must be controlled so that they are not able to move through each other, but are free to separate.

4.3. Application of Load, Boundary Conditions and Numerical Control

A roller support is provided at the slab end to nodes of the bottom flange of steel decking. Nodes at side face of concrete, at edges of steel decking along slab length, and at mid-span are given appropriate boundary conditions to simulate continuity of slab in lateral direction and half span model in the longitudinal direction [9, 11, 19]. Thus, model size and computational time have been reduced significantly. Load is specified by a prescribed maximum vertical displacement at concrete top fiber nodes where two equivalent line loads of equal magnitudes are incrementally applied to model to overcome convergence problems (Figure 9). Tolerance associated with this convergence criterion (CNVTOL command of ANSYS) and load step increments are varied in order to solve potential numerical problems. Newton-Raphson method is used as the incremental-iterative solution process. The convergence procedure is force-based and thus considered absolute.

4.4. Material Modeling of Concrete

For the concrete, the nonlinear material simulation of the ANSYS 11 software program is used. This includes both cracking and crushing failure modes. This observation is consistent with observation of slab behavior during test where concrete did not fail by crushing but rather separated by excessive tensile cracking due to slip.A similar observation is reported by Abdullah and Luttrell [19, 23]. The concrete elements in top most layer are assigned a linear elastic property where no cracking is allowed. This is done to avoid convergence problems where the model could become numerically unstable before reaching ultimate load if crack is allowed all the way up to the top most layer. The concrete properties applied in model are given in Table 2.

Table 2. Concrete properties used in the FE model.

Concrete properties	Values
Density	2500 kN/m^3
Characteristic cylinder strength of concrete	20.8 N/mm^2
Elastic modulus	21572 N/mm^2
Poisson ratio	0.2
Ultimate uniaxial tensile strength	2.08 N/mm^2

A Multilinear Isotropic Hardening (MISO) is used to model the stress-strain curve of concrete material as shown in Figure 10. The von Mises yield criteria uses along with the William and Warnke five parameter model to define the failure criterion [9, 11, 24]. The modulus of elasticity of concrete (E_c) is based on the equation [9, 25]:

$$E_c = 4730\sqrt{f_{ck}} \tag{1}$$

Compressive uniaxial stress-strain relationship for the concrete model is obtained using following equations to compute the multilinear isotropic stress-strain curve for concrete:

$$f = \frac{E_c}{1+\left(\frac{\varepsilon}{\varepsilon_0}\right)^2} \tag{2}$$

$$\varepsilon_0 = \frac{2f_{ck}}{E_c} \tag{3}$$

$$E_c = \frac{f}{\varepsilon} \tag{4}$$

In which f_{ck} is the characteristic cylinder strength of concrete, f is the stress at any strain ε, ε is the strain at stress f and ε_0 is the strain at the ultimate compressive strength f_{ck}.

Multilinear isotropic stress-strain implemented requires first point of curve to be defined by the user. It must satisfy Hooke's Law;

$$E = \frac{\sigma}{\varepsilon} \tag{5}$$

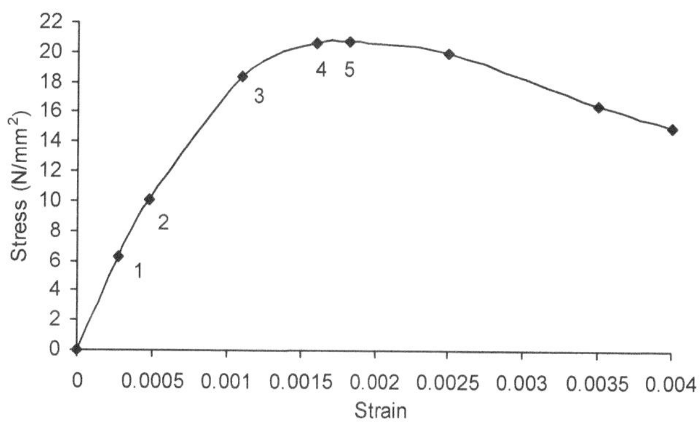

Figure 10. Uniaxial stress-strain curve for concrete.

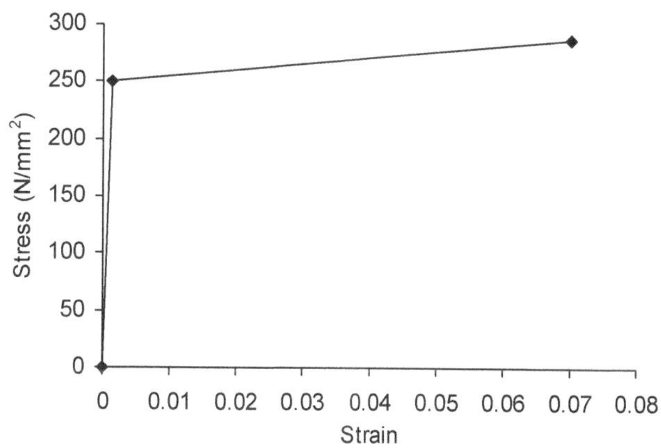

Figure 11. The elasto-plastic stress-strain curve for steel decking.

Multilinear curve is used to help with convergence of nonlinear solution algorithm. Point 1, defined as $0.30f_{ck}$, is calculated in linear range (Equation 4). Points 2, 3, and 4 are calculated from Equation 2 with ε_0 obtained from Equation 3.

Stresses are calculated for every value of Strain. Point 5 is defined at f_{ck} and ε_0 = 0.003 indicating traditional crushing strain for unconfined concrete.

4.5. Material Model for Profiled Steel Decking

The steel decking is modeled by shell element and is well suited for large strain/displacement during failure. The constitutive material law selected for steel decking is bilinear elastoplastic strain hardening (BISO) model using the von Mises yield criterion. The material model is determined for 0.8 mm thick plate and the characteristics are given in Table 3, and Figure 11. The material of steel is considered as homogenous and isotropic.

Table 3. *Steel decking properties used in the FE model.*

Steel decking properties	Values
Density	7850 kN/m^3
Elastic modulus	210000 N/mm^2
Yield stress	250 N/mm^2
Ultimate tensile strength	287.5 N/mm^2
Poisson ratio	0.3

Friction which develops in the steel-concrete interface is not deemed negligible; instead, a constant friction coefficient of 0.20 is considered [4, 11].

4.6. Material Model for Interface Contact Element

The horizontal shear interaction between the concrete and the steel deck is modeled by the contact element as discussed in section 4.2.

The COMBIN 40 is defined by two nodes, two spring constants K1 and K2 (Element stiffness = Force/Length), a damping coefficient C, a gap size GAP (length), and a limiting sliding force FSLIDE. The FSLIDE value represents absolute value of spring force that must be exceeded before sliding occurs.

The CONTACT 52 is defined by two nodes, two stiffnesses (K_n and K_s), an initial gap or interference (GAP), and initial element status (START). Orientation of the interface is defined by node locations. Only material property used is the interface coefficient of friction μ.

If rigid Coulomb friction model is selected in the ANSYS, $K_s$is not used, and elastic sticking capability is removed. This option is useful for displacement controlled problems where sliding dominates.

4.7. Simulation Results

In Figures 12 to 15, FE analysis results are presented for the specimen 16ST675. These results depict the behavior of the proposed simulation model.Maximum central load is set equal to 29 kN in agreement with the experimental outcome. Figure 12 shows deformation of composite slab under loading is depicted. Figure 13 and 14 shows deflection of profiled steel decking and composite slab at the time of the maximum vertical load is plotted. Longitudinal slip at the interface of the composite slab is shown in Figure 15 and is equal to 1.8 mm. Results of experimental (static and cyclic) and FE analysis are presented on the load-deflection curves as shown in Figure 16(a)-(f) and load-end slip curvesas shown in Figure17(a)-(f).

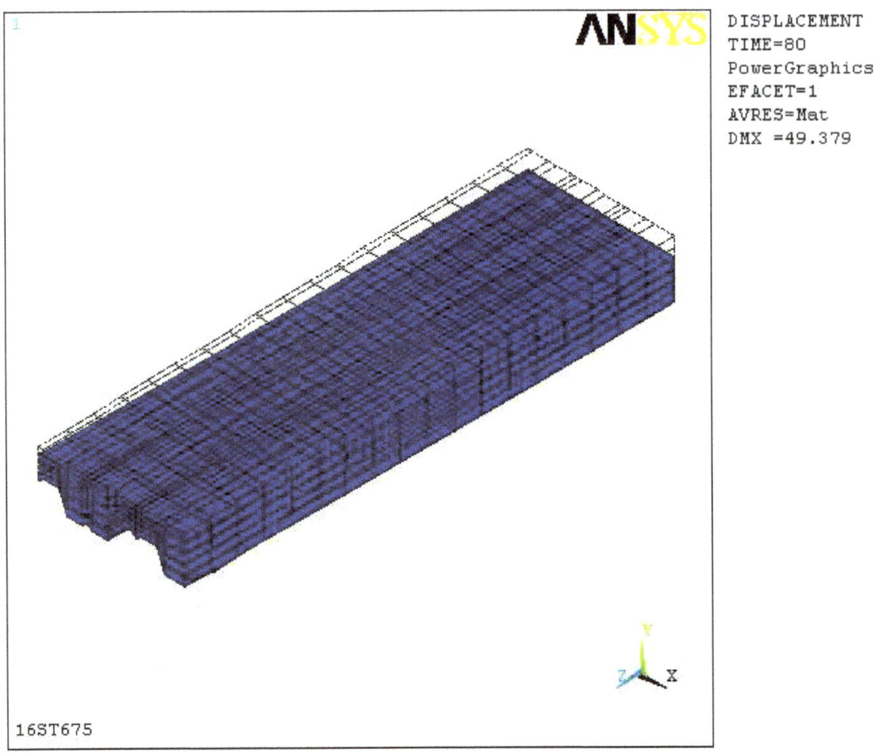

Figure 12. *Deformation of composite slab.*

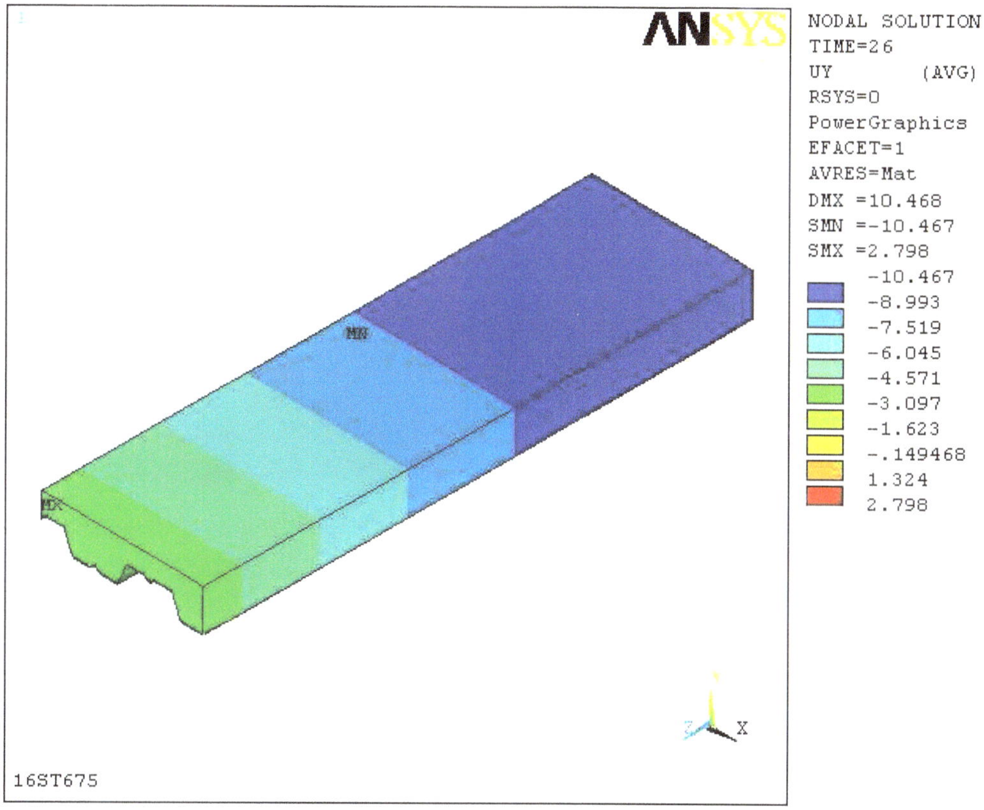

Figure 13. *Deflection (10.468 mm) of composite slab at load of 26 kN.*

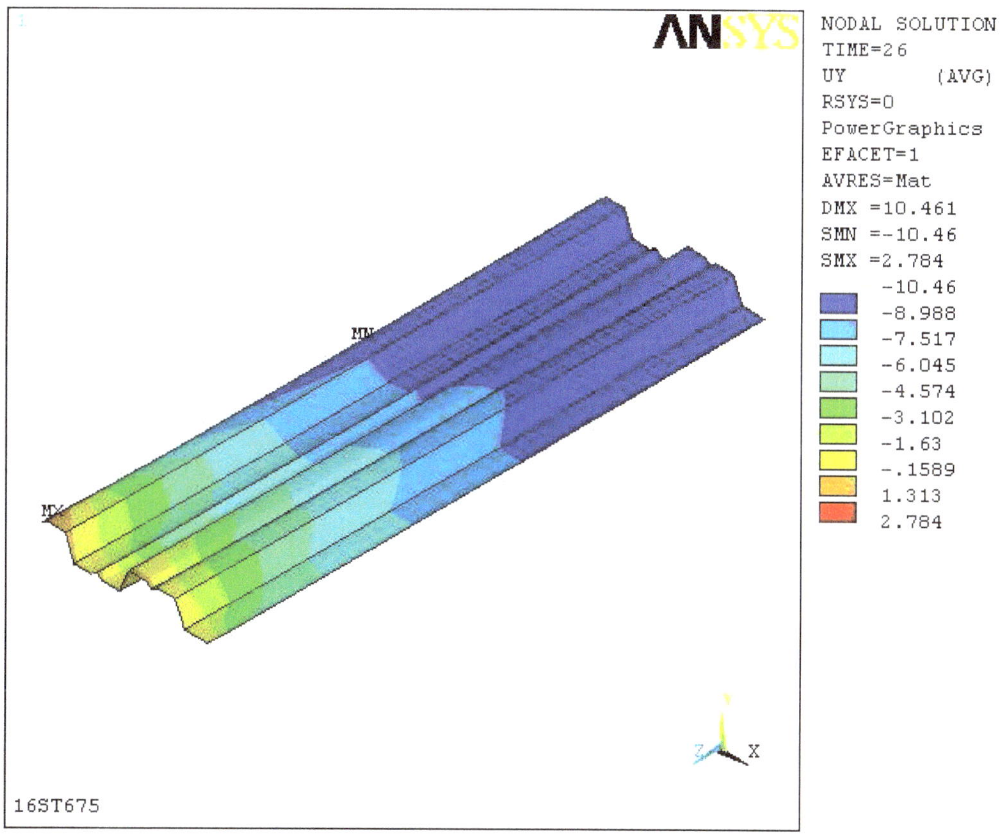

Figure 14. *Deflection (10.461 mm) of profiled.*

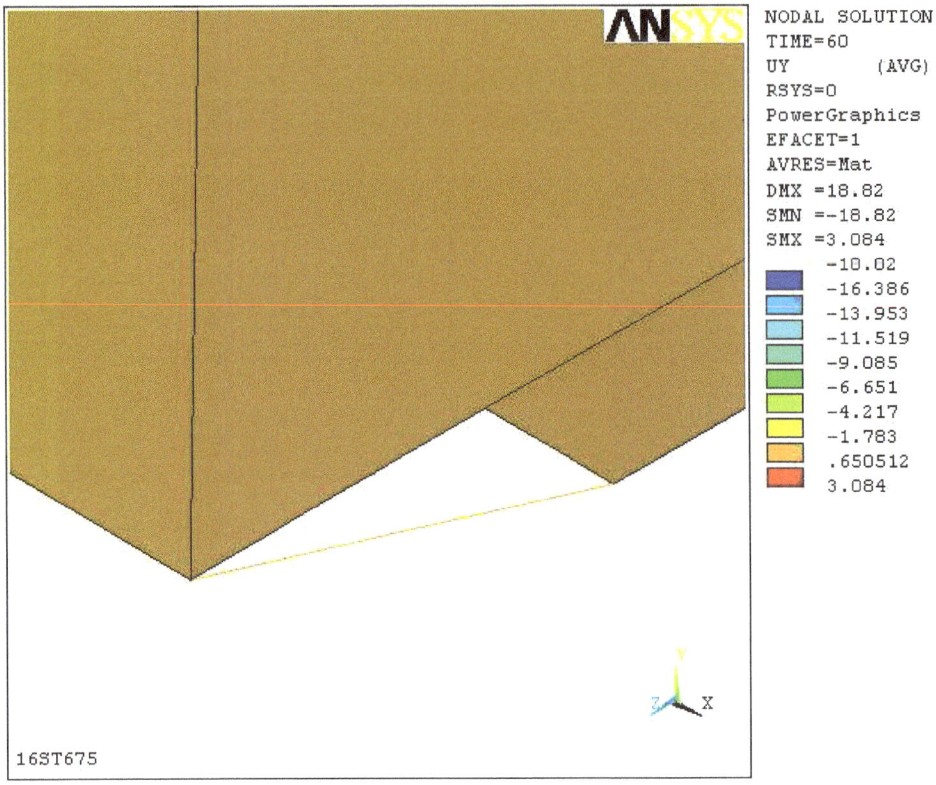

Figure 15. *Slip between concrete and steel decking at load of 26 kN.profiled steel decking.*

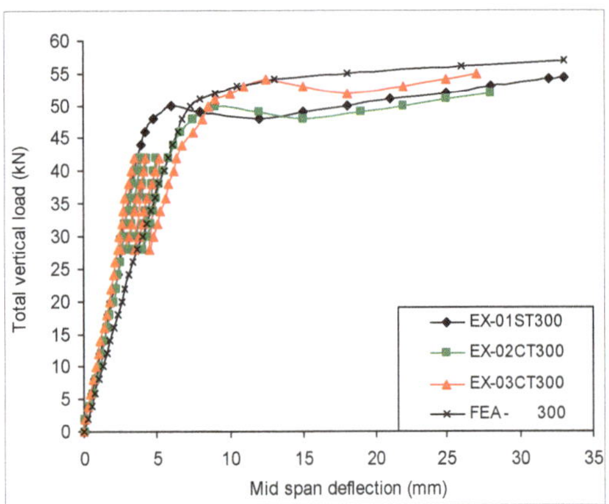

(a) shear span (L_s) = 300 mm.

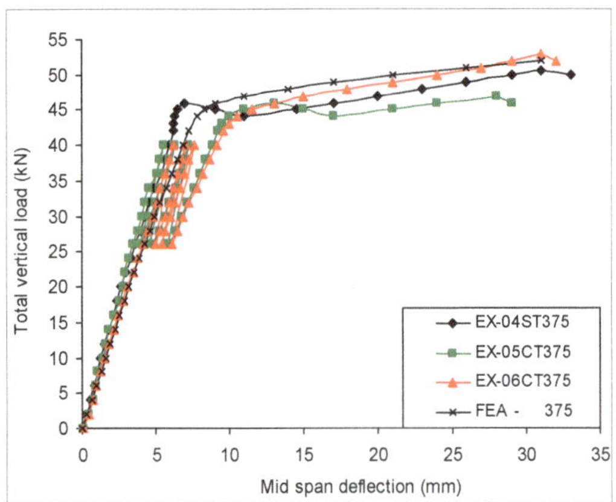

(b) shear span (L_s) = 375 mm.

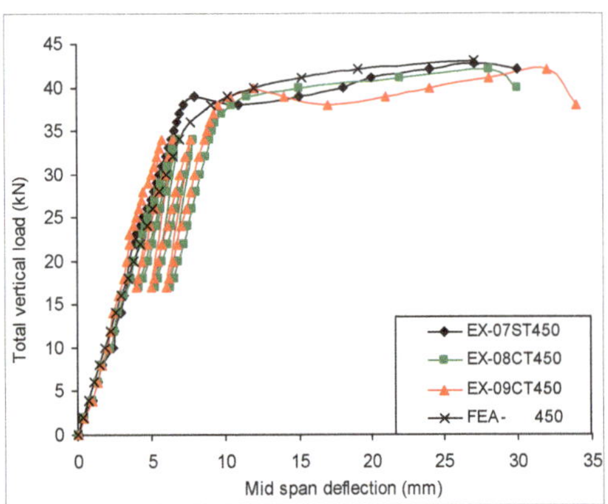

(c) shear span (Ls) = 450 mm.

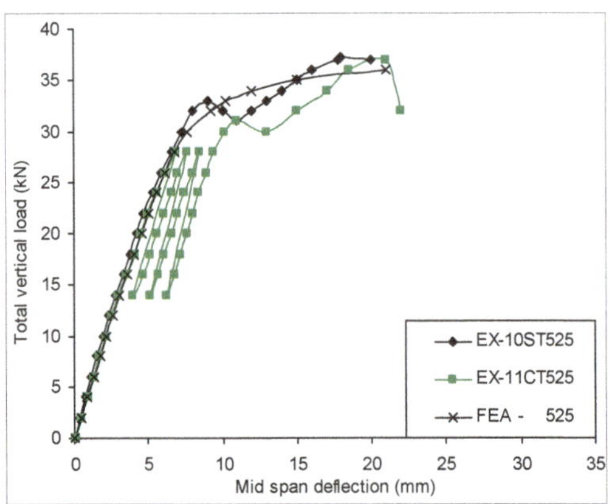

(d) shear span (Ls) = 525 mm.

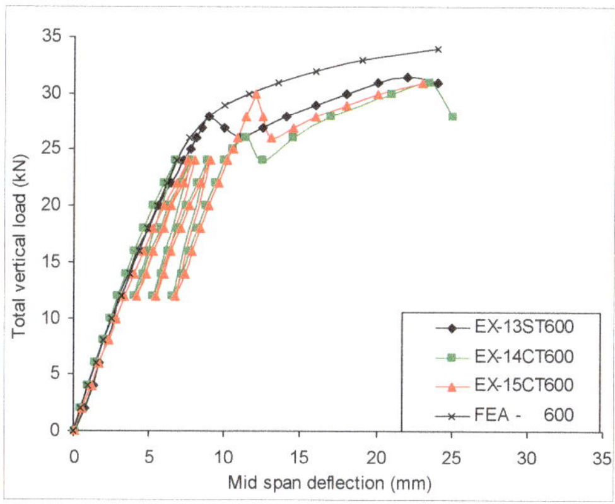

(e) shear span (Ls) = 600 mm.

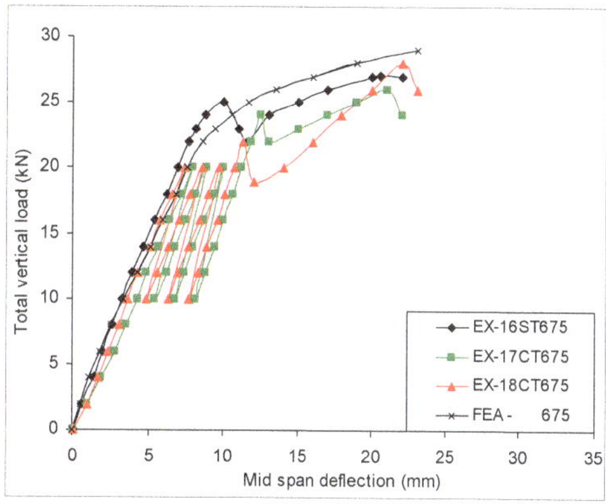

(f) shear span (Ls) = 675 mm.

Figure 16. *Experimental & FE analysis load-deflection curves.*

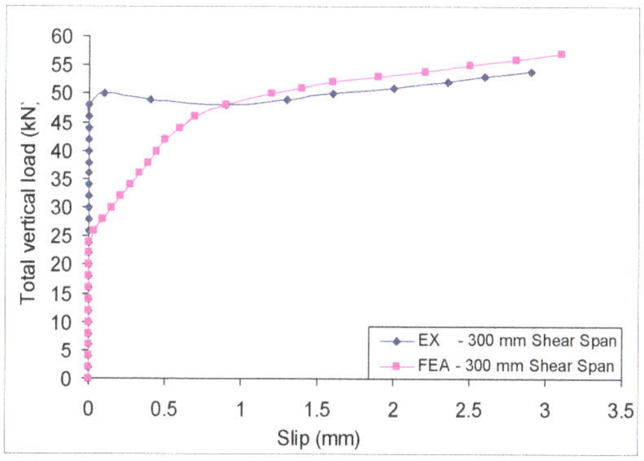

(a) shear span (Ls) = 300 mm.

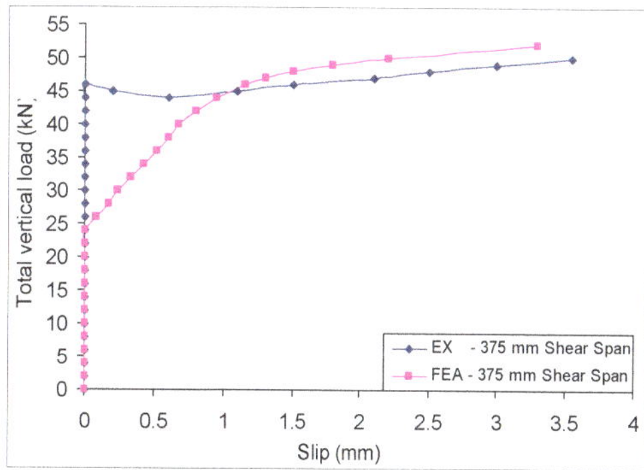

(b) shear span (Ls) = 375 mm.

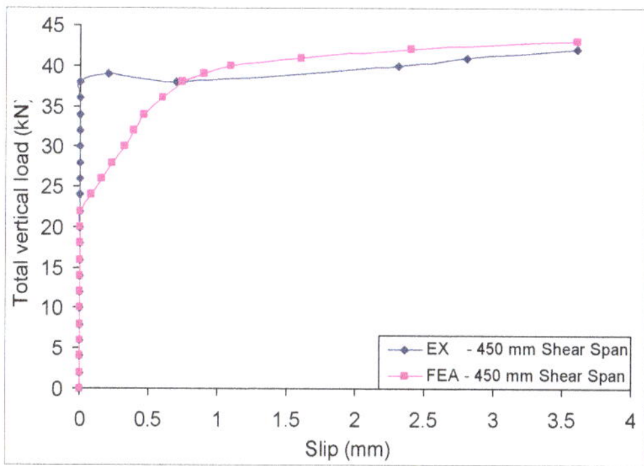

(c) shear span (Ls) = 450 mm.

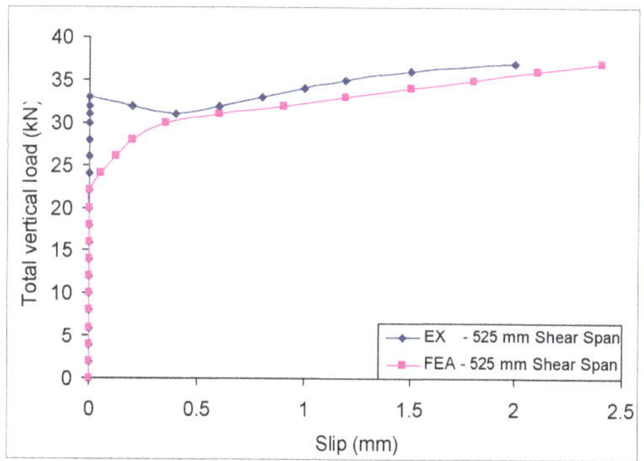

(d) shear span (Ls) = 525 mm.

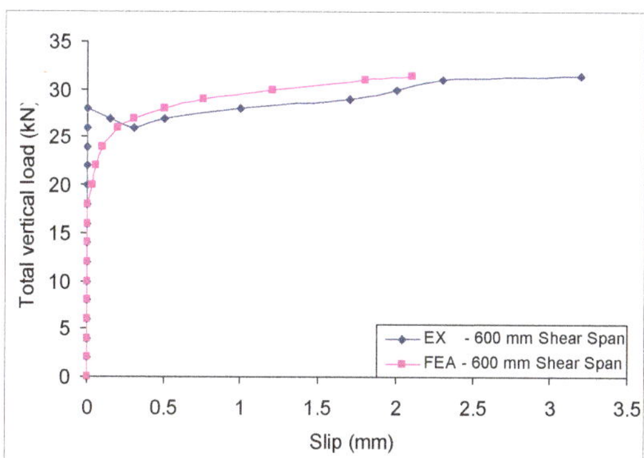

(e) shear span (Ls) = 600 mm.

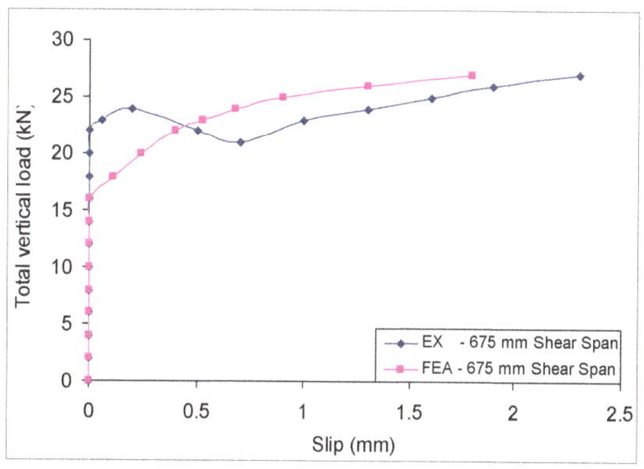

(f) shear span (Ls) = 675 mm.

Figure 17. *Experimental & FE analysis load-end slip curves.*

Table 4. *Experimental, FE analysis & Design load (EC4) results.*

Test No.	Test specimen ID No.	Experimental results				FE analysis results				EC4 results
		Failure load (kN)	Average failure load (kN)	Mid span deflection (mm)	Average mid span deflection (mm)	Average Slip (mm)	Load (kN)	Mid span deflection (mm)	Average Slip (mm)	Design Load (kN)
1	01ST300	54.30		33						
2	02CT300	52.25	53.85	28	29.33	2.9	57	33	3.1	55.32
3	03CT300	55.00		27						
4	04ST375	50.59		31						
5	05CT375	47.10	50.23	28	30	3.55	52	31	3.3	44.25
6	06CT375	53.00		31						
7	07ST450	42.65		27						
8	08CT450	42.00	42.18	28	29	3.6	43	27	3.6	36.88
9	09CT450	41.90		32						
10	10ST525	37.19		37						
11	11CT525	37.00	37.10	37	37	2.0	36	21	2.4	31.62
12	12CT525	-----		--						
13	13ST600	31.52		24						
14	14CT600	31.00	31.17	23.5	23.5	3.2	34	24	2.1	27.65
15	15CT600	31.00		23						
16	16ST675	27.10		27						
17	17CT675	26.00	27.00	26	27	2.3	29	23	1.8	24.57
18	18CT675	27.89		28						

5. Comparison Between Experimental and Simulation Results

The comparison between the experimental and the simulation results are listed below:

- Table 4 shows the comparison between experimental, simulation and EC4 results.
- The failure mode of all composite slab specimens is ductile.
- The load-deflection curve for static and cyclic cases of experiments is in linear range up to 75% to 80% of failure load and then turned in to the nonlinear range. This is due to slight decrease in bearing capacity as initial cracks are developed in concrete, progressing almost up to failure due to the mechanical bond between materials. Mechanical interlock originates from the interaction embossments of the profiled steel decking with the concrete. However, load-deflection curve produced by the FEanalysis is in close agreement with the experiment up to the linear range and is slightly less stiff, while in nonlinear range the curve slightly stiffer and seems deviates away from experimental result at the end of applied load (Figure 16). This difference can occur due to the effect of the embossment which play role in the experiment, but is not considered in the FE model.
- A parametric study has been carried out using FE model for different shear span lengths to examine the effect to modeling technique. It can be seen that, the FE models are approximately 6% higher ultimate loads than the experiments (Figure 16).
- The end slip at the interface of composite slab is monitored after 70% to 75% of experimental failure load, and then the rate of end slip is increased till the maximum load is achieved. In case of FE analysis, the end slip is monitored after 45% to 60% of the ultimate load, and then the rate of end slip is increased till the failure. It shall be noted that the corresponding end slip values from FE analysis did not differ appreciably with experimental results (Figure17).

6. Conclusions

In the paper, an experimental and simulation investigation of composite slabs with trapezoidal profiled steel decking is presented to investigate theultimate load, load-deflection behavior and load-slip behavior with different shear span lengths. Experimental study is based on Eurocode 4: Part 1.1. The three dimensional non-linear FE modeling and analysis of full-size slabs are carried out by considering the intermediate stiffeners on top and bottom flange of the profiled steel decking and interface contact elements using ANSYS 11 has been studied.

From the observation and experimental data presented above, the following conclusions can be summarized as follows:

- Behavior and strength of composite slab depends mainly on shear span. The ultimate failure load decreases from shorter to longer shear span and moves towards the midspan.
- Application of preliminary cyclic loading is carried out as per provisions in Eurocode 4. However, there is negligible effect of the cyclic loading on the load carrying capacity of the composite slabs as compared to static loading (Figure16).
- Failure modes of all experimental specimens are determined in accordance with the Eurocode 4

definition and exhibited a ductile failure.

- For shorter shear spans, strength of slab is governed by shear bond failure.
- For shorter to longer shear span, the behavior of slab is governed by shear to flexural failure respectively.

By the comparison of the experimental and simulation results it is found that:

- The ultimate load, load-deflection curve and load-end slip curve of the composite slabs produced by the FE analysis is in close agreement with the corresponding experimental curves and the slope of the curves is identical.
- The FE model proposed in this paper is accurately describes the interaction of concrete and profiled steel decking in a composite slab.
- The FE slip model also has capability to show the separation of concrete slabandsteel decking.

Acknowledgements

The authors thank Director, VNIT, Nagpur and Head of the Department, Department of Applied Mechanics, VNIT, Nagpur, for their kind support during experimental investigation. The authors express their grateful thanks to Institute for Steel Development and Growth (INSDAG), Kolkata, for sponsoring the research project.

References

[1] Andrade Vellasco. Standardized composite slab systems for building constructions. Journal of Constructional Steel Research 2004; 60:493-524.

[2] Makelainen P, Sum Y. The longitudinal shear behavior of a new steel sheeting profile forcomposite floor slabs. Journal of Constructional Steel Research 1999; 49:117-128.

[3] Porter ML. Ekberg CE. Design recommendations for steel deck floor slabs. J. Struct. Engg. ASCE. 1976; 102(ST11):2121-2136.

[4] Porter ML, Ekberg CE, Greimann LF, Elleby HA. Shear-bond analysis of steel-deck reinforced slabs. Journal of Structural Engineering, ASCE 1976; 102(12):2255-2268.

[5] Wright HD, Evans HR, Harding PW. The use of profiled steel sheeting in floor construction. Journal of Constructional Steel Research 1987; 7:279-295.

[6] Crisinel M, Marimon F. A new simplified method for the design of composite slabs. Journal of Constructional Steel Research 2004; 60:481-491.

[7] Mohan G. Upadhyay A. Kaushik SK. Simplified design of composite slabs using slip block test. Journal of Advanced Concrete Technology 2005; 3(3):403-412.

[8] Marimuthu V, Seetharaman S. Experimental studies on composite deck slabs to determine the shear-bond characteristic (m-k) values of the embossed profiled sheet. Journal of Constructional Steel Research 2007; 63:791-803.

[9] Baharom S. Xiao RY. Comparison between modeling of ribbed decking composite slabs with and without slip in finite element analysis. Proceedings of the eighth international conference on computational structures technology, Scotland, 2006; Paper 127:1-13.

[10] Eldip ME, Maaly HM, Beshay AW, Tolba MT. Modeling and analysis of two-way composite slabs. Journal of Constructional Steel Research 2009; 65:1236-1248.

[11] Tsalkatidis T, Avdelas A. The unilateral contact problem in composite slabs: experimental study and numerical treatment. Journal of Constructional Steel Research 2010; 66: 480-486.

[12] Chen S, Shi X. Shear bond mechanism of composite slabs – A universal FE approach. Journal of Constructional Steel Research 2011; 67: 1475-1484.

[13] EN 1994-1-1: Eurocode 4. Design of composite steel and concrete structures- General rules and rules for buildings. Brussels, 2004.

[14] BS 5950: Part 4. Structural use of steelwork in building: Code of practice for design of composite slab with profiled steel sheeting. London, 1994.

[15] Oehlers DJ, and Bradford MA. Composite steel and concrete structural members-Fundamental Behaviour. Pergamon Press, 1995.

[16] ASCE. Specification for the design and construction of composite slabs and commentary on specifications for design and construction of composite slabs. (ANSI/ASCE 2-84). New York, 1985.

[17] ANSYS Version 11.0, Reference Manual, USA.

[18] Chen S. Load carrying capacity of composite slabs with various end constraints. Journal of Constructional Steel Research 2003; 59:385-403.

[19] Abdullah, Easterling. Quasi-static analysis of composite slab. Malaysium Journal of Civil Engineering 2007; 19(2), 91-103.

[20] Abdullah Redzuan. Experimental evaluation and analytical modeling of shear bond in composite slab. PhD Dissertation, Virginia Polytechnic Institute and State University, Blacksburg, USA, 2004.

[21] Daniels BJ, Crisinel M. Composite slab behavior and strength analysis. Part I: calculation procedure. Journal of Structural Engineering, ASCE 1993; 119(1):16-35.

[22] Daniels BJ, Crisinel M. Composite slab behavior and strength analysis. Part II: comparisons with test results and parametric analysis. Journal of Structural Engineering, ASCE 1993; 119(1):36-49.

[23] Luttrell LD. Flexural strength of composite slabs. Composite steel structures-advances, design and construction, Elsevier, London, 1987; 106-115.

[24] Widjaja BR. Analysis and design of steel deck-concrete composite slabs. 1997, PhD Thesis USA, Virginia, 1997.

[25] Ellobody Ehab, Young Ben. Performance of shear connection in composite beams with profiled steel sheeting. Journal of Constructional Steel Research 2006; 62:682-694.

Procurement of Developmental Projects in Ghana: A Literature Review of the Designer-Led System Verses the Producer-Led Systems

Joseph Ignatius Teye Buertey[1,*], Emmanuel Dadadzogbor[2], Felix Atsrim[1]

[1]Department of Built Environment, Pentecost University College, Accra, Ghana

[2]School of Engineering, Accra Institute of Technology, Accra, Ghana

Email address:

jbuert@yahoo.co.uk (J. T. Buertey)

[*]Corresponding author

Abstract: Procurement is a very important aspect of the chain link process of infrastructural delivery. Through procurement choices in terms of project delivery and construction technology is agreed by key stakeholders. Irrespective of the concomitant challenges of the traditional Design Bid Build (DBB) method of project delivery commonly employed in Ghana, which has been noted for schedule delay, corruption, cost overrun and low quality; project stakeholders still employ this method in the delivery of over 90% of infrastructural project. The purpose of this review is to advance the reasons for the continuous use of the traditional method of procurement, factors hindering the adoption of other procurement methods and explore the possibility of stakeholders embracing other producer-led procurements. As a qualitative based study, the work was based on extensive desktop literature review from journal, periodicals, articles and previous related works. Findings from the study pointed out that the use of the DBB was basically due to familiarity by stakeholders, the small scale of projects executed in Ghana, the competence and capacity characteristics of the local contractor and the form of contract used in the delivery of construction projects. The study concludes that the construction industry in Ghana is not fully ready for a switch to producer-led system as an alternative to DBB since less than 10% of local contractors have the capacity. Again, most of the high profile firms ready for the adoption of DB are foreign-based firms; a switch would disadvantage the local contractor who lacks the necessary capacity. In conclusion, the Ghanaian contractor needs to build capacity in terms of human and technical skills to help position on the right pedestal.

Keywords: Local Contractor, Design and Build, Design Bid Build, Procurement, Designer-Led, Contractor Led, Relationship

1. Introduction

Since the inception of the architectural and quantity surveying practice in Ghana, the traditional system of procurement has been the main procurement method adopted in Ghana [1]. On the level of knowledge of the various procurement systems by contractors, consultants and clients, Kyei [2], ranked design and build second to the traditional system even though it is sparingly used. However, the Ghanaian construction industry continued to use the traditional system basically for reasons of familiarity with its procedures and wider applicability. Other major reasons for its use include size of firms and contracts involved, and because it is also the only procurement system with legal backing through the form of contract. Its advantages include competitive fairness and satisfactory public accountability, relative ease in quality checks, arranging and valuing changes and probably fully developed designs details prior to commencement of construction. The traditional system of procurement is however widely criticized for a number of weaknesses. The system is generally slower to start and open to abuse resulting in less certainty [3]. It commits the client to several contractual relationships and does not encourage coordination and integration amongst project delivery teams, it is adversarial, and tends to encourage the proliferation of small inexperience firms in the market. The concept does not bring the rich expertise of the contractor to bear on the project during the design stage [4]. It is important to note that with the current trends and requirements of the constructions industry in the world over and Ghana in particular, the traditional system of procurement may have to studied and reviewed as the main procurement option for delivering construction projects. This is particularly so since it lacked the 'character' to adequately address the challenges of today's construction needs as in early start of construction, more integration and collaborative working, early cost certainty, risk management and avoidance by clients, better buildability etc. It is therefore time the Ghanaian construction industry started reviewing the strength and weakness of the traditional procurement method vis-vis contemporary systems to enable a re-orientation towards modern contemporary procurement systems for the purpose of strategic positioning. For the purpose of this research the focus shall be a review of the key characteristics, advantages and disadvantages of the designer led and producer led systems [5].

Winter [6], classifies procurement into 3 broad categories: designer-led, management oriented and producer-led as displayed in figure 1 below. The designer-led system is primarily the traditional design and build system whilst the producer led system an amalgamation of integrated procurement systems and discretionary systems.

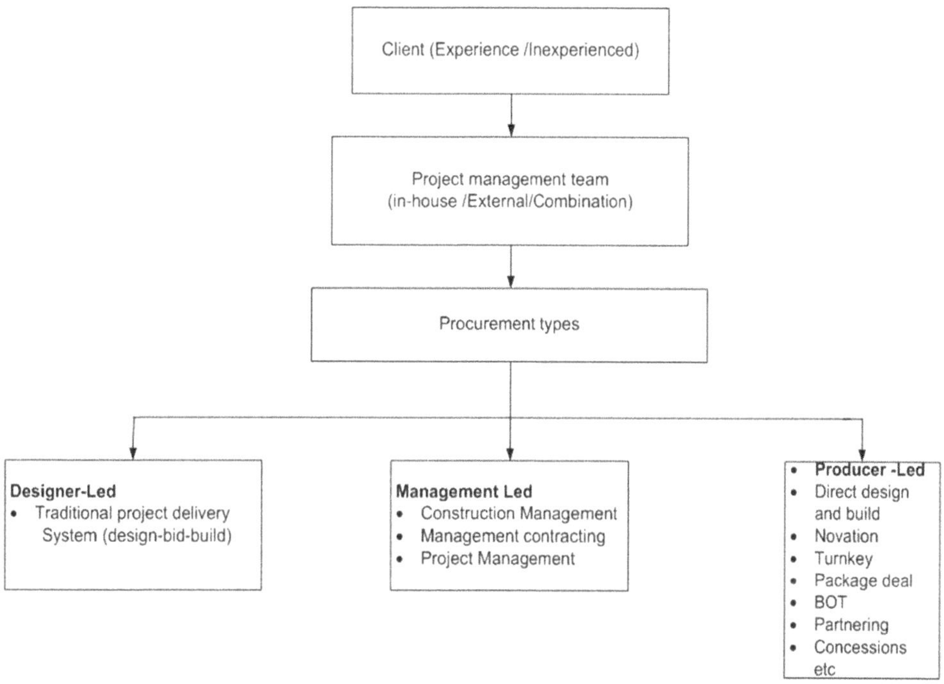

Fig. 1. *Classification of Procurement System. Source [6].*

2. Design and Build Procurement System

According to [5], Design and Build (D&B) is an "arrangement where one contracting organisation takes sole responsibility, normally on a lump sum fixed price basis, for the bespoke design and construction of a client's project". This means that the responsibility for design and construction rests with one organisation. [2], describes D&B as a procurement system where the client manages the design and construction process by using a single point of contact, i.e. the main builder, usually referred to as the Design-Builder. This is displayed in figure 2 below. In their book entitled Procurement, Tendering and Contract Administration, [2], referred to D&B as an arrangement in which the contractor undertakes both to design and to construct a project for a single contract sum. In addition, [9], described the system as a method of procurement that enables one building contractor, or a construction company, to take the full responsibility and to carry the sole liability for the design and construction of the building.

In agreement to the above definitions, the Construction Round Table (1995) in "*Thinking about Building: the Business Round Table Ltd.*" sees the procurement approach as design combined with construction. This source stated that, D&B is the arrangement where the finished building is bought from a single contractor who is responsible for its design and construction. In effect, apart from the construction role, D&B essentially combines all the fundamental tasks in the construction project design, production and the management in one single package. Various authors and construction professionals give several definitions and explanations of D&B, but the simplest idea of the system that is contained in all definitions and descriptions is that the contractor takes the responsibility for both the design and construction for a lump sum [10] as shown in figure 2. The system is generally a procurement approach that encompasses variants such as novation, develop and construct ('turnkey'), and package deals, which are dealt with in much more detail later.

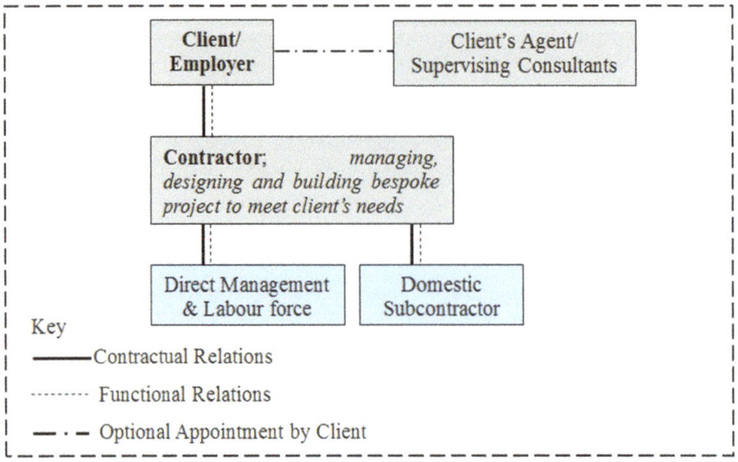

Fig. 2. *Contractual and Functional Relationships in Design & Build Source: [7].*

Once the main contractor or the contracting organisation takes sole responsibility for both the design and construction, the contractual and functional relationship between the construction team becomes simpler when compared with most other procurement systems. The contractual and functional relationship between the various members of the project team is illustrated in figure 2 above.

2.1. Evolution of Design and Build

Design and Build (D&B) is believed to be the oldest procurement system that is still in use, preceding the emergence of architecture as a profession in the middle of the eighteenth century. During these early times, it was first the client, then the architect and finally master builder who was solely responsible for both the design and construction of most building projects. Hughes [11] indicated that, the emergence of the traditional general system of procurement in the nineteenth century was as a result of separation of the responsibility for fabrication from that of design, and which saw the dominance of the traditional system of procurement. The period between the middle of the eighteenth century and the middle of the twentieth century, which marked the renaissance of architectural profession, saw the 'traditional' general system of procurement dominate all other forms of procurement. However, in the late 1960's and early 1970's, or the period after the Second World War, D&B begun to re-emerge from its period of dormancy but only initially to answer the needs of ambitious targets set by the UK government for the public housing sector [7]. At about the same time, there was a greater use of D&B system for industrial and commercial projects in the USA, and gradually, following the lead from across the Atlantic, private-sector clients began to adopt the approach which was vigorously marketed by contractors. These developments therefore, culminated in significant increases in the level of popularity and usage, of D&B system, in UK and across Europe from the 1970's to date. The D&B began to emerge as the dominant procurement method in the construction industry in the 1970's and 1980's. It was used for more and different types of industrial buildings, hotels and repetitive housing schemes. The growth of D&B and its market share are further discussed below.

2.2. Features of Design and Build

Design and Build (D&B) is characterized by certain essential features that are unique to this type of procurement. These features according to [12], are best dealt with in terms of how the client/employer describes his requirements (employer's requirements), how the contractor proposes to achieve them (contractor's proposal), the pricing mechanism, and the roles and responsibilities within the project delivery process.

2.2.1. Employers' Requirements and Contractors Proposals

The process of D&B commences with a client (employer) approaching a contractor with a set of requirements known contractually as Employer's Requirements, defining what he (employer) wants. The contractor responds with a proposal, also known contractually as Contractor's Proposals, which include production and design works [12]. The extent of design work carried out by the contractor usually depends on the level of preparatory designs commissioned by the employer. The contractor's design work ranges from detailing the employer's brief to the full design process with proposals of sketch schemes and information on fabrication [11]. Upon the agreement of the employer's requirements and the contractor's proposals, the contract can be entered into and the work executed. At this point, the contractor assumes total responsibility for undertaking the outlined design and for the construction of the bespoke project, including the integration and co-ordination of the entirety of the process. The contractor is equally responsible for appointing consultants (if he does not have the necessary skill within his/her firm). According to [3], the client may also wish to appoint his/her own consultants to monitor various aspects of the project, even though this is not always the case.

2.2.2. The Price Mechanism

"One of the commonest features normally present in most D&B contracts is a guaranteed maximum price (GMP), which helps to reassure the employer that he/she is not signing a blank cheque" [12]. In the Joint Contract Tribunal, The Design and Build Contract, the price for this system is governed by what is referred to as Contract Sum Analysis (CSA). The CSA differs from the traditional bill of quantities (BoQ) in nature, and its form is not determined by the contract. It is prepared in any form appropriate to the project but most are parallels to the BoQ, i.e. calculations of stage payments. In the Engineering and Construction Contracts (The New Engineering Form of Contracts) 3rd Edition, 2005, option clause A (Priced contract with activity schedule) give a useful guide on pricing of D&B [12].

2.2.3. Roles and Responsibilities of Parties Under D&B

The D&B form of procurement also differs from other forms, in that the arrangement exhibits a lack of an independent certification role in the contract. Under the procurement system, there is no contract administrator to settle differences between parties, neither is there an independent quantity surveyor responsible for preparing the basis upon which contractors tender. [11], stated that "*this changes some of the basic assumptions about the roles which are required on construction projects*". According to [8], the DB does not provide for the appointment of an architect or a quantity surveyor by the employer, instead an Employer's Agent is appointed who acts on behalf of the employer and receives or issues applications, instructions, consents, notices, requests or statements in accordance with the conditions.

In the D&B system of procurement, the contractor is responsible for 'everything' (single-point responsibility). Clients are quite attracted to this single point responsibility, especially those who may not be interested in distinguishing the difference between a design fault and a workmanship fault [12]. Once the contractor is responsible for the design under the contract, he bears the same professional liability as a consultant designer. The contractor is therefore expected to exercise reasonable skill and care expected of a competent designer. Consequently, greater risk in the project is transferred to the contractor owing to the single point of responsibility that rests in him/her [3]. Due to the numerous advantages and inherent flexibility of the D&B procurement system, it is applied to a wide range of buildings. The common characteristics of the projects to which D&B is ideally used, are seen in the nature of the employer's requirements regarding risk apportionment, the nature of the client's experience and the availability of construction firms suitable to undertake project on D&B basis [12]. These features are considered under the following areas;

- The client's familiarity with the construction
- The relative importance of client's priorities (time, cost, function, quality, value for money etc)
- The technical complexity of the project
- The need to make variations to the requirements as work progresses
- The pattern of responsibility and communication
- The need for an early start and completion

2.2.4. The Processes Involved in D&B

After identifying the need for a building, the client states his/her requirements adequately in terms of physical design needs and the intended use of the building. According to [14] from figure 3 below, a selected number of building contractors are invited to submit their proposals together with their estimated cost. The system invokes design competition among contractors, which is absent in other procurement systems [9] and permits the optimization of design and production costs.

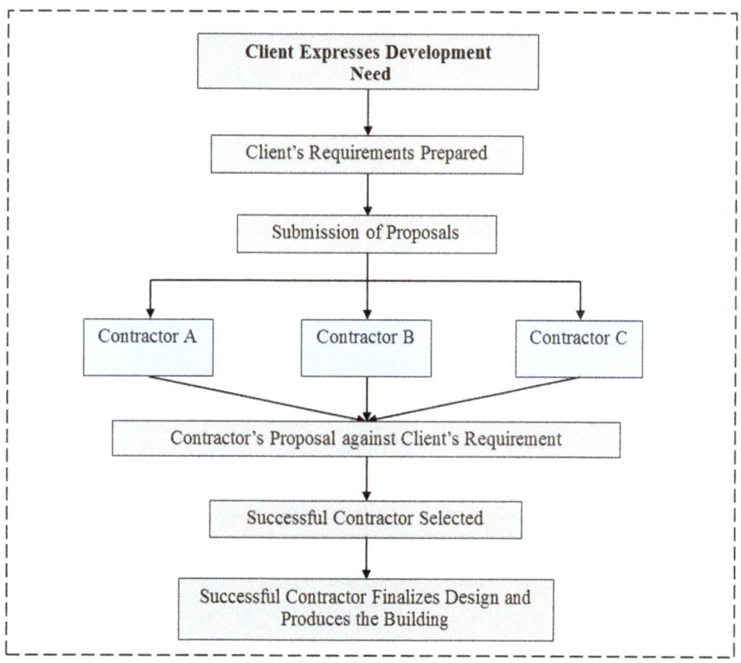

Fig. 3. *Design and Build (D&B) Process [14].*

The D&B system is suitable for standard buildings: industrialized such as factories and warehouses, office buildings/complexes, residential flats and complexes, educational and/or institutional buildings and hotels. On large complex or specialist projects, D&B companies may employ, or appoint a designer from consultancy firms. In such cases, the appointed designer's responsibility is to the D&B Company and not directly to the employer. However, the client normally appoints an agent to look after his/her interest, and to ensure that the contractor's proposal receives planning approvals [9]. [14] describes the process involved in D&B procurement system as illustrated in figure 3 above.

3. Development of Design and Build in UK and Europe

Since its emergence in the 1970's and 1980's, D&B has experienced a continuous growth in market share as it is used for more and different kinds of buildings. Over the years, this method has greatly sharpened and oriented the UK construction sector with numerous advantages. D&B was trumpeted as the ideal way to avoid delays and cut down on costly claims and litigations. It is claimed, by the UK construction industry, to have produced measurable cost and time benefits in the construction of industrial buildings, hotels and repetitive housing schemes. In the USA as well as the UK, D&B experienced a rapid growth in market share from less than 10% in the early 1980's to 23% by 1990, and was further up to more than 30% by the early part of 2000. According to [15], figures published in the Royal Institution of Chartered Surveyors (RICS) "*Contracts in use survey*" (1994) indicated that in 1993, D&B accounted for 35% (by value) of contracts. Currently, D&B's market share in the UK construction industry stands at more than 43% [15]. In the USA, the increase in the number of D&B projects has been astounding. D&B has risen from 15% in 1990, to 35% by 1999. Compared with the traditional procurement system, D&B exceeds construction speed by 12% and total project delivery speed by 30%. Customers of D&B reckoned that the single point of responsibility, a guaranteed maximum price and avoidance of design and construction risks, are the most important reasons for using Design and Build [16].

The D&B system increases the likelihood that the building will be constructed within the owner's budget. Projects procured under D&B are 50% more likely to be completed on time, and on budget [16]. The system makes for better co-ordination amongst the various professionals involved in the construction project. According to Rimmer [17] of Slough Estates, "*the biggest surprise is that D&B performs well on more complicated, hi-tech buildings rather than simple ones*". The present trend, coupled with the numerous advantages (that are discussed later in section 3.6) that D&B offers, is an indication that, the system could soon dominate all construction procurements especially in the non-public sector [15]. Design and Build (D&B) system of procurement encompasses four main variants, stated earlier in sections above.

The four variants discussed here include turnkey, develop and construct, novation and package deals. It is however important to note that, like all other procurement systems, many other less common variations of the method may be in use within the construction industry.

3.1. 'Turnkey' Method

According to [9], the D&B approach is referred to as "turnkey" when it includes the complete equipping and/or staffing and commissioning of the building. Under the turnkey system, one organization, generally a contractor, is responsible for the total project from design through to the point where the key is inserted in the lock, turned and the facility is immediately operational [7]. It is stated that the responsibility of the contractor under this variant often extends to include the installation and commissioning of the client's process equipment and sometimes the identification and purchase of the site, recruitment and training of management and operatives as well as arrangement for funding of the project. The operation of the project could equally be taken up by the contractor as part of his/her responsibility under this variant, especially through a private finance initiative (PFI). A typical contractual relationship between the client and the contractor is illustrated in figure 4 below.

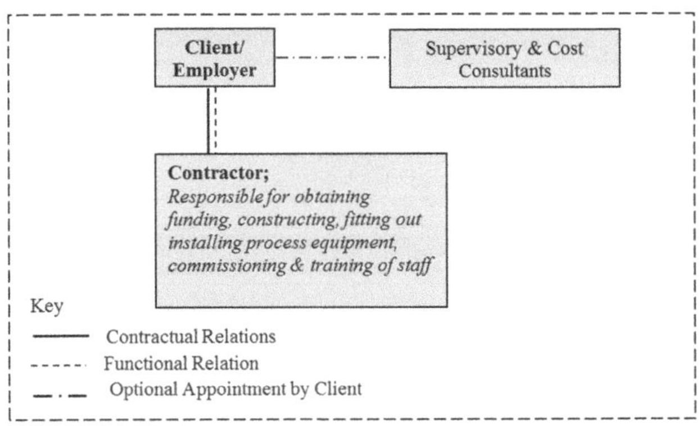

Fig. 4. *Contractual and Functional Relationships in 'Turnkey' Source: [7].*

The 'turnkey' method according to [7] "was pioneered in USA in the 1900s, where it has been extensively used since that time by the private sector, for the construction of process plans, oil refineries, power stations and other complex production facilities". The system has been used sparingly in UK as the amount of work carried out in the industrial and commercial sector appears to be small by comparison to the USA. However, the introduction of the PFI by the UK government in the early 1980's, resurrected the concept which is being used to build and operate major public projects such as the Channel Tunnel, the Dartford river crossing and more recently, various hospitals, prisons, with private finance [3]. From a construction viewpoint, this arrangement echoes all the features of D&B, and provides the client the advantage of being able to take over a fully operational facility. Where PFI is used, the scheme reduces public-sector capital expenditure in the short term while establishing a commercially viable development in the future. These benefits notwithstanding, [7] asserts that the cost to the client of using the 'turnkey' method can be higher than other more conventional procurement systems.

3.2. Develop & Construct

Under this variant, the client appoints a consultant to design the building to a certain stage and then invites tenders from contractors to develop and complete the design either using the client's consultants, or their own designers and to construct the building [3]. With this approach, the client provides his consultant with a detailed brief, which in some cases he has helped to formulate and from which conceptual drawings/sketch designs and a site layout, including the disposition of individual buildings and their plan forms are prepared [7]. The contractor develops the conceptual design, produces detailed drawings, chooses and specifies materials and submits these proposals with a bid in the same way as with design and build proper. The figure 5 below illustrates the relationship between the construction team under a Develop and Construct arrangement.

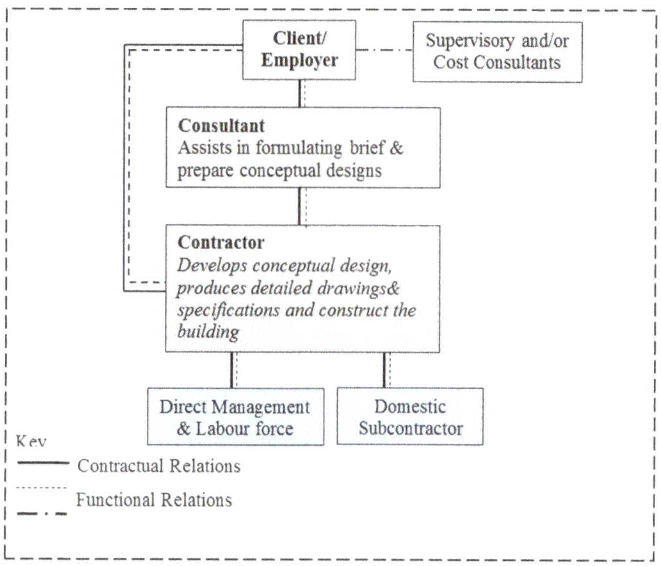

Fig. 5. *Contractual and Functional Relationships in Develop & Construct Source: [7].*

The most appropriate circumstances under which develop and construct form of procurement operates is where the client desires to determine the detailed concept of a project before inviting competitive tenders and yet still requires a single organization eventually to take responsibility for the detailed design and execution of the project. Develop and construct variant differs from the rest in the extent to which the design of the project has been developed by the client before inviting tenders [7]. In most cases, the design will be developed at least up to outline planning stage and may, in sensitive planning locations, be taken to the point where full planning approval could be obtained. The method is said to be frequently used where the client:

- Sees advantages in using a consistently retained consultant with previous experience of similar types of projects
- Employs his/her own in-house consultant
- May wish to limit knowledge of his/her intentions to a trusted few
- Wishes to minimize the differences, so often experienced at tender stage, among normal, individual design and build submissions.

The main benefits associated with this approach as espoused by [10] are: full integration between design and construction through collaborative working, overlapping of the design and procurement without risk of un-priced design development, the reduced need for an additional shadow design team especially where the original team is "novated" and progressive co-ordination. "*The basic criticism for develop and construct lies in possible dispute owing to the involvement of both the design consultants and the contractor in the design of the project*" [7]. Once again, all aspects and characteristics of Develop and Construct echo those of the other forms of D&B which are together classified as "integrated methods". The form of contract for this variant is same as those previously described for the parent design and build approach.

3.3. Novation

Novation is one of the common arrangements under D&B contracts that emerged in recent time, and takes the form of an extended develop and construct. By this approach, the client appoints an architect to design/develop an outline, and the contractor is then made to contract the (same) architect to further develop the details of the outline and to construct the bespoke project [3]. According to [18], "*novation is a form of design and construct agreement in which the client initially employs the consultant to carry out design and documentation to an extent that the client needs and intent are clearly identified and documented (pre-novation contract). On the basis of these documents, tenders are invited and a building contractor selected. The client then novate the consultant's agreement to the contractor (post-novation contract), who then takes responsibility for the project completion"*. Masterman [7], noted that the tender documentation will normally contain details of the client's consultants and the proposed novated procedures, together with a requirement that the eventual winner of the contract will have to accept responsibility for the total design of the project, including the initial work carried out under the client's auspices. The novated arrangement should enable the design of the project to proceed more smoothly from the pre-contract to the post-contract stage. *"Once the contract is novated, the consultants will no longer be available to provide advice to the client on the detailed design, and it is therefore likely that he will need to employ new consultants to examine the final scheme and to confirm or otherwise its suitability"*.

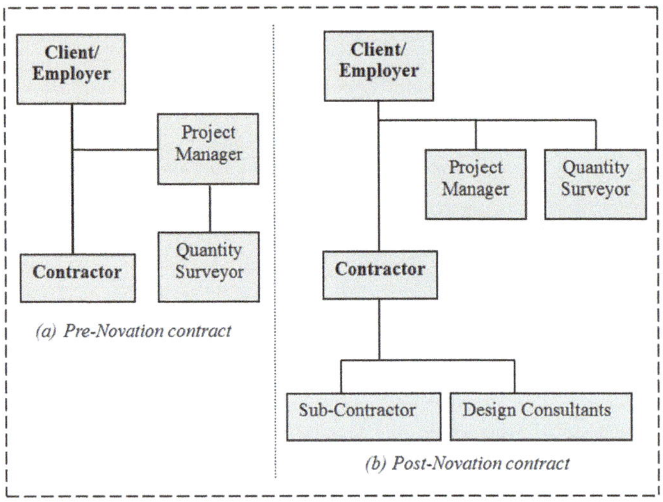

Fig. 6. *Contractual Arrangements in Novation Source: [18].*

Maintaining the same design consultants through all stages of the process which seeks to ensure that design standards are consistent throughout the pre-contract and post-contract phases of the project is reckoned as the main benefit of this system. Siddiqui [18] identified novation to offer additional advantages in several respects; certainty of time and cost, single point responsibility, better transfer of risks, better buildability, less adversarial relationship, easier co-ordination, less bidding competition etc. Figure 6 above, illustrate the contractual relationships between the project team under novation procurement. Like any other construction procurement system, novation cannot be a panacea to all the problems confronting the industry and is indeed beset with certain flaws. It was observed that the main of novation are less control over design quality and less flexibility. As observed by [7], problems can arise when the contractor is compelled to adjust to some predetermined consultants rather than being able to choose his own designers whom he may have successful working relationship with on projects of such nature. This kind of 'forced' arrangement may well produce a less than happy team and productive team compared to the results of a design and build arrangement

According to [7], "while many successful projects have been carried out using this method of procurement, there appears to be no general consensus among any of the participants as to its suitability for use on the majority of design and build projects". However, the University of Reading (1996) report found that novation led to the worst possible outcome for D&B projects, where just about 28% of client's quality expectations were met [19].

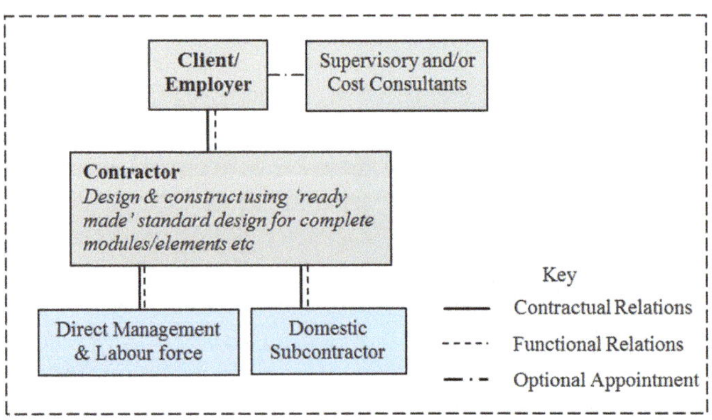

Fig. 7. *Contractual and Functional Relationships in 'Package Deal' Source: [7].*

3.4. Package Deals

The package deal system is reckoned to be the antecedent of D&B method of procurement. The original concept was to give clients the opportunity to purchase a complete 'package', virtually off the shelf, in order to satisfy speedily their building needs at an economic price [7]. The unique features of a package deal are that it uses proprietary building systems to produce scheme. The most fundamental difference between package deals and the other variants is that whereas other variants provide a custom-made design solution to suit the client's specific requirements, package deal uses proprietary systems for the delivery of the scheme. The majority of package deal contractors by their very nature, employ own in-house designers and can thus be categorized as pure Design Builders. These contractors are therefore better disposed to perform well, particularly as a team, and to deliver with speed. It is argued that some of the products of this method lack aesthetic appeal. However, once *"the potential client is often able to see actual examples of the contract's product before reaching a decision this potential difficulty can often be avoided"* [7]. Figure 7 above illustrates package deal system and the various relationships between the members of the project team.

The package deals contracts also replicate most of the characteristics of the D&B system. The forms of contract used are likely to be contractor drafted rather than any of the recognized standard forms and care must be taken by the client if this type of contract is to be used. One outstanding advantage of the system is that the client is usually able to see actual examples of the product in real situations, and be able to assess their practical and aesthetic appeal before entering into the contract (purchase). Also, if the client's requirements are flexibility, the method becomes an attractive proposition particularly as the probable reduction in the design, approval and construction stages of the project can lead to savings in time and cost. The main criticism of this approach lies in the fact that package dealers are unable to satisfy fully the needs and criteria of the majority of clients because they provide an adapted standard product. Some serious structural failures have occurred amongst some of these proprietary systems. The product of this procurement system have also suffered from other less serious defects as a result of poor design and detailing [5]

3.5. Engineering Procurement and Construction

In recent times, the design and build has been branded into an Engineering Procurement and Construction (EPC) for huge engineering projects. The features of the DB and EPC and similar: In both cases, the contractor carries out the engineering design for the project, procures all equipment and materials necessary and construct to deliver a functioning facility. During the execution of an EPC, the client usually procures the services of a Project Management Consultancy team to manage the project on behalf the client. Usually the team that does the Front End Engineering Design (FEED); a preliminary design which forms the basis for the bidding and selection of the EPC contractor. It must be however emphasized that the EPC is different from EPCM (Engineering, Procurement and Construction Management), which is a core professional service contract.

Generally, the EPC contractor has to execute and commission the project within agreed time, lump sum budget and according to the agreed deliverables. This places the risk of time and schedule on the EPC contractor. It is particularly associated with developing and operating facility eg hotel, mine, power plant, water station etc and functions as a turnkey contract when it is expected to commission the facility.

The EPC contract provides a single point of responsibility, provides a fixed time and a fixed cost. In the EPC, procurement is the responsibility of the contractor, who will as well guarantee the achievement of certain performance standards, efficiency and reliability of the facility.

4. Benefits/Advantages of Design and Build

The D&B procurement system, which encompasses amongst others, the variants discussed earlier provides a wide range of benefits both to the client and the contractor. Some of the benefits (advantages) associated with D&B are stated below.

(1) Hackett [8] identified early certainty of contract price as a major advantage, especially where the JCT form of contract is used. According to the USA's Legislative Analyst's Office (LAO) Report (2005), D&B system offers price certainty because the employer (agency) specifies what he is willing to pay for a building before proposals are solicited from D&B contractors. Interested contractors respond with configurations, material specification and methods of fabrication that they are willing to provide for the specified price. [7] observed that once the client's requirements are accurately specified, certainty of final project cost could be achieved. It has also been established that a feature sometimes present in D&B deals is the guaranteed maximum price (GMP), and where this is so, the client has a feeling of reassurance that he is not "signing a blank cheque" [12].

(2) D&B systems increase the likelihood that the building will be constructed within the owner's budget. Projects procured under D&B are 50% more likely to be completed on time, and on budget [16]. Contractors often can provide better prices and information regarding construction methods, than architects. The contractor is able to conduct a value engineering and constructability analysis from the start.

(3) Another important benefit associated with the D&B system of procurement as put forward by [7] is that of single point responsibility. *"The single point of contact between the client and the contractor that is unique to D&B system means that the client has the advantage of*

dealing with a single organisation that is responsible for all aspects of the project. Expressing a similar viewpoint [19], stated that direct contact between the client and the contractor is one of the numerous advantages claimed for the use of a D&B arrangement. In support of this claim, [18] indicated that the responsibility for the design, construction and the required performance of the building lies entirely with one party, the contractor.

(4) D&B involves the contractor at an early stage of the development process, and this results in a greater degree of co-ordination between the members of the team. Given this co-ordination and single point responsibility, variations during construction tend to be fewer and risk of post-contract price escalations are reduced [8]. [11] argues that the D&B process increases the opportunities for harnessing the benefit of the contractor's experience during the design stages of the project. Some professionals and stakeholders in the construction industry recognize that the involvement of the building contractor in the design process from the beginning provides helpful insights on the construction materials and methods used, and that can make the design more efficient and less costly to construct. Good teamwork and cohesion between the various experts/specialists to function as a unit, engenders buildability [3].

(5) Furthermore, [19] noted possible reductions in the overall timescale of the project as an advantage unique to this type of procurement system. D&B imposes a discipline on the employer to define the brief fully at an early stage, making it possible to overlap the design with construction and thus leading to shorter project durations [8]. Adding to this, [7] noticed that the integration under D&B, enabled overlapping of the design with the construction and improved the communication between the client and contractor.

(6) One other benefit of using the D&B system of procurement in the UK and Europe as observed by [19], is that it offers little/no claims for possible delays due to lack of drawn information. In supporting this assertion, [14], stated that D&B was trumpeted as the ideal way to avoid delays and cut down on costly claims and litigations. Because the designer and builder are part of the same D&B entity, and especially with the employer not being the guarantor of the completeness and accuracy of the work of the architect/engineer, the employer may avoid conflicts and disputes that can arise between the architect, or the engineer, and the main contractor.

(7) In An Analysis of the JCT D&B Contract in UK, [11] suggests that one of the strengths of D&B is that the contractor's proposals will normally include design solutions to problems posed in the employer's requirements. By this, contractors are not only competing on price as in most other procurement systems, but also on any other criteria important to the client. The method through such an approach motivates the use of innovation and creativity while ensuring buildability.

5. Disadvantages of Design and Build Procurement

Notwithstanding the above benefits that the construction industry stands to enjoy, and indeed has enjoyed through the use of D&B, the procurement system is criticized for a number of shortcomings.

(1) The greatest difficulty with the D&B system is that the employer is required to provide a great deal of project information/briefing with all necessary operational considerations at the outset of the development, usually referred to as Employer's Requirements. [11] stated that *"one of the biggest disadvantages of D&B is that the brief (in terms of the employer's requirements) must be clear and unambiguous at a very early stage"*. Inadequate and hazy briefs which are unable to communicate precisely to the contractor the client's need, and may engender difficulties in evaluating proposals and tender submissions [11].

(2) With the D&B system of procurement, the design and construction work generally is awarded based on subjective criteria such as experience, qualifications, and best value for money. Where several tenders are invited, comparison can be difficult as the end product in each case is different. The final decision is most often than not influenced by subjective judgment [8]. In support of this criticism, [3] indicated that there is a relative difficulty in comparing tenders/bids under D&B since the process is often characterized by the subjective opinion of the evaluator. Ashworth [20] noticed that D&B is found to be unwieldy where it is necessary to provide competition between building firms, as it poses difficulty in evaluating a project across a wide range of attributes.

(3) D&B is also criticized for its limited assurance on quality control. [7] asserts that *"although well-designed and aesthetically pleasing buildings can be obtained using D&B, the client's control of this aspect of the project is less when compared to other methods of procurements"*. There is always a risk regarding quality and standards as observed by [8]. The view expressed suggests that where the client's brief is not precise and the specifications offered by the contractor are vague, there is likelihood for the builder to reduce standards. Postulating further that the benefits attributed to the system often amount to a mere shifting of the risk from the employer to the contractor. Private sector clients for instance, believed they are escaping having to pay design fees, and that *"the major item that suffers from this is usually overall quality"*.

(4) D&B is a very rigid method of procurement and despite some level of tolerance for variations; it does not lend itself to the developing requirements and ideas

by the client. This means that any variation required by the employer, especially after signing the contract, can be expensive and difficult to evaluate. [11] did state that a client who wishes to reserve the right to make extensive alterations to the requirements during the construction and fabrication process, should not use the D&B system of procurement.

(5) Limited access for small contractors as mentioned before is another shortfall of the system of procurement. Because D&B contracts mostly are awarded based on qualification and experience, it tends to lockout small newly established contractors who may not have the range of experience possessed by large, long-established firms. As a result, access to D&B contracts, especially the large contracts, may be limited for these smaller contractors. The presence of fewer firms in the D&B market is reckoned to a minimal extent, as a disincentive to the use of the system [3].

6. Challenges Inherent in the Adoption of D&B (Firms and Products in a Design and Build Market)

The manner in which D&B contracts are organized can profoundly affect the quality of the final product. [7], indicated that the recommendation to use the system must therefore be based not only on the type of building required, but also upon the client's expectations in terms of programme, cost in construction, cost in use, risk level, level of specification and quality of design.

The D&B systems or contracts are suitable for a range of standard buildings:

- Industrialized buildings such as factories and warehouses where an early return on capital outweighs design excellence considerations.
- Building using proprietary systems where the manufacturer of the system might well become the main contractor, examples include repetitive housing or low cost hotels, office buildings, residential flats and complexes, educational and/or institutional buildings and hostels.
- On complex/specialized projects for which some contractors are specialists.

By its very nature, D&B requires a well-established and highly experienced firm to undertake such projects. The firm, or organisation, should be capable of employing architects, engineers and/or quantity surveyors either as in-house members, or on fee consultancy basis. According to [3] very few specialist construction organisations undertake D&B contracts, since most firms in the construction industry are smaller or medium-sized and may not have the resources to engage the services of professionals. The single point responsibility, coupled with the fixed prices, imply that much risk is borne by the D&B contractor than would have been under general contracting. This risk carries a premium according to [12] and it is to be expected that the D&B contractor would add this factor to a tender in order to allow for the extra risk. Figure 8 below illustrates the level of risk (speculative) that the contractor bears under D&B contract in relation to the level of risk borne under other forms of procurements [5].

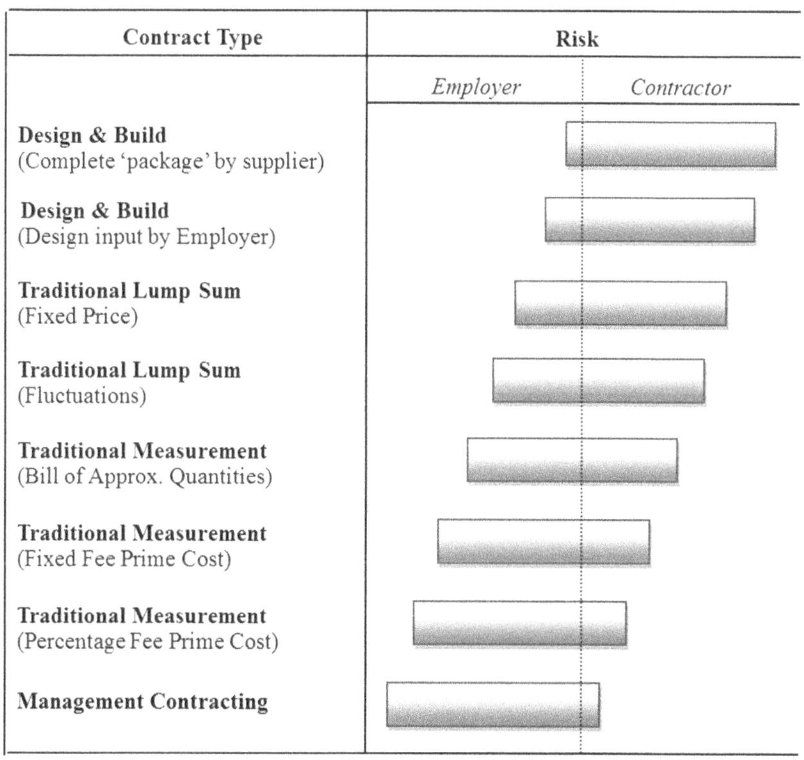

Fig. 8. *Speculative risk by Contract Type Source: [3].*

7. Design-Bid-Build Procurement System

The traditional system of procurement, also called the conventional system, involves discrete design development, tender and contract award and the construction delivery phase. Each phase in theory is separate and distinct as depicted in figure 9 below [21]. The procurement process begins when a client commissions an architect to design a facility; on completion of the design and documentation, contractors are invited to tender for the work. The client subsequently enters into a contract with the successful tenderer to execute the construction process. This process requires that design development is close to 100% complete before tenders can be invited. In practice, however, there are many design issues left incomplete and unresolved requiring amendment or further detailing during the construction process. Open tendering and pre-qualification are the two methods adopted during the tendering process in the above process [21]. The above system has as well 'cast roles in stones' for all parties, making it difficult to negotiate outside the risk of the contract. Another major challenge is that the contractor is isolated during the design process, without any input into the management and buildability of the project. The above has high cost implications since the rich experience and advice of the contractor on specification and cost is not brought to bear in this procurement process. Research has shown that the above results in high contract claims from the contractor. The designers and employers have highlighted the poor design coordination, subsequent design changes to make design detail workable. The implication is that in the adoption of the above procurement process, a client can only be absorbed off the vagaries of this challenge through the incorporation of higher design contingencies and allowances [22].

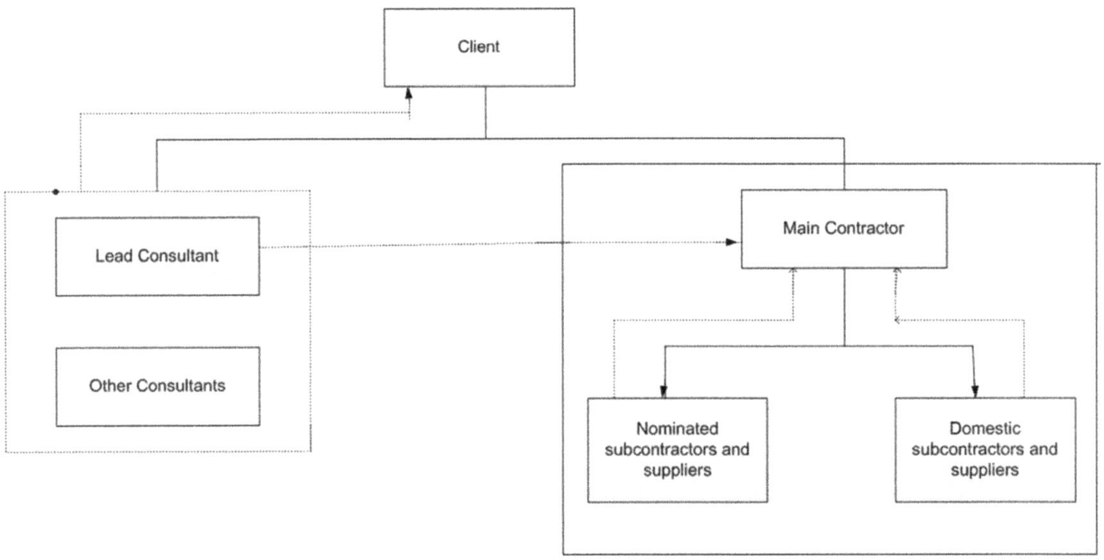

Fig. 9. Traditional Procurement System Source: [22].

8. Factors Promoting the Use of DBB in Ghana

As stated in [9], a number of factors tend to encourage the continuous use of the traditional system of procurement even when it has become clear that the increasing complexities of today's construction industry requires the adoption and use of other contemporary methods as alternatives, or to be used alongside the DBB system.

8.1. Long Usage

Long usage of the design-bid-build (DBB) system of procurement has made it become more of an in-house kind of practice and the players of the industry have accustomed themselves to its procedures. [23] noticed that the system dominates the Ghanaian building industry largely because it is well established with wide applicability and simple procedures. [3] asserted that one of the advantages of DBB that accounted for its continuous use is the fact that its procedures are well known in the construction industry. This popularity and long-standing reputation of the procurement system in the Ghanaian construction sector makes it difficult to introduce new and contemporary procurement systems.

8.2. Government as Major Employer

DBB system of procurement has been preferred to the more contemporary approaches especially in Southern Ghana as a result of the fact that for a very long time, central government, through its Ministries, Departments and Agencies including assemblies have been the major employer (client) in the construction industry [24]. Usually, the agency first award an architectural and/or engineering firm (consultant) a contract to design the project based on subjective criteria of qualifications and experience. This contract generally accounts for a relatively small portion of total costs of the projects, between 5% and 10% [25].

After detailed project plans and drawings including estimates are completed, then funding/budgeting is sought and subsequently a contractor is selected to execute the project. In almost all cases, contracts for construction work are awarded on the basis of competitive bidding [2]. It is this process from single source clientele who basically uses the traditional method for all its procurements of construction projects, significantly accounted for enrooting the system in the construction industry. The effect is that the use of other contemporary systems of procurement is indirectly discouraged.

8.3. Form of Contract

It is also reckoned that the DBB system of procurement will continue to dominate the Ghanaian construction industry largely because of the statutory backing that it receives as against other alternative procurement approaches. The standard 'form of contract' that is used for building contracts is the Ghana government's Articles of Agreement and Conditions of Contract for Building Works *(The Pink Form),* Edition 7 being the latest [23]. There is also the Conditions of Contract for Works of Civil Engineering Constructions (FIDIC) [26], 4th Edition (1992). These documents have been the main forms establishing the responsibilities of the parties to a building and civil projects respectively, and are used for general contracting. Despite several amendments to the *Pink Form* up to the current edition (7th Edition), the document did not state clearly the specific roles of the employer and the other members of the project team especially where alternative methods of procurement is used. Therefore, the players in the construction industry 'stayed' with the traditional form of procurement. Notwithstanding this, a limited number of projects have been delivered in Ghana using contemporary procurement arrangements such as design and build (D&B), management contracting and construction management [24]. Other legislation, like the Public Procurement Act (Act 663) of 2003 [27], which seeks to guide public procurement procedures in public entities in particular, is however mute on any other form of procurement apart from the common practice of activities that are explained under the traditional procurement approach.

8.4. Kind and Magnitude of Projects Involved

[28] Identified the size and/or magnitude of construction contracts in Ghana as a significant factor promoting the continuous use of DBB system of procurement. Until quite recently, the kind of designs and structures put up were so simple, ranging from single to a few storey heights buildings such as school buildings, residencies like Social Security and National Insurance Trust (SSNIT) Housing Schemes, Low Cost Housing Schemes, office accommodations etc. [24]. With particular reference to Southern Ghana, most projects are single storey with straight and simple geometry that do not require extensive engineering or multi-faceted technical applications. This therefore does not encourage the use of newer/alternative procurement options such as D&B which is said to perform well on complicated and hi-tech projects [17], and which is the focus of this research. It is recognised that the traditional system of procurement (DBB) works best where the kind of projects involved are simple, and size or magnitude smaller. Complex and hi-tech projects will definitely require greater integration and co-ordination of all facets of the project to which modern procurement systems such as D&B is best at delivering.

8.5. Size and Nature of Firms in the Construction Market

Furthermore, the sizes of the firms in Ghana's construction business contribute significantly to the reliance on DBB system of procurement. Most of the registered construction organisations (firms) in Ghana are relatively small in size with single proprietorship [2]. Most of these firms are registered with the registrar general as limited companies and hence have the legal mandate to operate as construction firms in the sector. However, the fact that they are sole proprietorships imposes a limitation on their capital base [3]. The limitation in size of the construction firms hampers their ability to employ professionals either as in-house members or on fee basis to constitute one organisation for purposes of undertaking projects on D&B basis. It must be mentioned however, that there are a few well established construction firms dotted around the country, especially in Southern Ghana, which are well resourced and have the experience and capability to deliver projects as D&B companies. [24] identified a number of projects that has been delivered by some of these firms using D&B approach. What is yet to be established is the *"form of contract"* under which these arrangements, or the terms of engagement were.

9. Advantages of the DBB Procurement System

The traditional procurement system (DBB), which is the dominant method of procurement in Ghana's construction industry, possesses a number of advantages or benefits. Perhaps these are the reasons for which it dominates other procurement methods in the market. These advantages, as identified by various practitioners, researchers and stakeholders in the construction businesses, included the following.

9.1. Wide Applicability

The system is said to be widely applicable, well understood and with well-established and clearly defined roles for the parties involved [23]. It is the most common approach for public projects and government contracts where there is the need to comply with state procurement statutes [2]. According to [3], the traditional system of procurement is one whose procedures are well known, largely due to its use over a long period of time.

9.2. Fully Defined and Detailed Project Designs

DBB system of procurement makes for fully defined and detailed project designs. Under this approach, the facility that the client or employer wants, is completely defined by detailed working drawings and specifications before bids are solicited [9]. This means there is little uncertainty about what is desired and what the contractor is required to deliver. DBB thus ensures that there is certainty in the projects' design. The processes under the procurement system are a series of end-on activities and each activity/phase must be completed before the subsequent one is undertaken [1]. The sequence of these phases which include brief, design, tendering and construction is illustrated in the diagram (figure 10) below.

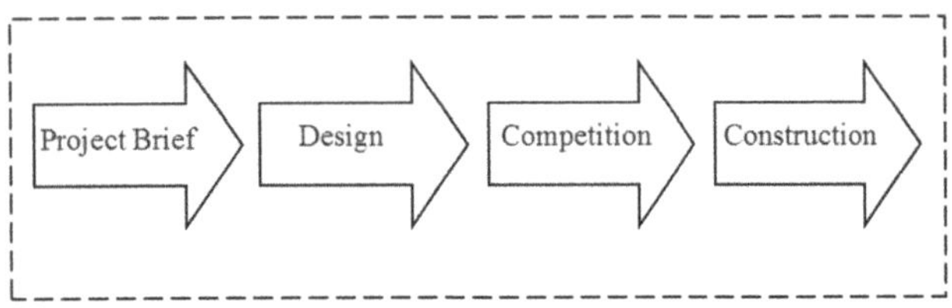

Fig. 10. Sequence of Project Delivery under DBB Source: [3].

9.3. Objectivity of Award

One of the greatest advantages of the traditional system of procurement is its objectivity in the award of contracts. Unlike D&B and others, DBB procurement method gives tenderers an opportunity to price their tenders on a common basis. In Ghana, the contract is mostly awarded to the bidder who offers to construct the building for the *lowest evaluated price.* It must be said that the bids are usually subjected to detailed scrutiny and proper evaluation by contract administrators and quantity surveyors (these consultants acting on behalf of the client). The winner most often than not emerges from an objective process. Furthermore, the competition motivates bidders to offer the lowest but economically viable price possible since most contracts are awarded on the criteria of lowest price amongst other factors. The detailing of the working drawings and specifications eliminates the need for high contingencies which most often than not culminate to increase bids/tenders.

9.4. Competitive Fairness and Satisfactory Public Accountability

In addition to the above, the DBB procurement system offers competitive fairness and satisfactory public accountability [3]. Ghana in particular, the Public Procurement Act of 2003 (Act 663) [24], which is the blueprint for procurement procedures in all public entities, categorically requires that procurement activities be carried out in a competition [24]. The procedure as described has the tendency to reduce the opportunity for bias and inappropriate influence in awarding the contract, though this does not necessarily mean that it eliminates completely potential abuses and improprieties. The contracts for the consultancy practices themselves are awarded based on competitive criteria of experience, qualifications and competencies, though this is often subjective under certain circumstances. The traditional system therefore is practiced in line with required public procedures and supported by statute.

9.5. Better Assurance on Quality Control

[24] indicated that DBB system of procurement provide relatively better assurance on quality control. The detail working drawings and specifications provide the basis for controlling quality. Since the drawings and specifications are the basis of the contract, they form the yardstick for comparing quality standards such as workmanship, materials and technology. In Ghana's construction industry, the use of a traditional bill of quantities (BoQ) is an integral part of the contract. The BoQ makes it relatively easy to arrange and value changes as the system is characterized by unlimited variations. In support of this feature of DBB, [3] observed that the traditional system makes the valuation of variations relatively easy to calculate.

9.6. Access for Small/New Contractors

Furthermore, DBB system of procurement to some extend provides good access for small contractors. Perhaps, the factor that proliferate a lot of small construction firms in the Ghanaian construction industry especially in Northern Ghana. [2] suggests that by awarding contracts based on price, the traditional process provides the best opportunity for qualified small and new contractors to obtain contracts which otherwise would have been difficult under arrangements such as Design and Build. Small and newly established contractors may be able to carry out certain types of work at a lower cost than large competitors because of lower overhead and less mechanisation.

10. Problems Associated With DBB in Ghana

The use of the traditional system of procurement in Ghana is not without difficulties. A number of researchers and practitioners do reckon that notwithstanding the benefits enumerated, the system is characterised by the following disadvantages.

10.1. Lack of Inherent Buildability

According to [23], under the DBB system of procurement, there is absence of input into the project design by the contractor, and that this may limit the constructability of the project delivery. The builder is not usually known until after the design work has been completed, tendered for, and a contract awarded. This means the design rightly does not incorporate inputs from the contractor on materials and methods (constructability) that could improve the building's design, functionality, and cost. [7] observed that, during the design process, the designers work in isolation, far removed from the contractor who will carry out the project construction. This makes for limited opportunities for ensuring better co-operation and buildability.

10.2. Adversarial Relations

Design-Bid-Build (DBB) procurement system as practiced in Ghana tends to promote adversarial relationships among the project team rather than co-operation and co-ordination [2]. Errors and omissions in the working drawings and specifications prepared by the consultants, constitutes a major source of conflict in a DBB procurement system. [23], observed that under the traditional system in Ghana, the owner is generally exposed to contractor claims, over-design and constructability issues since he (the employer) accepts liability for design in the contract with the contractor. Under the system, the client hires the consultants directly, whilst the law holds the employer to be the guarantor of the completeness and accuracy of the consultant's work. This often draws the client into disputes between the designer and the builder. [1] reckoned that one of the major disadvantages of DBB system of procurement has been the attitude of "them and us" that tend to develop between consultants and contractors. This often results in rigid lines of communication and creates a low level of flexibility in the relationships [1]

10.3. Requires Increased Oversight

Under DBB, it is claimed that the contractors usually pursue a least-cost approach to completing projects. According to [23], this requires increased oversight and quality review by the employer. To this extent, most employers are 'compelled' to engage additional technical staff (Site Agent, Supervisors) on the project to ensure that the construction and project delivery is in line with agreed standards.

10.4. Multiple Contractual Relations

According to [7], the traditional system "involves the client in a number of differing relationships with several organisations, and many inexperienced customers are dismayed at the complexity of the process and at the size and cost of employing the design team itself". [2] recognizes that the traditional systems 'pushes' the client into several contractual relations, which are illustrated in figure 7 above. This tends to be a disadvantage in the sense that in the event of problems, the architect accuses the contractor of faulty construction, and the contractor blames the architect for faulty design and so on, resulting in "finger pointing".

10.5. Slow in Starting on Site

DBB system of procurement is also criticized for being generally slower compared to others such as design-build and management contracting. This is attributed to the separated and sequential nature of the processes involved. [2] asserted that DBB is time-consuming since all design work must be completed before contracts are solicited. According to [3] the traditional procurement method is slow to start on site and hence does not allow for parallel working. The fact that the process of designing, tendering/bidding and building is far removed and separated (illustrated in fig. 2 and fig. 3 above), gives room for delays that otherwise could have been avoided. This weakness, however, may not necessarily be a disadvantage in the case of larger projects because procurement extreme care and professionalism needs to be exercised in such projects to ensure minimal errors.

10.6. Proliferation of Miniature Construction Firms

The traditional procurement system's advantage of creating opportunities for small contractors is nulled with the proliferation of miniature and small size construction firms. This is particularly so in the case of Southern Ghana where most contractors lack the necessary experience and capacity to execute construction contracts to the exactness of standards and specifications. This might be the reason for abandoned and uncompleted building projects all over the country, as many contractors lack the resources, both finances and professionals, to apply appropriate techniques to specific contracts.

10.7. Unwieldy Variations

DBB system is also characterized by huge and unwieldy variations especially in the case of public clients. The introduction of changes tends to throw construction programs over-board, compelling protracted extensions and associated problems. This also acts as a disincentive to contractors to save cost, and adopt best management practice, thereby exposing clients to avoidable claims and controllable cost and time overruns.

11. Future of DBB System of Procurement in Ghana

In the wake of continuous changes in clients' demands and characteristics, some of which are given below, the Ghanaian construction industry may have to review the use of the traditional system as the main route for the procurement and/or delivery of construction contracts.

- Movement of corporate businesses and clients into the construction market
- Clients becoming increasingly aware of alternative procurement options and are now demanding better value for money
- Early knowledge of project cost and its certainty for planning purposes
- Increased interest in collaborative working and joint ventures
- Risk avoidance and/or transfer
- Early start and completion of projects
- Better co-ordination of the construction delivery process

There is evidence that major projects have been delivered using other procurement routes [24]. For instance, Brunei International Student Hostel at Kwame Nkrumah University of Science and Technology-Kumasi was constructed under

BOOT by a group of lecturers of the university. Major Banks such as Barclays, Stanbic, Cal-Bank, National Investment Bank, and many others are constructing new branches in other parts of the country as quickly as possible. Private universities and secondary institutions are delving into construction and requiring multi-storey complexes that could provide multi-purpose functions.

The successful execution of these projects demands a high level of expertise and integration of project teams, which require a more collaborative, and co-ordinated procurement arrangement rather than the mundane conventional procurement system (traditional). The presence of a good number of well established construction firms such as Taysec Construction Ltd., P. W. Ghanem Ltd., Consar Construction Ltd., Myturn Construction Ltd. some of which are located in Northern Ghana suggest that these firms may be able to undertake projects on D&B bases. The challenge however is that, most of these contractors are foreign-based contractors. These construction firms are so developed that they established branches in other regions of the country, with in-house professionals with vast experience in construction, thus they are capable of undertaking projects on the basis of D&B given the opportunity and appropriate forms of contract. With the recent discovery of oil, there exists a host of opportunities in several sectors including the construction sector in Ghana. The construction industry in particular will soon be tasked to supply physical infrastructure to meet the growing/development needs of the country. There is also the Northern (Savannah) Accelerated Development Programme which seeks to inject massive capital investment in the Northern regions in order to bridge the development gap between the North and the South of Ghana. This obviously will include the construction of ultra-modern facilities and edifices and hence place a huge responsibility on the construction industry to deliver all this within the near future.

12. Conclusions and Recommendations

Based on the discussions above, it is clear the future of designer led procurement system has a daunting future since by its very nature; it may not be able to respond adequately to address the future demands of the industry. There is therefore the need for the construction industry in Ghana to brace itself in readiness to vigorously adopt modern contemporary procurement systems. It was revealed in the study by [5] that on the average 91% of projects procured in Ghana used the traditional procurement method with 2% using design and build and the remaining 7% for the other finance based procurement routes. In a related study by [5], 92% of professionals in Ghana hold the opinion that less than 10% of the local contractor is ready for a switch to emerging producer-led procurement routes. This finding supports the assertion by [3] that, very few construction firms have the expertise to embark on design and build. The observation that many Ghanaian contractors lack the expertise to undertake design at the same time the construction meant that on the basis of contractors' capacity and capabilities alone, the adoption and use of D&B on a large scale at present may not be feasible.

By implication, though the design and build strategy may be adopted for isolated projects, it can be concluded that Ghana may not be ready as a nation to fully embrace the producer-led design and build for majority of its projects. Thus we risk switching to design and build procurement system to favour the foreign-based contractors at a disadvantage of the local contractor. It is thus recommended that the Ghanaian origin local contractor must position himself strategically through capacity building by developing their technical, professional, financial and human capacity base to position them for the future in order to be found competitive when such procurement strategies are to be adopted.

References

[1] Osei-Tutu, E. (1999) *Construction Procurement Decision in Ghana.* Unpublished MSc Thesis, Department of Building Technology, KNUST-Kumasi

[2] Kyei, K. A. (2009) *Procurement Systems and Project Success in Ghanaian construction industry.* Unpublished Masters Thesis, KNUST-Kumasi

[3] Coles, D. (2009) *Lecture Notes on Project Evaluation and Development.* HAN University

[4] Buertey, J. T. I., Asare S. K. (2014), Public Private Partnership in Ghana: A Panacea to the Infrastructural Deficit? International Journal of Construction Engineering and Management 2014, 3(5): 135-143

[5] Buertey, J. Dadadzogbor E. (2015). *The Design and build Procurement system: An elixir of Project failure?* (1st Edition) Lambert Publications.

[6] Murdoch, J., and Hughes, W. (2008) *Construction Contracts, Law and Management.* (4th Edition). London and NY: Taylor & Francis.

[7] Masterman, J. W. E. (2008) *Introduction to Building Procurement Systems.* (2nd Edition). London: Taylor & Francis

[8] Hackett, M., Robinson, I., and Statham, G. (2007) *Procurement, Tendering & Contract Administration.* Oxford: Blackwell Publishing

[9] Abudey, W. (2005) *The Use and Effectiveness of Construction Management as a Procurement System in the Delivery of Building Projects on Ghanaian Construction Industry.* Unpublished BSc. Thesis, Department of Building Technology. Kwame Nkrumah University of Science and Technology (KNUST)-Kumasi

[10] Rowlinson, S. (2006) *Procurement: Develop and Construct.* The Building, February 8, pg 68

[11] Hughes, W. P. (1992) *An Analysis of the JCT D&B Contracts.* Chartered Institute of Building (CIOB) Construction Paper, No. 6

[12] Murdoch, J., and Hughes, W. (2008) *Construction Contracts, Law and Management.* (4th Edition). London and NY: Taylor & Francis.

[13] Engineering and Construction Contracts, (2005) *The New Engineering Form of Contracts.* (3rd Edition) (NEC3). Office of Government Commerce, UK

[14] Kwakye, A. A. (1997) Construction Project Administration in Practice. Addison Wesley

[15] Hoggs, K. and Morledge, R. (1995) *Risks and Design & Build and build: keeping a meaningful perspective,* CSM, Monthly, May, 1995, Vol 14, No. 7. 32-33

[16] Chevin, D. (1996) *Keeping up Appearances.* The Building, June 14, page 38-41

[17] Rimmer, B. (1996) *Keeping up Appearances.* The Building, June 14, pg 38-41

[18] Siddiqui W. A. (1996) *Novation: and its comparison with common forms of procurement.* (CIOB) Construction Paper, issue № 60

[19] Minogue, A. (1999) *Clash Points.* The Building, (October, 29)

[20] Ashworth, A. (2008) *Pre-Contract Studies, Development Economics, Tendering and Estimating.* (3rd Edition). Oxford: Blackwell Publishing

[21] Winch, G. M. (2000). Institute reform in the British construction: Partnering and private finance *Building research and information*, 28, 141-155

[22] Winter, C. H. H. (2002). On the appropriateness of contractor-led procurement-an investigation of the circumstance and consequences. *Von der Fukult Fur Wirtschaftswissschaften der Technischen Universitat Bergakademic*

[23] Adu-Poku, P. O. (2006) *Project Management as an Alternative to the Traditional Procurement Method in Project Delivery in Ghana.* Unpublished BSc. Thesis, Kwame Nkrumah University of Science and Technology (KNUST) - Kumasi

[24] Ameyaw, C. (2009) *Design and Build verses Design-Bid-Build Method in Ghana.* Unpublished MSc. Thesis, Department of Building Technology. Kwame Nkrumah University of Science and Technology (KNUST) - Kumasi

[25] Articles of Agreement and Conditions of Contract for Building Works (2000). 7th Edition. (*The Pink Form)*

[26] Ghana Government's Public Procurement Act, Act 663 (2003)

[27] Kojo, M. A. (2005) *Procurement of Contracts in the Informal Sector of the Construction Industry of Ghana.* Unpublished BSc. Thesis, Department of Building Technology, (KNUST) - Kumasi

Characteristics and Preparation Method of Sulfur Extended Asphalt Mixtures

Vahid Rezvani, Hassan Saghi

Department of Civil Engineering, Hakim Sabzevari University, Sabzevar, Iran

Email address:
vahidrezvani2002@gmail.com (V. Rezvani), h.saghi@hsu.ac.ir (H. Saghi)

Abstract: Huge amount of sulfur production as a side product in petrochemical industries and its low application in country has been a challenging issue. In this paper, application of sulfur as a partial substitute for bitumen in process of asphalt mixture production is investigated. According to the results, the characteristics of the sulfur extended asphalt in comparison with the conventional asphalt made from pure bitumen, and the impact of sulfur addition on asphalt performance are expressed and the optimized formula for making the asphalt mixture is obtained. In this content, the modified asphalt was produced in mass production scale and its performance was studied. In this paper, the method of sulfur addition to pure bitumen and then its mixture with aggregate and finally optimized condition of sulfur modified asphalt production regarding pollution and environmental issues are also proposed.

Keywords: Sulfur Extended Asphalt, Sulfur Asphalt Concrete, Sulfur, Bitumen, Asphalt

1. Introduction

Sulfur recovery from fossil fuel for prevention of air pollution and corrosion in devices has led to production of huge amount of sulfur in the related industries. Sulfur production is particularly significant in gas and oil refining industries and many of them have no use. On the other hand, exportation of this material has its own problems in some of refineries and in most cases is not economically efficient. Predictions show that annual sulfur production in Iran and rest of the world is increasing due to environmental limitation and gradual acetifying of hydrocarbon resources. Therefore studies have been conducted on application of excess sulfur; among them is its application in road construction.

2. Related Study

Application of sulfur in road construction has almost hundred-year history. The first approach was in the 1970s following the energy crisis which led to construction of tens of experimental sulfur modified asphalts throughout the world. First sulfur extended asphalts were made in USA and Canada through the 1970s [1, 2]. Then by 2002, about 4200 km of sulfur asphalts were made and tested in more than 200 countries [3, 4]. Among these countries, USA, England, France, Poland, Spain, Finland, Belgium, Norway, Iraq, Saudi Arabia could be mentioned [5-7]. Experimental construction of sulfur asphalts decreased in the 1980s however, at the end of th 1990, it increased again. The reason is the change in relative cost of sulfur/bitumen in this period of time which provided the economic justification for producing sulfur extended asphalt. In Iran, research studies were made from the 1970s. The results of performance investigation of sulfur asphalt all around the world shows that it has equivalent or better performance in comparison with the ordinary ones.

3. Theory

3.1. Sulfur Properties

Pure sulfur is a yellow solid which melts at 119°C and its viscosity suddenly increases at 160°C. At high temperatures, sulfur produces toxic gases such as SO_2. At specific temperature and conditions, in combination with water, this material produces sulfuric acid which is highly corrosive.

3.2. Sulfur Role in Chemical Structure of Bitumen and Its Properties

The reaction between sulfur and bitumen depends on temperature and sulfur concentration. At its melting point, it has octet ring form and at higher temperatures it forms polymer and gets the form of 2-radical chain. Radicals could react with bitumen in two ways; forming carbon-sulfur bond or absorbing hydrogen and consequently dehydrogenization occurs. The probability of all these reactions depend on temperature therefore, the product would be different. At 240°C, sulfur acts like Oxygen and the main reaction will be oxidation and H_2S would be released. While at 140 °C, the dominant reaction is entrance of sulfur atoms into aromatic naften molecules of bitumen. As a result, the amount of aromatic naften group of bitumen decrease and percentage of polar aromatic groups of bitumen increases. Consequently, the bitumen gets softer and more adhesive. It must be noted that at 140°C, all the added sulfur do not participate in reaction, part of it will remain in solution form and the rest will float in bitumen. The percentage of the sulfur participating in reaction depends on sulfur concentration, aromatic naften part of bitumen and as a result the type of the bitumen. Therefore, addition of sulfur to bitumen at 140°C has the following stages:

Sulfur entrance into chemical structure of bitumen, sulfur solvation in bitumen, suspension of excess sulfur in the form of colloidal particles and finally, sulfur tendency to precipitate.

4. Impact of Sulfur Addition to Asphalt Mixtures

4.1. Asphalt Concrete

As the majority of warm asphalt used all around the world is asphalt concrete type, application of sulfur in asphalt concrete has gained more attention, in a way that most of the implemented sulfur asphalts in different countries are the sulfur extended asphalt concrete type, as a results most of the reports are on the effect of sulfur addition on the performance characteristics of asphalt concrete. The following results have been obtained from the previous researches:

- Enhancement of some of mechanical properties such as increase in Marshal stability, increase of hardness, increase of resistance against rutting and decrease of transverse cracks in sulfur modified asphalt [13, 17, 18]
- Increase in durability of sulfur asphalt in comparison with ordinary asphalt concrete [2]
- Ease of application and improvement of workability after the addition of sulfur to asphalt [2]
- Decrease of energy costs in asphalt production by reduction of mixing temperature [2].
- Higher resistance against petrol solvents [12]

4.2. Sand Asphalt

As sand asphalt has lower quality and strength in comparison with ordinary ones, they have no application in asphalt layers of high traffic paths. The results show that addition of sulfur to sand asphalt will increase its strength significantly in a way that it could be used in the asphalt layers of high traffic roads. Moreover, by application of sulfur, a proper mixture could be make out of desert sand and bitumen which has sufficient strength for use in major roads. In some studies the mixture including 82% sand, 6% bitumen and 12% sulfur is mentioned. Due to high amount of sulfur, this mixture hardens quickly therefore it could be used for road maintenance and let the cars to pass the road quickly after the repairing procedure [14].

5. Sulfur Extended Asphalt Preparation Methods

Sulfur could be added to the asphalt mixture in different stages of preparation; following methods are mentioned in studies:

5.1. Previous Mixing of Sulfur and Bitumen

Sulfur added bitumen is added to hot aggregate in mixer as follows:

Mixing of hot bitumen and melted sulfur is done by application of colloidal mill, vane stirrer, mixing in the path or in weighing container. Each method has its own advantages and disadvantages.

Mixing of hot bitumen and solid sulfur in a way that the solid sulfur enters melted bitumen and after contact with bitumen it melts and properly distributed in bitumen by powerful stirrers. Finally sulfur added bitumen would be conducted to mixer [15].

5.2. Direct Addition of Sulfur to the Mixer of Asphalt Plant

In this method, solid or liquid sulfur, after being weighed, is directly injected to mixer; in this way, the containers needed for liquid sulfur or premixing of sulfur and bitumen (which leads to production of H_2S) would be eliminated, therefore the emission would be reduced and the work would be done more easily [16].

According to studies, direct mixing is more popular due to less emission. In this method, sulfur would be transported by conveyor belt to the top of mixer, and after direct weighing; it would be poured on mixture of aggregate and hot bitumen and properly mixed.

6. Experimental Investigations

Before the production of sulfur asphalt in industrial scale, it is needed to obtain the technical parameters such as temperature, optimized percent of bitumen and sulfur, method of mixing, design of asphalt mixture and so on. In this regard, numerous mixtures were prepared and the results are presented here. In order to achieve to a uniform and homogeneous mixture, the method of mixing melted bitumen (60/70 pure bitumen of Isfahan refinery) and solid sulfur was selected. It must be noted that the mixture of bitumen and sulfur should be enough in a way that it could be uniformly distributed and any crystal bump of sulfur should be prevented. Sulfur added bitumen is relatively unstable and should be considered as an intermediate product rather than a product suitable for storing. Microscopic investigations showed that the amount of dissolved sulfur in bitumen is independent of bitumen grade and it was not crystallized by passing of time.

6.1. Experimental Results of Sulfur Addition to Bitumen

Results showed in table 1 are related to the mixtures prepared in this paper:

Table 1. *Physical characteristics of pure and sulfur added bitumen.*

No.	Pure Bitumen 60/70 (%)	Sulfur (%)	Softening Point (°C)	Weight loss due to heating (w %)	Penetration in 25°C (0.1 mm)	Ductility in 4°C (cm)	Ductility in 15°C (cm)	Ductility in 25°C (cm)	Relative Specific Gravity in 25°C	Flash point (°C)	Fraass breaking point (°C)
1	100	0	49.5	0.01	65	0	96	+150	1.015	339	-11
2	85	15	48	0.71	92	2	31	+100	1.103	215	-12
3	75	25	46.5	0.82	93	2	25	39	1.143	205	-12
4	65	35	48	0.91	94	1	24	36	1.100	201	-10
5	55	45	48	0.98	93	1.5	19	24	1.316	191	-7

RTFO						
Viscosity in 60 °C (poise)	Viscosity in 135 °C (cSt)	Viscosity in 98.9 °C (cSt)	Ductility in 25 °C (cm)	Softening point (°C)	Penetration in 25 °C (0.1 mm)	Weight loss due to heating (w%)
1372	330	2422	100	54	39	0.04
908	133	1125	68	55	41	4.1
1119	154	1371	33	56	41.50	4.57
2188	232	2851	24	57	39	4.84
764	174	1091	21	58	37.5	4.59

6.1.1. Penetration Grade

As it was mentioned before, by addition of sulfur, bitumen would get softer and more penetrative. Addition of 5% sulfur will result in significant increase in its penetration, but by gradual increase of sulfur to 45%, no significant increase in penetration would be observed. High increase in sulfur content would lead to lower penetration, even less than the basic pure bitumen.

6.1.2. Softening Point

Due to conversion of aromatic naften part of the bitumen into polar aromatic and constant ratio of asphalteness to saturated hydrocarbon in the procedure of adding sulfur to bitumen, no significant change could be observed in softening point.

6.1.3. Ductility

By increasing the ratio of sulfur to bitumen, ductility would be reduced in high temperatures 15, 25 °C), but in lower temperatures, ductility will improve. It is noted, by increasing of sulfur content, significant changes in the rate of this feature are not observed.

6.1.4. Effect of Temperature on Sulfur Added Bitumen

Results of weight reduction due to heating indicate that higher percent of sulfur in bitumen will result in more weight dissipation. As a result, reduction in penetration of sulfur added bitumen after the experiment would be more than the pure bitumen. This trend was also observed in RTFO experiment.

6.1.5. Viscosity

By addition of sulfur, viscosity also decreases, but by reach of sulfur content to 35%, it will increase and approaches to value of pure bitumen, by further increase of sulfur, viscosity will decrease again.

6.2. Asphalt Concrete

According to experimental researches, in order to avoid emission of harmful gases such as H_2S, the temperature of sulfur added bitumen would be never more than 150 °C. This would make no problem in preparation procedure of sulfur asphalt and it is only due to improvement of rheological properties of bitumen. The next stages of sulfur addition to asphalt concrete such as the method of sulfur addition, determination of the best condition of mixing, and obtaining the best practical formula were investigated for sulfur added bitumen with different percent of sulfur. After comparing the results with pure bitumen, the data were analyzed and at the next stage, the effect of sulfur on adhesive properties of bitumen to mineral aggregate, in presence of water, was studied. As viscosity of sulfur added bitumen is less than pure one, the temperature of mixing and compacting the sulfur extended asphalt mixture would be less than pure one.

In this study, the temperature of mixing is 140-145 °C, and the temperature of compacting is 130-135 °C. As specific gravity of sulfur is twice the bitumen; the following equation could be applied for obtaining weight percentage of sulfur added bitumen:

$$SEA\% = \frac{SGSA\ /\ SGA\ *\ PA\ *\ 100}{SGSA\ /\ SGA\ *\ PA\ +\ (100 - PA)}$$

Where, SG_{SA} is specific gravity of sulfur added bitumen, SG_A is specific gravity of pure bitumen, P_A is weight percentage of bitumen in conventional mix design and $SEA\%$ donates the required weight percentage of sulfur added bitumen.

6.3. Preparation of Asphalt Mixture

The design of asphalt mixture with 60/70 pure bitumen and sulfur added bitumen were made in lab. Similar materials were used for preparation of them whose properties are as follows.

Table 2. *Gradation and specification.*

Percent Passing						
Gradation (½ inch)	**¾ inch**	**½ inch**	**No. 4**	**No. 8**	**No. 50**	**No. 200**
ASTM-D3515 (Specification range)	100	90-100	44-74	28-58	5-12	2-10
ASTM-D3515 (Specification midpoint)	100	95	59	43	9.5	6
Aggregate Gradation A	100	15	-	-	-	-
Aggregate Gradation B	100	100	-	-	-	-
Aggregate Gradation C	100	100	90	10	-	-
Aggregate Gradation D	100	100	100	90	6	2
Mineral filler	100	100	100	100	80	35
Theoretical results of gradation	100	91.5	68	48	10.4	4.3
Experimental results of gradation	100	91	69	48	10	4

The results of Marshal Mix Design calculation of the samples made from 60/70 pure bitumen are presented in table 3.

Table 3. *Calculation of Marshal Mix Design for samples made from 60/70 pure bitumen.*

Bitument content (%)	Stability value (kg)	Flow value (mm)	Bulk specify gravity (gr/cm3)	Voids in Mineral Aggregate (%)	Air Voids (%)
4.0	1092	2.3	2.328	14.7	5.7
4.50	1115	2.5	2.341	14.6	4.4
5.0	1069	2.8	2.338	15.2	3.8
5.5	1022	3.0	2.335	16.5	3.7
6.0	999	3.3	2.324	16.6	3.6

Optimum conditions	
Bitument content (%)	4.6
Stability value (kg)	1110
Flow value (mm)	2.5
Bulk specify gravity (gr/cm3)	2.341
Voids in Mineral Aggregate (%)	14.8
Air Voids (%)	4.2

Similar calculations were conducted for the samples made from sulfur added bitumen with different content of sulfur (25,35 and 45 %) which are presented in tables 4, 5, and 6.

Table 4. *Calculation of Marshal Mix Design for samples made from sulfuric bitumen with 25% sulfur content.*

Bitument content (%)	Stability value (kg)	Flow value (mm)	Bulk specify gravity (gr/cm3)	Voids in Mineral Aggregate (%)	Air Voids (%)
4.0	929	2.0	2.356	14.1	6.4
4.50	1208	2.5	2.376	13.8	5.1
5.0	1277	2.8	2.386	13.9	4.2
5.5	1138	2.9	2.380	14.6	4.0
6.0	1092	2.3	2.375	15.2	3.8

Optimum conditions	
Bitument content (%)	5.6
Stability value (kg)	1260
Flow value (mm)	2.7
Bulk specify gravity (gr/cm3)	2.386
Voids in Mineral Aggregate (%)	14
Air Voids (%)	4.1

Table 5. *Calculation of Marshal Mix Design for samples made from sulfuric bitumen with 35% sulfur content.*

Bitument content (%)	Stability value (kg)	Flow value (mm)	Bulk specify gravity (gr/cm3)	Voids in Mineral Aggregate (%)	Air Voids (%)
4.0	1069	2.4	2.357	14.0	6.4
4.50	1092	2.5	2.380	13.7	5.0
5.0	1208	2.7	2.389	13.8	4.3
5.5	1162	2.8	2.385	14.4	4.0
6.0	1045	3.0	2.381	15.0	3.8

Optimum conditions	
Bitument content (%)	5.6
Stability value (kg)	1200
Flow value (mm)	2.6
Bulk specify gravity (gr/cm3)	2.390
Voids in Mineral Aggregate (%)	13.9
Air Voids (%)	4.2

Table 6. Calculation of Marshal Mix Design for samples made from sulfuric bitumen with 45% sulfur content.

Bitument content (%)	Stability value (kg)	Flow value (mm)	Bulk specify gravity (gr/cm3)	Voids in Mineral Aggregate (%)	Air Voids (%)
4.0	1208	2.3	2.360	13.9	6.3
4.50	1231	2.4	2.385	13.5	5.2
5.0	1245	2.5	2.391	13.7	4.3
5.5	1162	2.9	2.388	14.3	4.1
6.0	1138	3.0	2.386	14.8	3.9

Optimum conditions	
Bitument content (%)	5.7
Stability value (kg)	1250
Flow value (mm)	2.6
Bulk specify gravity (gr/cm3)	2.392
Voids in Mineral Aggregate (%)	13.8
Air Voids (%)	4.2

6.4. Adhesion to Mineral Aggregate

One of the properties of sulfur added bitumen is the increase of adhesion to mineral aggregate in presence of water. In this content, according to AASHTO method, 95 gr of stones with specific grading was covered with 5 gr of bitumen and then floated in water. At 60 °C, and in specific time intervals, the bitumen cover of aggregate was investigated. As it is observed from table 7, sulfur added bitumen has better adhesion properties in comparison with pure one.

Table 7. Adhesion of pure and sulfur added bitumen to aggregate at presence of water.

No.	Bitumen type	Bitumen (%)	Sulfur (%)	Percentage of being washed		
				After 1 h	After 24 h	After 48 h
1	60/70	100	0	60	65	70
2	60/70	75	25	25	45	50
3	60/70	65	35	20	35	40
4	60/70	55	45	5	10	15

7. Field Operation

According to the results, bitumen with 35% sulfur was prepared at industrial scale and mixed with aggregate to prepare sulfur extended asphalt mixture. Then it was laid down in one of the sub roads of Mashhad-Ghoochan road beside ordinary asphalt pavement. All the materials and road conditions were considered similar with ordinary asphalt pavement, only the optimized percent of bitumen, mixing and compaction temperature were different. Fortunately, due to close care of temperature and safety points, no emissions were observed in location of preparation, transportation or implementation. This piece of road would be evaluated in different seasons in future.

8. Discussion and Conclusion

The results of the study could be summarized as follows:

- In spite of bitumen properties modification by addition of sulfur, however, the main motif for application of sulfur is economic reason. Therefore, when the price of sulfur is lower than bitumen, this sounds reasonable.
- Addition of sulfur to pure bitumen will result in modification of some properties of bitumen such as increasing penetration, viscosity reduction and adhesion improvement. And the modified asphalt concrete has higher tension strength.
- Due to viscosity reduction, temperature of mixing and compacting is lower, which leads to energy saving.
- In order to reduce the emission, technical and safety points especially temperature control should be performed [19].
- No emission was observed in this study as the temperature was closely controlled.
- Preparation and implementation of sulfur extended asphalt mixture is possible with current equipment and facilities and no modification is needed, only some consideration should be mentioned for addition of sulfur.

References

[1] Filippis P.D., "Sulfur–Extended Asphalt, Reaction Kinetics of H2S Evolution", Fuel, Vol. 77, No.5, 1998.

[2] Weber H., "Sulfur- Asphalt & its Global Aspects", Sulfur Institute, 19th Sulfur Phosphate Symposium, 2003.

[3] Mills D.R., "Field Trial with Sulfur- Extended – Asphalt (SEA) Binders", Sulfur Research & Development, Vol.3, No.3, 1980.

[4] Nazarbegi Ali Ehsan, "sulfur application in asphalt mixture for roads", research report of oil industry research center 1998.

[5] Strickland D., "Performance Properties of Paving Mixtures Made with Modified Sulfur Pellets", International Symposium on Antennas & Propagation (ISAP), Taipie, Taiwan, 2008.

[6] Asphalt Institute, "State-of-Knowledge on the Use of Sulfur–Extended Asphalt (SEA)", 2007.

[7] John S.C., "Review of Relationships between Modified Asphalt Properties and Pavement Performance", SHRP-A-631, 1993.

[8] Love G.D., "Sulfur-Paving Material of the Future", Sulfur Research & Development, Vol.3, No.3, 1980.

[9] Deme B.H., "Mix-Paver-New Approach to Sulfur Paving", Sulfur Research & Development, Vol.3, No.3, 1980.

[10] Alikhani, "sulfur use in asphalt and its safety", road development organization, 1999.

[11] Petrossi V., "Reactions & Technological Properties of Sulfur - Treated Asphalt", Industrial Engineering of Chemical Production Research and Development, Vol.11, No.2, 1972.

[12] Miller L.J., "A Technical Review of Sulfur– Association Asphalt Paving Technologists", Vol. 42, 1981.

[13] Tabatabaei A.M., "sulfur application in asphalt pavement", journal of technical faculty, No 46, 1984.

[14] Parrett F., "Sulfur Extended Asphalt for Highways", Civil Engineering, Vol. 28, 1984.

[15] "Solid Sulfur- Extended Asphalt Composition and Method and Apparatus therefore", U.S. Patent, No.4, 1982.

[16] Taylor J.A., Kennepohl G.J., "Environmental Aspects of Paving with Sulfur-Asphalt", Can. Tech. of Asphalt Assoc., 1980.

[17] Bissada A., "Rheological Characteristics of Sulfur Extended Asphalts used for Road Pavements", Sulfur Research & Development, Vol.9, 1985.

[18] Kandhal P.S., "Evaluation of Sulfur Extended Asphalt Binders in Bituminous Paving Mixtures", Association of Asphalt Paving Technologists, Vol.51, 1982.

[19] Liddle W. A., Mahoney J.R., "Sulfur Extended Asphalt Pavement Evaluation for the Baker River Highway", FHWA Report, 1983.

Effect of ground motions on nonlinear seismic behavior of corroded buried gas pipeline

S. Mohammad S. Kolbadi[1,*], Nemat Hassani[2], S. Mahdi S. Kolbadi[3], M. R. Shiravand[2]

[1]Department of Civil Engineering, Shahid Beheshti University of Technology, Tehran, Iran
[2]Department of Civil Engineering, Shahid Beheshti University, Tehran, Iran
[3]Department of Civil Engineering, K. N. Toosi University of Technology, Tehran, Iran

Email address:
mohamadkolbadi.pwut@gmail.com (S. M. S.Kolbadi), nemathasani@yahoo.com (Nemat Hassani), mahdi_kolbadi@sina.kntu.ac.ir (S. Mahdi S. Kolbadi), shiravand@aut.ac.ir (M. R. Shiravand)

Abstract: Recent earthquakes have damaged many lifeline structures. Buried gas pipelines are also no exceptions. Sometimes, theses pipes are subjected to a thinning of the wall thickness due to corrosion. Therefore, it is important to evaluate the strength of the pipes undergoing local wall thinning to maintain the integrity of the piping systems. The main purpose of this study is to understand failure aspects of buried pipeline caused by earthquake and degradation of metal due to corrosion through FEM. In dynamic analysis, corroded pipeline approaches to critical criteria (reaching the yield stress) in much less time. In this case, if the corroded pipe thickness be half, the yielding angle is reduced to a quarter.

Keywords: Buried Pipeline, Corrosion, Fault, Landslide, Yielding Pressure, Seismic Response

1. Introduction

The damage or disruption of buried pipelines due to earthquakes may severely affect civil lifeline structures since it may cause fires, economic losses, and disable of lifeline networks. Furthermore, if they are exposed to corrosion, material strength is severely reduced. As a matter of fact the damage will be propagated.

Corrosion is one of the most critical degrading mechanisms of structural strength. Moreover, corrosion of most metals is often an inevitable process. When localized, pit or uniform corrosion occurs, strength reduction can be difficult to establish because of the complicated effects of uneven surfaces and uneven material properties on the stress fields and failure modes of the structure. In order to accurately estimate pipeline structure life, it is important to understand and evaluate corrosion effects on overall ductile fracture, bending capacity and the corresponding local strength [1].

Based on the severity and intensity of corrosion, the material loss can be converted into an overall reduction in plate thickness and a corresponding decrease in the strength of the structure. Although the concept of an overall thickness reduction is simple to apply, it may be inaccurate as some important local phenomena may be overlooked [1].

Gas pipelines, are types of structures that are highly susceptible to corrosion. As the geometry features of such in- and out-of-water structures, corrosion effects on shell structures are of special interest in studying the corrosion effects on pipelines. For pipelines, internal and external pressures, and bending moments are regarded as the most important loading conditions. Pipelines sustain huge external pressures from water and soil, internal pressures from transmitted oil or gas and bending moments due to thermal expansion and curved lay-out. When these pipes are corroded from either external or internal (or both), burst and bending capacities are decreased; thus, the structural integrity is significantly affected. Therefore, guidelines are needed for the estimation and calculation of these reductions [1].

2. Pipe Specimen

The material of pipe specimens are carbon steel pipes for high pressure service, called 'STS370', which are commonly used in piping systems. The carbon steel of STS370 is similar to that of ASME A333 Gr.6[3].The inside pressure of those pipes are 250 Psi[2] andelasto-plastic behavior of the steel material is defined as follows (Figure 1):

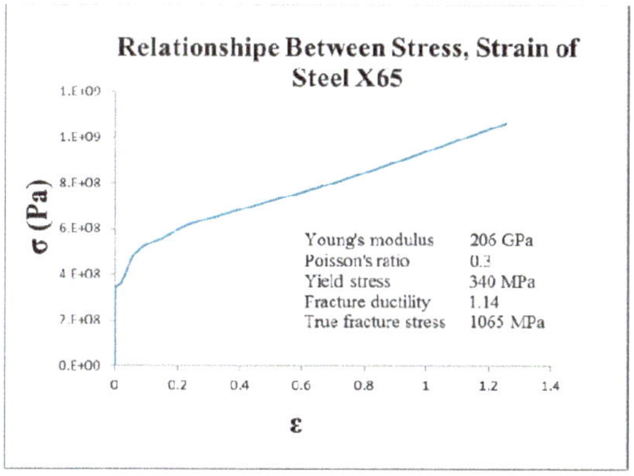

Figure 1. Stress-strain curve for FE analysis [3].

Since the chemical process ofcorrosion caused a huge reduction of the stiffness and the fracture behaviors of pipes with wall thinning on the outside of the pipes are almost identical to those of pipes with wall thinning on the insidesurface [4], thus; corrosion effect is expressed by local wall thinning roundly.

The corroded pipe geometric can be seen in the figure2.

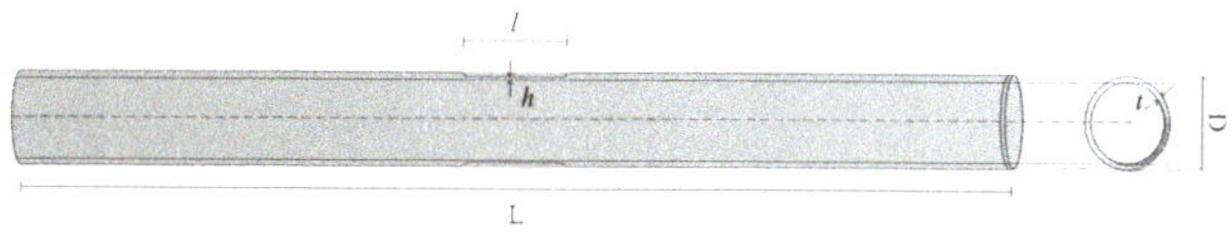

Figure 2. Shape and dimensions of a pipe specimen with circumferentially local wall thinning

Table 1. Geometric dimension of the pipe[2,3]

α=h/t	*l*	*h*	*t*	*D*	*L*
Corrosion ratio	Length of corrosion	Thickness of corroded pipe	Thickness of sound pipe	External diameter	Pipe's length
0 ~ 1	*L*	Variable	14.3mm	1m	30m

The magnitude of αvaried from 0 to 1. If α be closed to 1, it means the pipe is sound and if αtended to 0, it demonstrates there is so corrosion on pipe's wall.

3. Properties of Buried Pipeline

3.1. Geometric Properties

The model geometry, due to the impact of soil-structure interaction is intended as figure 3.

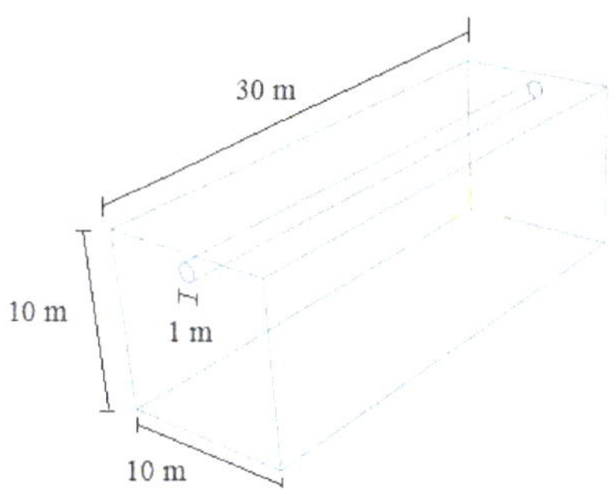

Figure 3. Geometric properties of buried pipeline

Figure 4. Buried pipeline meshed- 3D plane

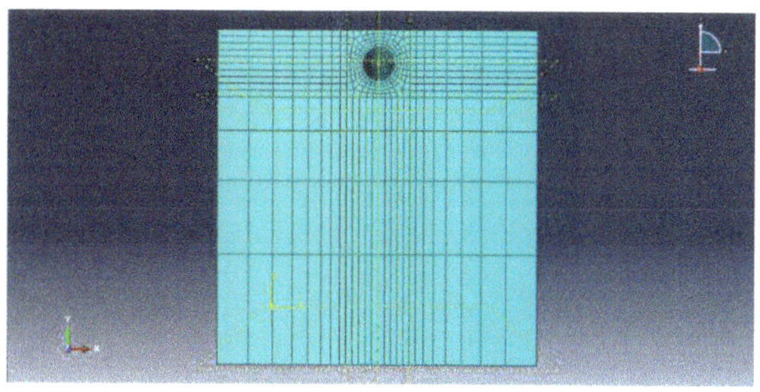

Figure 5. Buried pipeline meshed- 2D plane

3.2. Soil Properties

Table 2. Soil properties[5]

Cap Model		
Elastic Modulus	E	5 MPa
Poison's Ratio	υ	0.2
Unsaturated Density	ρ	1923 Kg/m^3
Cohesion	C'	5.1 KPa
	d	30 KPa
Internal Friction Angel	ϕ'	30 Degree
	β	50.2 Degree

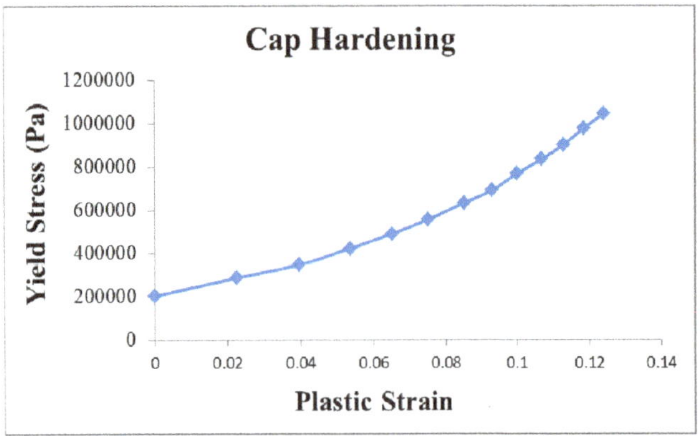

Figure 6. *Cap hardening curve of cap model[5].*

4. Dynamic Analysis

The behavior of buried pipelines is depended to the ground motion closely in both lateral and longitudinal directions. Therefore, examining the relative displacements between buried pipelines and ground are suitable for estimating failure aspects of buried pipelines caused by earthquake because the behavior of buried pipelines is mostly governed by relative ground movement from earthquake [6].

Two major influences of the earthquake on buried pipes is:1. drift between soil and pipe in longitudinal direction (Figure7, 8);2. bending pipe caused by earth movement (Figure9).Therefore both factors may impact to the stress caused by corrosion.

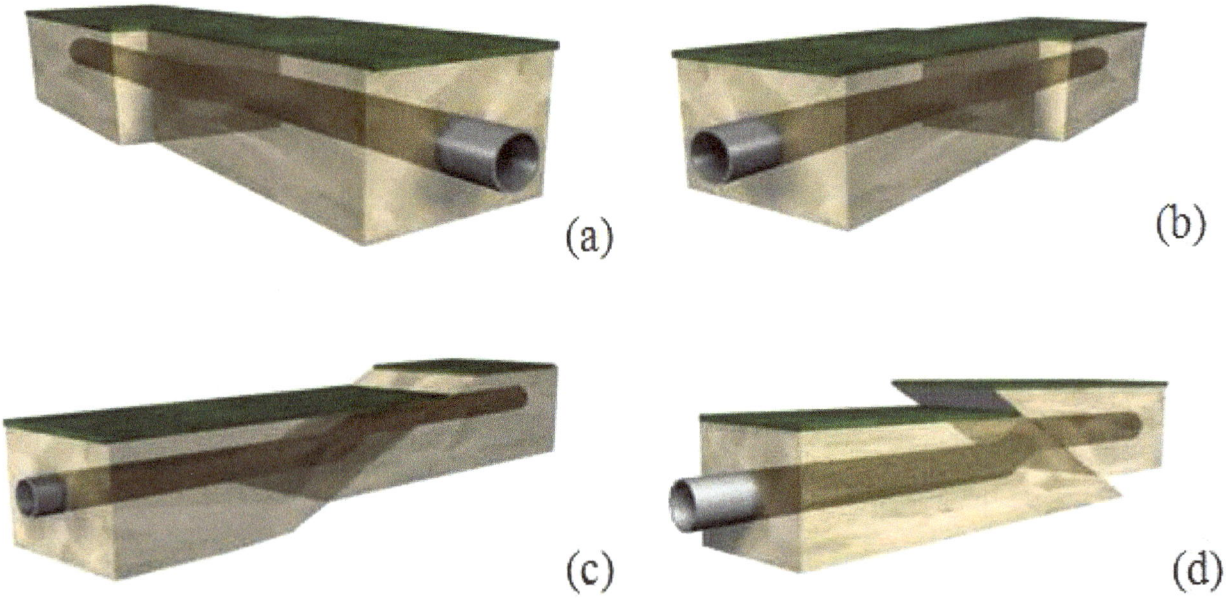

Figure 7. *Type of buried pipeline-fault crossing (a) Pipeline subjected to left lateral fault movement, (b) Pipeline subjected to right lateral fault movement, (c) Pipeline subjected to normal dip slip fault movement and (d) Pipeline subjected to reverse dip slip movement*

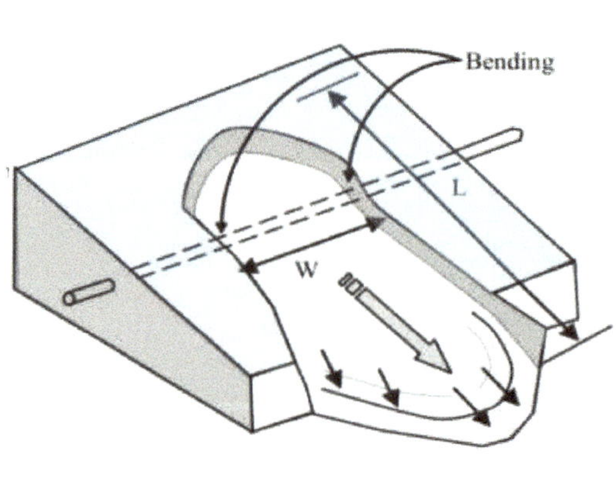

Figure 8. *Transverse Pipe-PGD crossing*

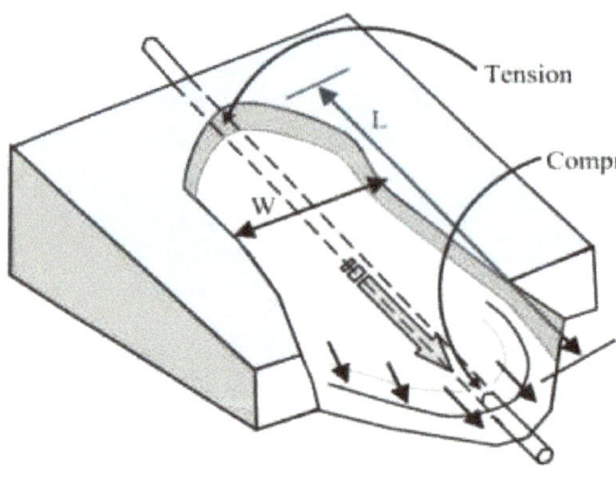

Figure 9. *Longitudinal and Pipe-PGD crossing*

4.1. Pipe's Bending

At first the impact of ground motion that causes the bending of pipe is studied. This phenomenon happens where, fault and pipe are crossover. In this regard, a distinct drift were applied to the soil in the perpendicular of pipe`sdirectionfor describe to the corroded pipe crossing fault and it was compared with sound pipe. Ground motion had 10 seconds duration and 10 cm drift.

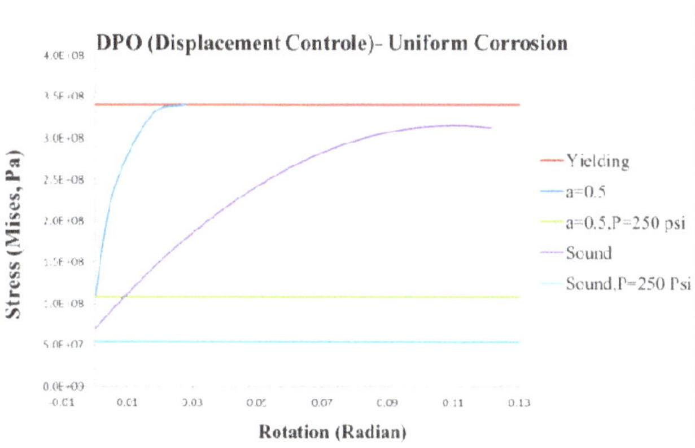

Figure 10. The effect of bending caused movement of fault on the corroded pipe

As it can be observed the curves are performed by movement control. The corroded pipes with a $\alpha = 0.5$ was chosen. Corroded pipe reached to the yielding level at lower rotation angel compared to the sound pipe. The yielding rotation angle of sound pipe is 4 times more than of corroded pipe approximately.

4.2. Pipes Tensile

If the ground be parallel to the pipe`s direction, the tensile stresses will be create. Thus to understand the influence of the drift between the pipe and the soil, it is important to evaluate this phenomenon that occur in landslide usually. In this case, the effect of soil-structure interaction is very important. Whatever the involvement of soil and pipe be too much, tensile stresses are more in pipe. So,when the pipe will be corrode,the pipe would be much weaker.

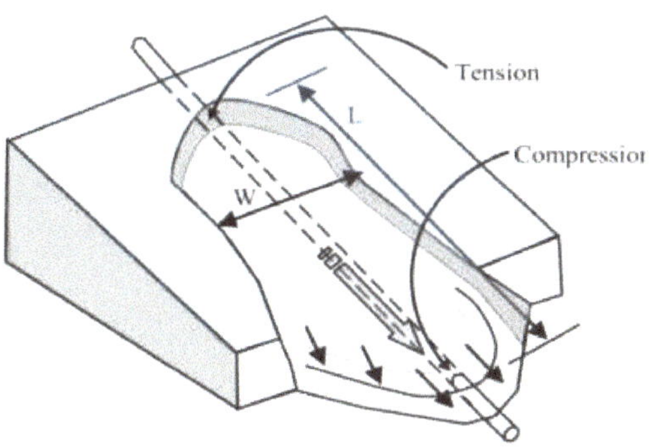

Figure 11. Longitudinal and Pipe-PGD crossing

It is possible to divide soils into two groups, which are sandy soils and cohesive soils. Sandy soil is often called frictional soil or drained soil and cohesive soil is normally classified as undrained soil. These two groups of soils have different characteristics and can cause different failure modes for buried structures.

The interaction effects were applied by contact element that it is contain two types of transmissionforces:

- Normal
- Tangential

The tangential behavior with considering the friction coefficient (F) between the pipe and soil was definedusing Columbus theory [5] that provides the equivalent stiffness of the soil and the pipe. For cohesive soils and coarse-grained soil,F=1 and F=0.1 was considered respectively and the results were compared to sound and corroded pipes.

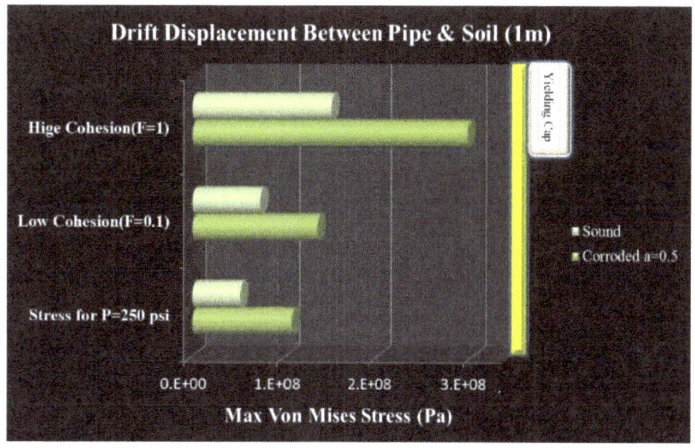

Figure 12. Effect of the drift displacement between pipe and soil on the corroded pipe

It can be seen that the stress values incorrodedpipe is more than of the sound. With increasing the friction between the soil and the pipe, maximum stress increases.

5. Conclusions and Discussions

Thestudieswere performed on the effect ofcorrosionontheseismicresponses of pipes causedby the ground motions. Since the buried pipes confine with surrounding soil, the dynamicresponse of pipeaffected by the ground movements. Themass of thepipeisvery small compared to thesurroundingsoil, so equal stiffness between soil and pipe be considerable. Thus; it seems reasonable to use displacement control instead of acceleration-time in historical analysis.

In the first study, corrodedpipelineapproaches to critical criteria(reaching the yield stress) in muchlesstime. In thiscase, ifthe corrodedpipethickness be half, the yielding angle is reducedto a quarter.

In the other case, formodeling the landslidephenomena, a falsedrift(about100cm) was apply betweenthe pipeand soil. During thisinterval, the tension stresscontourshowedvarying amountsthat were highly dependent on the surrounding soil type.Therefore, forbetter comparison, the mean of the valueswasused. This comparison was madebetween thesound pipeand thecorroded pipe with $\alpha = 0.5$. The important parameterinthisphenomena is the equivalentstiffnessbetween thepipe and thesoil.

To express the equivalent stiffness, the simple and practical theory of Columbuswas used. The result shows, whateverthe equivalent stiffness between the pipe and the soil be less, the tension will less in both sound and corroded pipes. Average stresses in corroded pipe is 2 times the sound pipein all of the modes (Without motion, movement of cohesive soil, and movement of granular soil) approximately.

The most critical state occurs when the soil is cohesive. In this state, the corroded pipe stress values are very close to the yielding stress. Therefore, it is better to use, coarse and granular soil around the pipe when the pipes are buried.

References

[1] Weiwei Yu. Corrosion Effects on the Ductile Fracture, Strength and Reliability of Membranes, Plates and Shells, Dissertation, 2009, 1.

[2] A. Aghasi. Investigating the performance of the pressure gas pipe on explosions, PWUT University, 2012, 46.

[3] M. Ono, K. Woo Nam, K. Takahashi, K. Ando. Effect of Local Wall Thinning on Fracture Behavior of Straight Pipe, Nuclear Engineering and Design, 2002, 1

[4] K. Miyazaki, S. Kanno, M. Ishiwata, K. Hasegawa, S. H. Ahn, K. Ando, Nuclear Engineering and Design, 1999, 191

[5] A.S. Daryan, H. Bahrampoor, H. Arabzade. Complete Manual of ABAQUS. Angeizeh Publishing Co- ISBN 978-964-7517-44-7, 2011.

[6] LEON, R. L. & WANG, M. (1978) Performance of underground pipelines in earthquake. IN DIVISION, A. G. E. (Ed.) Earthquake engineering and soil dynamics. New York, American Society of Civil Engineers.

20

Reconstruction of urban worn-out textures and reduction in risk taking (case study: Shushtar City)

Fatemeh Monjezi[1], Hasan Beik Mohammadi[2]

[1]Department of Geography, Payame Nor University of Shushtar, Shushtar, Iran
[2]Research Centre of Shakhes Pajoh, University of Isfahan , Isfahan, Iran

Email address:
Fatememonjezi1@yahoo. com (F. Monjezi), Beikmohammadi_h@yahoo.com (H. B. Mohammadi)

Abstract: Worn – out textures for a number of reasons such as lack of technical and engineering criteria in the construction of the buildings, non – efficient communication network, lack of urban installations and facilities, are more susceptible to natural hazards. The aim of this research is to reduce risk- taking of old and worn – out textures of Shushtar City in the process of reconstruction of this texture. Tools of gathering information in this research are the desk studies, observation, using of the maps, documents and queries. SWOT strategic model has been employed and the SPSS software has been used for the data analysis. The results of this research show that in this texture , the low quality of the buildings , construction material , narrow width of passages, inability of the inhabitants in reconstruction and renovation of old and worn – out textures , makes it necessary to seek a remedy for elimination of risk-taking.

Keywords: Worn – Out Textures, Risk-Taking, Urban Planning, Reconstruction, Shushtar City

---

1. Introduction

Human efforts to confront natural disasters and calamities in modern era possess 3 stages and approaches. The first approach based on offering emergency aids and rescue and relief actions, began at early 20 century and continued till the Second World War and during which rescue aids were offered to the injured. The second approach developed after the World War II and in the atmosphere of victory of scientific and technical dialogue, and schools of growth and development and Behaviorism, following the war in America – Europe and then all over the world. And by incorporating scientific and engineering outcomes in building and development of built environment such as buildings and roads and urban construction became a priority and remained the dominant and fundamental approach till the 1980s. The third approach , considering the inadequacy of scientific and Engineering approach in confronting disasters and calamities and not enough taking into account human factors and variables in preventing disaster, gradually a new approach named crisis management approach began to form at the end of 1980s and early 1990s which based its emphasis on the reduction of individual and social human environment risk-taking against natural disasters and calamities and now is the dominant and widespread approach at the international level and in many countries of the world (1).

Old and worn-out textures are one of the many types of urban textures which, due to worn-out fabric and inappropriate possession and existing vulnerable infrastructure possess local and spatial instability (4).

The aim of urban reconstruction is the improvement of economic, spatial, natural and physical environments within and around the city. In the process of urban reconstruction, government organization, local authorities, regional development organizations, urban development companies, private developers and other business investors are involved (11).

Renovation means reconstruction of buildings and urban spaces, which through achievements or processes, the signs of weariness, destruction and lack of prosperity Disappear (5).

2. Problem Presentation

Natural disasters have influenced the lives of millions of people in the world and challenged the development of the nations. Despite many efforts to reduce risk-taking and negative impact of disasters, the risk-taking of many disasters is increasing. This increase in risk-taking is due to the increase in people and assets encountered with hazards. For example through rapid economic and urban growth in the bay area's subject to storms or earthquake prone cities (8).

Therefore the goals of this research are to define indices of weariness determination, of the urban texture as well as .
analysis and articulation of optimum guidelines for renovation to reduce risk-taking in this area.

As disqualified materials with low durability are used in the construction of almost all existing buildings, the renovation in these textures could be an appropriate mean to reduce risk taking in this area.

The present research methodology is a combination of descriptive, documentary and analytical methods.

3. Theoretical Bases

Being worn-out means, disfiguration of fabric components of the city from their main shape and movement towards fabric and functional destruction of the city (6, 9).

The main criteria for the establishment of the process of weariness of urban texture are:

1-Micro-residential blocks, indicating tightness and intensity of the blocks, more than 50% of blocks possess less than 120 square meters.

2-fabric instability, inappropriate structural system and lack of strength in buildings.

3- Inscrutability, indicating lack of access to and appropriate network of passages-Alleys less than 6 m in width (7).

Hyogo's framework has 3steering goals and is the most integrated and comprehensive global framework in disaster risk reduction and management. The steering goals of Hyogo's framework are:

(A)-Integration of the most influential considerations of disaster risk reduction into the policies and sustainable development programs.

(B)-Development and strengthening of institutions, mechanisms and capacities in all levels of local communities with the aim of helping empowering confronting risks.

(C)- Introducing reduction of risk taking approaches in the design and implementation of preparedness, respond and reconstruction of hazard destructed areas (8).

Main schools in the scope of urban improvement and renovation are listed in order in the Table 1:

***Table 1.** Theories of improvement of worn-out textures*

Cultural orientation	Priority of spiritual needs over material ones: improvement and renovation through revival of the past cultural values.
Progressive orientation	During renovation and reconstruction of the old cities, nothing is preserved except the roads
Ecological school	Recession of old texture because of social and economic groups and segregations due to social injustice
Functional orientation	Improvement and renovation for the aim of economic and functional development of the city
Structural orientation	Collective like and systemic attitude towards city: improvement of fabric and spatial structures in collections and not single buildings.
Modernism	Extreme inclination for preservation of old monuments that even required the renovation and improvement of dead and non-historic buildings.
Post modernism	Thinks more about social relationships: provoking the sense of locality, revival of what that possesses local and special value.
Humanism	Considering people and their partnership in improvement and renovation of urban space and people-oriented intervention.

Source: (1)

4. Museum-Like Attitude towards Urban Texture

This view is based on the preservation of cultural heritage and does not accept intervention in the old and historical textures, unless for their preservation. This attitude believes that the old texture should be preserved like things existing in a museum. This view has sheer emphasis on this that by least intervention in the buildings and urban elements should be preserved. In this attitude preservation of identity and heritage of ancestors are prioritized over requirements of contemporary life. In consequence the inhabitants of these textures , because of not applying the required changes in today's everyday life and shortages and insufficiencies of urban infrastructures which have roots in such insight, have been forced to leave the texture and after collapse of social texture, follows the destruction of fabric texture (3) .

5. Location of Study Area

Shushtar City has an area of 2436 square kilometers, and is located between $48^{\circ},35'$ and $49^{\circ},12'$ from Greenwich Meridian and between $31^{\circ},36'$ and $32^{\circ},26'$ from Equator, at the north of Khuzestan Province. Its population is about 111200 (according to census 1390) and is 57th city of Iran according to the population .And the fourth in Khuzestan Province after cities of Ahwaz, Dezful and Abadan. (2).

***Table 2.** SWOT Matrix*

SWOT Matrix	Strengths(S)	Weaknesses(W)
Opportunities(O)	Strategies that, based on the strengths, are founded to take advantage from opportunities (SO).	Strategies that based on diminishing weaknesses are founded to take advantage from opportunities (WO).
Threats(T)	Strategies that based on the strengths are founded to combat the threats (ST).	Strategies that based on the weaknesses, are founded to combat threats (WT)

The old and worn-out texture of Shushtar has 176 ha. area which is located almost at the center of the city and the four boundaries of it are: from the north , Shahid Sherafat Street, from the south, Shahid Rejaei Street, from the east , Shahid

Shariati and Imam, Khomeini Streets and from the west , Taleghani Street. Considering the main business and economic activities being concentrated in this area which include all the shopping centers of the city like Bazaar and insurance offices and banks, wholesalers and retailers…., the concentration and over crowdedness is well visible in this area (same source ,6).

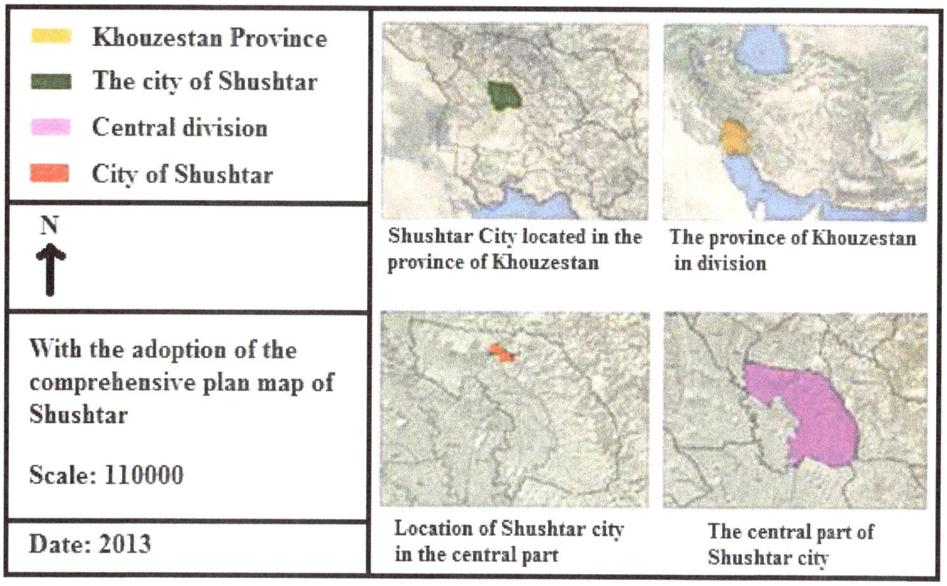

Fig. 1. Political divisions of Khuzestan Province

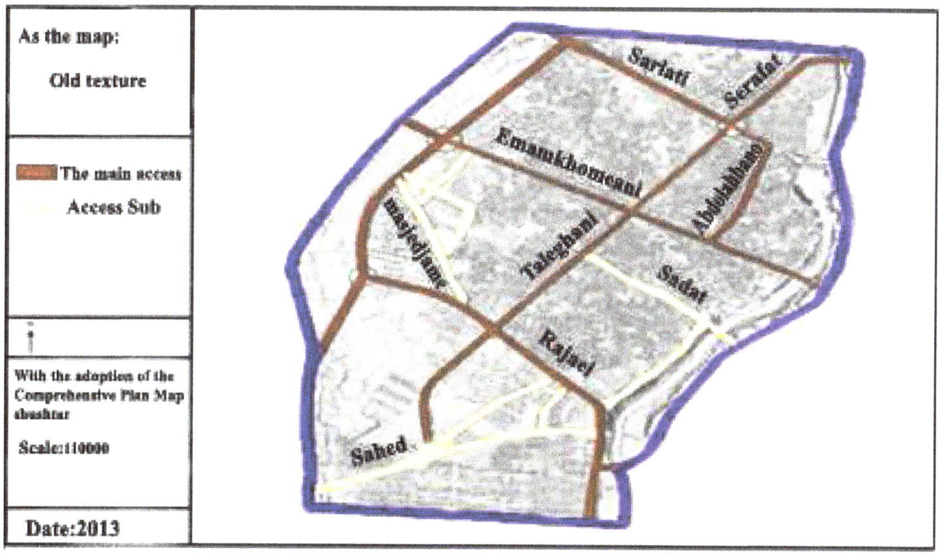

Fig. 2. Map of worn-out part of the Shushtar City

In the designed query, 4 general questions are asked which include, gender, education, age, career position. In this section we will analyze the data related to these questions.

1-The results of investigation into the quality of the buildings show that only 8.82% of buildings are recently built and live able and nearly 70% are completely worn or should be destroyed.

2-The investigation into the quality of construction material used in residential units of the worn texture, is among the indices which enables to analyze the fabric situation of the texture. The results and field studies show that 7.64% of houses of worn-out texture possess brick and steel material, 52.35% possess brick, wood and mud, 34.1% possess brick, steel, wood, mud and 5.88% possess reinforced concrete material. The oldness of texture depends to a great degree on the age of the buildings. One of the characteristics of the old textures is the presence of old buildings. The results of the field studies show that 5.88% of the residential units , in the worn-out texture of Shushtar City do have an age less than 5 years,7.64% an age of 5-10 years,14.7% an age of 10-15 years ,20% an age of 15-20 years and the last 51.76% of buildings possess an age more than 20 years.

3-In investigation of the main problems of the neighborhood and based on the field studies, it is observed that 36.47% of people mentioned shortage of services andeducation,12.94% mentioned social and cultural problems, 22.35% mentioned economic problems,7.64% mentioned environmental problems as the main obstacles.

4-In investigation of the kind of problems of the neighborhood, in improvement and reconstruction and based on the field studies, we can conclude that 24.11% have mentioned resistance to natural hazards , 45.29% have mentioned narrowness of passages and lack of access to material for improvement tasks, 8.23% have mentioned the age of buildings and juridical problems, 4.7% have mentioned the smallness of land parcels and atlast 17.64% have mentioned the lack of willingness for improvement and renovation as the main factors for not to achieve improvement tasks.

6. SWOT Analysis

From this model view, appropriate strategies, maximize the strengths and opportunities and minimize weaknesses and threats. At this stage, based on the information from previous studies and investigations and by considering theories and views of experts and test assumptions, the guidelines and general and specific policies in every index for the purpose of reconstruction of the worn-out texture of Shushtar City, to reduce disaster risk, will be presented.

Table 3. *Analysis of internal factors in the Strengths*

Strengths	weaknesses	opportunities	threats
Oldness of residence In the neighborhood. Presence of historical And precious buildings in the neighborhood. Political and social History. Presence of social institutions in the neighborhoods. Low rates of rent and price of houses. Possessing willingness of partnership and cooperation in improvements and renovations of neighborhood. Low noise pollution in the neighborhood. Private ownership of real estates. Existing low rise buildings. Existing local councils.	Intense weariness in the fabric texture of the neighborhood. Low income and economic level of the employees. Population evasion. Lack of strength of buildings against natural disasters. Narrow alleys and passages. Low strength of buildings against natural risks. Non-standard material used in most the buildings.	Economic saving through reconstruction of this part. Existing laws related to broadening of the passages. Possibility of reinstatement of space and fabric. Possibility of utilizing patterns of precious architectural buildings in the new designs. Ease of possessing buildings because of weariness and low strength of them, low intensity and low price. Determination of authorities to intervene for reconstruction. Existing multitude of global and internal experiences in reinstatement and revival of worn textures.	Continuation of Population evasion. Weak authority of local management. Lack of incentive practices Intense population concentration during day and it's evacuation at night. Not considering people's partnership. Intense reduction in social security. Increase in poverty and un-employment rates among residents of the texture. Non uniformity between new constructions and the old structure.

Table 4. *Analysis of internal factors in the Strengths*

Strengths	Weight	Ranking	Score
Oldness of residence In the neighborhood	0.05	3	0.15
Presence of historical And precious buildings in the neighborhood.	0.05	3	0.15
Existing Social institutions of the neighborhood.	0.06	2	0.12
Existing local councils.	0.06	3	0.18
Possessing willingness of partnership and cooperation in improvements and renovations of neighborhood.	0.07	4	0.28
Low noise pollution in the neighborhood	0.03	2	0.06
Existing low rise buildings	0.04	2	0.08
Private ownership of real estates.	0.05	2	0.10
Political and social History	0.06	4	0.24
Low rates of rent and price of houses.	0.03	2	0.06

Table 5. *Analysis of internal factors in the Weaknesses*

Weaknesses	Weight	Ranking	Score
Intense weariness in the fabric texture of the neighborhood.	0.08	4	0.32
Low income and economic level of the employees.	0.06	4	0.24
Population evasion.	0.07	3	0.21
Lack of strength of buildings against natural disasters.	0.07	3	0.21
Narrow alleys and passages.	0.06	3	0.18
Low strength of buildings against natural risks.	0.07	2	0.14
Non-standard material used in most the buildings	0.05	2	0.10

Table 6. *Analysis of external factors in the Opportunities*

Opportunities	Weight	Ranking	Score
Economic saving through reconstruction of this part.	0.09	4	0.36
Existing laws related to broadening of the passages.	0.08	4	0.32
Possibility of reinstatement of space and fabric	0.07	3	0.21
Possibility of utilizing patterns of precious architectural buildings in the new designs.	0.06	2	0.12
Ease of possessing buildings because of weariness and low strength of them, low intensity and low price.	0.04	2	0.08
Determination of authorities to intervene for reconstruction.	0.05	2	0.10
Existing multitude of global and internal experiences in reinstatement and revival of worn textures.	0.05	2	0.10

Table 7. *Analysis of external factors in the Threats*

Threats	Weight	Ranking	Score
Continuation of Population evasion.	0.08	4	0.32
Weak authority of local management.	0.06	4	0.24
Lack of incentive practices	0.06	3	0.18
Intense population concentration during day and it's evacuation at night.	0.05	2	0.10
Not considering people's partnership.	0.04	2	0.08
Intense reduction in social security.	0.06	2	0.12
Increase in poverty and un-employment rates among residents of the texture.	0.06	3	0.18
Non uniformity between new constructions and the old structure..	0.04	3	0.12

7. Discussion and Conclusion

Among the most important consequences of rapid growth of urbanization in Iran , in addition to inappropriate urban infrastructures for the residents and urban emigrants, is the formation of irregular and problematic intra-urban textures and ultimately the high factor of risk-taking, which has had many challenges .This problem alongside the issue of land, becoming a scarce economic resource in recent decade, has caused that optimum use of old and worn-out textures (dead and non used urban spaces) ,which is one the main sources of land supply, gains special importance. Establishment of one institution or strong team for intervention management in worn-out textures and also management of risks is recommended. Based on the surveys done among residents of the texture, concerning its reconstruction, one could remark that: The plan of reconstruction of old and worn-out textures of Shushtar City as many other urban projects face financial problems. Neglect of authorities toward this texture during many years and also lack of serious measures in the implementation of the project, has caused distrust of the residents. Renovation and reconstruction of old texture to reduce risk-taking in them, is possible only through the cooperation and partnership of the people. Therefore reconstruction of these worn-out textures should be done by the municipality and cooperation of Cooperative-company, and this will be realized only when the municipality possesses adequate credits for these issues. What that could be derived from the referred discussions are as follows:

8. The Results of Research Assumption

Considering that low quality and nondurable material is used in the most of the existing buildings, Renovation in these textures could be a good practice for reducing risk-taking in theses region. As nearly 70% of buildings are completely worn-out, 52.35% of houses have brick, wood and mud material, 51.76% of buildings are more than 20 years old, 36.47% of people have problems to achieve construction and 45.29% is the rate for narrowness of passages and lack of access to material for improvement, therefore one could conclude that the assumption that "reconstruction in the worn-out texture of Shustar City could be a good practice to reduce risk-taking" is confirmed.

9. Offering Appropriate Concepts and Recommendations in Reconstruction of Worn-Out Texture of Shushtar City

Real partnership of people in the projects and plans related to the revival and reinstatement in different dimensions, retrofitting of buildings within the texture and reducing the natural hazards risks, attempts to strengthening and restoration of buildings and historical monuments, pavement of the bed of passages with good and durable material , mitigation in reconstruction charges for buildings with non durable material(wood and mud), providing discounts in taxes for buildings' owners in order to begin the improvement and renovation tasks. Loans granted with low interest rates and with long terms to owners because of their financial in abilities to renovate their houses, attempts to eliminate obstacles, and solving problems and laws that have become obstacles to give speed to renovation and reconstruction of these textures, attraction of investors to accomplish improvement of textures, reconstruction and improvement in these neighborhoods to give neighborhood identity and reduce their risk taking against accidents, emphasis and supervision of urban management to widen passages within the textures, encouraging activities by municipalities, related organizations and people to reconstruct worn-out textures and reduction of risk taking in them, applying encouraging effective policies in regions that need retrofitting through the support, protection, maintenance, observation, strengthening and repair, establishment of one mediator organization to coordinate among the components of reconstruction cycle(people, banking system, government and municipality) and facilitation of reconstruction, identification and prioritizing of buildings with respect to different kinds of instabilities with the help of municipality and partnership of people.

References

[1] Shahidi, Akram, (2009), Social consequences of development and modernization of space for residents and local businesses Tabarestan central city of Mashhad. M.S Thesis, , Earth Sciences Faculty, Univ. Shahid Beheshti.

[2] Master Plan of Shshtar City, 2012, Tarh va Tadvin Consulting Engineers, Housing and Urban Development Organization of Khuzestan Province.

[3] Shamaaei, E, Poor Ahmad, Ahmad, Urban improvement and renovation from the view point of Geography, Tehran Univ. publications, 2006.

[4] Peter Dickens (2004), Urban Sociology, Community, local gathering and its nature, Mashhad, Astane Ghodse Razavi.

[5] Habibi et al. (2008), improvement and renovation of old urban textures, Kurdistan Univ. Publications.

[6] Zebardast,Esfandiar(2006), Application of AHP in Planning, Fine Arts Quarterly, No. 1 ,pg(4-56).

[7] Rules and regulations of the city,(2006), Housing and Urban Development Ministry 2006, Supreme Council of Planning and Architecture Secretariat

[8] Mohammad Farhadi Consultant to the Red Crescent Population Manager, Chief Consultant of U.N in Risk taking management, 2013, Quarterly Journal of Rescue and Relief, 4th year, No. 3, 108-110.

[9] Shamaaei, Ali & Poor Ahmad, (2005), an analysis of urban policies and programs of improvements and renovations in national development programs, Quarterly of Geographical Research, No. 48, 112.

[10] Ha seong, Kyu (2001), developing a Community-Based Approach to Urban Redevelopment, GeoJournal

[11] John Diamond, Joyce Liddle, Alan Southern and Philip Ose,(2010), Urban Regeneration Management International Perspectives, Routledge

Effective Factors in Causing Leakage in Water Supply Systems and Urban Water Distribution Networks

Hassan Saghi*, Abbas Ansari Aval

Department of Civil Engineering, Hakim Sabzevari University, Sabzevar, Iran

Email address:
h.saghi@hsu.ac.ir (H. Saghi), aaabas1192@gmail.com (A. A. Aval)

Abstract: Water has a vital role in the country's economic development, since most part of it is dry and desert. Whereas, water has been cheap and easy accessing, the authorities of the urban water distribution organizations have not paid so much attention to the decreasing of the unaccounted for water (U.F.W) as a solution for saving water, the decreasing of the cost of preparing water and the increasing of their efficiency or income. In recent years, the U.F.W has been paid attention because of the shortage of resources and the cost increasing of water preparation. Therefore, along with preparing water from new resources, the water wasting in different stages of transmission, refinement and distribution of water should be prevented as a long term programing. In this article, the effective elements in causing leak have been explained. The results show that the most leakage of water deals with the parameters such as pressure, quality, diameter and age of the pipes.

Keywords: Economic Development, Urban Water Distribution, Water Leakage

1. Introduction

In recent years, the international communities and the dveloped countries have paid a special attention to the shortage of the water resources and finding new solutions for it, especially to prevent water wasting. The U.F.W and leakage in the water supply system and the urban water distribution networks have been paid attention in many countries. In recent three decades good experiences in theoretical and executions in decreasing of the water leakage have been obtained through the scientific and systematic planning. Fortunately, in Iran, this subject has been dealt with for more than one decade. The first step was preparing a national development plan for upgrading the urban water distribution in the second development plan of the country. The second step has been done by controlling offices of the U.F.W in the water and waste companies of the country in which the study of decreasing the water wasting has been paid.

The unaccounted water has many privileges; some of them are mentioned here:

- Decreasing the investment cost of the constructing of the installations for water resources and the development of water supply system
- Decreasing the entering of the pollution through the leak points
- Increasing the efficiency of the water and waste companies in technical engineering, performance and subscription sections
- Increasing the income of the water and waste company.
- Obtaining the consent of the subscribers.

Decreasing the loss of water in water supply system network involves knowing the amount of loss, causes, methods and priority of these factors. The next stage includes the location detection and determining the position of the most hazardous leakage points and the repairing and rebuilding the network.

2. The Effective Factors in Causing Leak

The accidents which happen in the water supply system networks are usually high. For instance, about one million accidents have happened in water supply system networks in the country in 1998. This damage costs about 20% of the total income of the Water and Waste Company. This cost has been for repairing, rebuilding and improving of the network. Over 30% of these accidents have happened on the pipelines of the water supply system networks. Many factors are involved in the leakage detection of the water in the network.

Some of these effective factors will be paid to in this article.

2.1. The Material of the Pipes

One of the effective factors in water leakage is the material of the pipes. An experiment on the pipes in Chaloos showed that the asbestos pipes have the least ruins and the galvanized pipes have the most efficiency [1]. This is shown in Fig. 1. The experience showed that in leakage detection methods, the material of the pipes has been one of the most important parameters in the causing leakage in the networks. The results of the studies, which have been done by Jing and Zhi-Hong, showed that the material of the pipes is the most important factor in causing leaks. Among different materials of pipes, iron pipes have the most leakage accidents and asbestos pipes have the least.

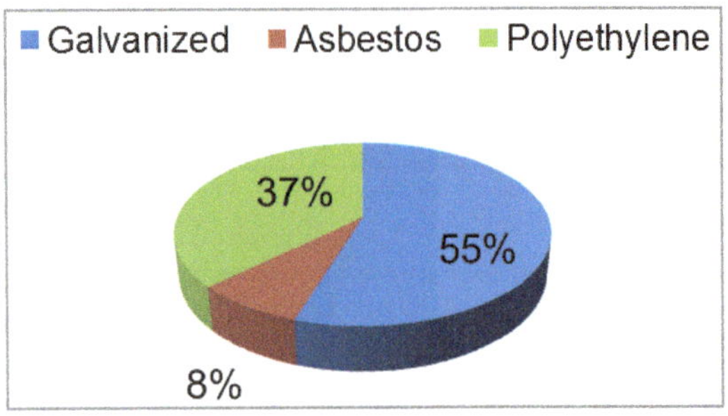

Fig. 1. Accidents percentages for various materials of pipes [1].

2.2. The Age of the Pipes

The age of pipes has a direct or indirect effect on the amount of leakage. So a program can be planned with a time table for the efficiency and rebuilding of the pipes with different qualities, so that the loss of water will decrease. On the basis of the experiments, which have been done in Chaloos, the suggested time for galvanized pipe is 15 years and for polyethylene is about 16 years [1]. The studies of recent years show that with the increasing of the pipes age, the loss can reach to 50% of the entered water into the networks [2].

2.3. The Diameter of the Pipes

The diameter of the pipes is one the effective element in the rate of the accidents and the loss. The hydraulic pressure decreases when the diameter increases, so the loss comes down. Although, using high diameter pipes involves more cost, this investment seems to decrease the loss of leaking water in a long range program.

The more diameter of the pipe, the thicker of the shell the pipe has. So the resistance of the pipe increases against the outside or inside pressure and the destructive elements of the environment. For instance, on the basis of a research, which has been done in Netherlands, in every 100 Km with 150 mm pipe diameter, 25 cracks are expected annually [3].

2.4. The Pressure

Another hydraulic parameter, which determines the performance condition and the servicing of urban water supply system network, is pressure with a high importance. High pressure causes the increasing of leak, and the loss of consuming water and the number of the accidents in the network. On the other hand, low pressure in the network causes the inability of complete supplying water or result in providing unsuitable water. The pressure also is effective on the stability and the structure of the system. With high pressure, the probability of the accidents and the crack of the system increase and the stability performance and suitable servicing of the network endangered. If the changing of the pressure happens according to the prescribed standards, it will be appropriate for a network user. So that the different aims of the network which are usually contradictory on the basis of the pressure reaches to a balance. Therefore, knowing the conditions of the changing pressure has a high importance. The loss is sometimes because of the network pressure conditions. In other words, the cracks open with the high pressure and close with the low pressure. The amount of loss in a point has a direct relation with the amount of pressure on that point so the amount of loss changes according to the changes in pressure.

The pressure parameter is the most important one among the different parameters; this element is paid attention so much in the new methods of leakage detection.

2.5. The Movement of the Soil Around the Pipe

The movement of the soil may happen because of sliding of the ground or shaking of the earthquake. This movement brings a lot of pressure on the pipe and can cause cracks in the pipe, gaps in the joints or the breaking of the pipes.

2.6. The Movement of the Soil Around the Pipe

The corrosion of the environmental soil and water for all of the metal structures is inevitable (especially about the buried pipes and joints). This corrosion in the motor pumps pipes and taps can happen through cavitation or chemical corrosion of the water supply system pipes make corrosion by water from inside and solid from outside. Both of them happen due to the existence of unsuitable components and salts in water and soil. This corrosion happens through the forming the galvanic cells or concentration cells, so that a part of the pipe takes the role of anode and other part of pipe the role of cathode. So the currency of the metal happens from anode to cathode therefore the metals which has anode role is ruined.

The iron or steel pipes have been used in the water supply system networks for more than five centuries. The corrosion usually happens over time. The corrosion of the pipes has an important role in the quality of the water. Some of the corrosions of the pipes are as follow [4]:

- The water takes color, smell and taste.
- It prepares a good background for the growth of microbes.
- It attracts and gathers the arsenic materials

2.7. Hit

Most part of the water supply system network goes under

the streets of the cities. In the cases of digging by the other organizations such as gas, electricity, communication and may damage the pipes.

2.8. *Using Unsuitable Materials for the Basis and Coverage of the Pipes*

While installing the pipes, the surrounding of them should be made so that a good basis is prepared for the setting of the pipes and the extra tensions does not transmit to the pipes. If unsuitable materials (such as big seed materials or those that their summit is remarkable) are used, unbalance pressures are applied to the pipes and damage them.

2.9. *Incorrect Installation*

Installation by unprofessional workers or using incorrect and non-standardized techniques may result in incomplete sealing of the joints. In this case there are some leaks in the joints from the beginning and this would increase over time.

2.10. *Water Hammer Hit*

When water is pumped directly into the water supply system network a water hammer hit may cause accidents and breaks the pipes and joints. Wherever the pipe is weak because of corrosion this hit becomes stronger. This hit makes some quicks waves and passes away fast, so it may cause different dangers and damages in the pipelines.

Generally, this water hammer hit in the pipeline projects has an important role and ignoring it results in very large damages in the equipment. In order to analyze the water hammer hit, many types of software, on the basis of different mathematical methods and digits are provided. Three software, Hytran, AFT impulse and Hammer are used for analyzing the water hammer hit in the pipeline network. The outputs of AFT impulse were the most realistic and Hammer's result was far from the reality and expectations. It was showed through Hammer's program that in spite of the installation of the controlling devices to decrease the minimal pressure, not only the minimal pressure did not change but the maximal pressure increased. But AFT impulse and Hytran had a good compiling with each other and with the practical results.

Other performances against water hammer hit were done by Daneshfaraz and Moradi [5]. They used a tank of air with 3 m3 capacity with an air compressor and a one-way tap in a pump station. They also put an air tap and an empty tap along the pipeline, with using protection equipment, they succeeded to minimize the positive pressure and to omit the negative pressure [5].

2.11. *Clogging the Components Joints and Pipes*

Clogging happens when some sediment and objectives enter into the pipes and their components. This makes a condition in which the weak parts break.

2.12. *The Effect of Temperature*

When the temperature is high, the resistance of the pipes, especially the pipes with artificial material such as plastic, polyethylene components, comes down very much. When the depth is not suitable or the coverage is not enough, the low temperature causes freezing. This freezing makes some tension inside the pipes.

2.13. *The Depth of Placement of the Pipes*

The depth of placement of pipes in the ground can have a role in leaking. The studies which have been done by Jing and Zi-Hong showed that the depth of placement has an effect on leaking. But the results show that other factors such as the material and the diameter of pipe and the pressure have more importance [6].

2.14. *The Pressure of the Cars on the Surface of Roads and Ground*

Another factor of leaking is the pressure of the passing cars on road and ground where the pipeline buried. Although in new methods of leakage detections, this factor mostly is not considered, the studies which were done by lee and his colleagues mention that this factor has a special importance. This parameter should be considered in the future studies [7].

2.15. *Corrosive Waters*

The corrosive waters cause corrosion and weaken the pipes. Corrosion happens with tiny holes and developed over time.

2.16. *The Movement of Soil*

The clay soil contracts or expands according to its moisture. The resulted movement may cause loosing of the joints and local cracks on the pipelines.

2.17. *The Damages to the Pipes by the Other Service Organizations*

The performances which are done by other service organizations such as gas and electricity may cause some leakages, unsuitable joints and non-standard piping.

2.18. *Poor Quality Connections and Nonstandard Plumbing*

The loss of water is usually very much when unsuitable joints are used or incorrect piping is performed. Controlling these materials and performances can prevent loss of water, remarkably.

2.19. *The Summit of Ground*

So much using of the underground water in some areas causes the ground summit in which the pipe may crack and leak.

2.20. *Climate Conditions*

The climate conditions of the area which compose of the maximum and minimum air temperature, the quality of soil, raining, freezing and so on, have a remarkable effect on the breaking of the pipe.

3. The Priority of Effective Factors in Causing Leakage

The priority of the factors is needed in this stage. Some new methods of leakage detection through all over the world and the parameters which are effective in causing leakage are referred in the table 1.

Table 1. *Main parameters considered for leakage.*

Researcher	Reference	Main pipe parameters considered in causing leakage
Jing and Zhi-Hong	[6]	Material, age, pressure, diameter, the depth of the placement of the pipe
Li et al.	[7]	Material, age, pressure, diameter
Perez et al.	[8]	pressure
Marunga et al.	[9]	pressure
Nicolini et al.	[10]	pressure
Berardi et al.	[11]	Age, diameter, length
Araujo et al.	[12]	pressure

4. Discussion and Conclusion

As presented earlier, numerous factors have effective role in causing leakage. However, considering all of them is almost impossible in different methods of leak detections. Therefore, with considering some of the important factors, it is tried to do leak detection in the water supply network. With considering these methods, it can be easily found that the pressure, age, material and diameter of the pipe are taken as the main factors and other factors can be ignored. As shown in table 1, among the four parameters including material, age, diameter and pressure of the pipe, the pressure element is the most important one. This element is considered in the most methods of leakage detection. Therefore, studying more about this parameter can result in holistic view, collectivity and designing a more efficient water supply system network.

References

[1] Tabesh, M., Karimi, K. "Determination of the leak detection and the renewal time of urban water by analyzing the accident data," No. 5, pages 597 to 610, 2006.

[2] M. Schouten, and R. D. Halim, "Resolving strategy paradoxes of water loss reduction: A synthesis in Jakarta." J. Resources, Conservation and Recycling, vol. 54, pp. 1322-1330, 2010.

[3] Tabesh, M. Honari, H.r. "Analysis of accidents in the urban water distribution networks", water and environment magazine, NO. 50, pages 17-23, 2002.

[4] V. Chawla, P. G. Gurbuxani, and G. R. Bhagure, "Corrosion of Water Pipe: a Comprehensive Study of Deposits." J. Minerals and Materials Characterization and Engineering, PP. 479-492, 2012.

[5] Danshfraz, R., Moradi, N. "Evaluation of water hammer in pumping station and water pipeline using software AFT-Impulse", accepted at the first conference on Kerman water and sanitation, 2012.

[6] K. Jing, and Z. Zhi-Hong, "Time Prediction Model for Pipeline Leakage Based on Grey Relational Analysis." J. Physics Procedia, vol. 25, pp. 2019-2024, 2012.

[7] W. Li, W. Ling, S. Liu, J. Zhao, R. Liu, Q.Chen, Z. Qiang, and J. Qu, "Development of system for detection, early warning, and control of pipeline leakage in drinking water distribution: A case study." J. Environmental Science, vol. 23(11), pp. 1816-1822, 2011.

[8] R. Perez, V. Puig, J. Pascual, J. Quevedo, E. Landeros, and A. Peralta, "Methodology for leakage isolation using pressure sensitivity analysis in water distribution networks." J. Control Engineering Practice, vol. 19, pp. 1157-1167, 2011.

[9] A. Marunga, Z., Hoko, and E. Kaseke, "Pressure management as a leakage reduction and water demand management tool: The case of the City of Mutare, Zimbabwe." J. Physics and Chemistry of the Earth, vol. 31, pp. 763-770, 2006.

[10] M. Nicolini, C. Giacomello, and K. Ded, "Calibration and Optimal Leakage Management for a Real Water Distribution Network." J. Water Resources Planning and Management, pp. 134-142, 2011.

[11] L. Berardi, O., Giustolisi, and F. Primativo, "Exploiting multi-objective strategies for optimal rehabilitation planning." In Proceedings of Computer and Control in Water Industry (CCWI) - Water Management ChallengesinGlobal Changes-Ulaniki, Taylor and Francis Group, London, PP. 23-30, 2007.

[12] L. S. Araujo, H. Ramos, S. T. Coelho, "Pressure Control for Leakage Minimization in Water Distribution Systems Management." J. Water Resources Management, VOL. 20, pp. 133-149, 2006.

Research on Pore Water Pressure of Saturated Muck Soil Under Traffic Loading of Low Frequency

Luo Qiyang, Liu Yongjian, Zhou Yaoyao, Wu Jiansheng, Xie Zhikun

School of Civil and Transportation Engineering, Guangdong University of Technology, Guangzhou, China

Email address:
luoqiyang159@163.com (Luo Qiyang)

Abstract: In order to study the pore water pressure of saturated muck soil under traffic loading of low frequency in Pearl River Delta, the influence of loading frequency, cyclic stress and consolidation on pore water pressure are analyzed. The results show that the final pore water pressure is increased as the frequency decreases, but the increasing rate of pore water pressure is periodical. Moreover, isotropic consolidation can improve the effective stress of soil and slow down the development of pore water pressure, but it is not the key factor. The key factor of influencing the pore water pressure is the cyclic stress, the following is loading frequency, and the consolidation is the last one. In the end, considering cyclic stress, loading frequency and consolidation, a model of pore water pressure based on a math model is built up.

Keywords: Muck Soil, Low Frequency, Traffic Load, Pore Water Pressure, Pore Pressure Model

1. Introduction

Multiple urban agglomeration are lied in the east coast of China. The transportation network supports the development of the whole city, however, the high moisture content, void ratio and poor structural soft clay brings difficulties to the roads and subways construction. In addition, there are many factors affecting the development of dynamic pore water pressure of soft clay under traffic load, like cyclic stress, loading frequency, cyclic numbers, stress history and stress path, etc. There were some tests [1, 2] showed that when the loading frequency is limited, the lower the frequency is, the higher the cumulative plastic strain and pore pressure of soil develop. However, there were some scholars such as Ansal [3] showed that the influence of loading frequency only work at the beginning of the loading stage which can be ignored when the soil samples are closed to broken. Yasuhara [4] suggested that the influence of frequency on the acumulative plastic strain can be almost negligible. Zhang Ru [5] did a dynamic triaxial test on saturated clay soil, which showed that the influence of the frequency f on dynamic characteristics of soft soil worked periodically during the frequency f is from 0.1 Hz ~ 0.4 Hz. Wei Xinjiang [6] suggested that the pore water pressure develop slowly if the consideration degree become lower. Wang Jun [7] showed that the development of pore water pressure is not obvious when the loading frequency is less than 1 Hz. Wang Yuandong [8] suggested that the pore water pressure of reinforced soil decrease as the cell pressure increase. Ding Zhi [9] showed that the law of pore water pressure in un-drained condition is similar to the drained condition. Zhao Chunyan [10] built a model of accumulative pore water pressure considering cyclic numbers, deviatoric stress and the consolidation. The experiments above showed that there are many factors influencing the development of pore water pressure. This paper, firstly, summarize the previous research, and then analyze the dates by amounts of dynamic triaxial tests for the Pearl River Delta saturated muck soil under traffic loading of low frequency, which is to study the influence of the cyclic stress, loading frequency and the consolidation on the pore water pressure. In the end, a model of pore water pressure is established.

2. The Dynamic Triaxial Test of Saturated Muck Soil Under Traffic Loading of Low Frequency

2.1. Experiment Apparatus and Soil Sample

A dynamic test of saturated muck soil is carried out by GCTS SPAX-2000 (modified) static and dynamic true triaxial testing system which is developed and manufactured by United States. The system consists of seven sub-systems: (1) SPAX-100 true triaxial pressure chamber; (2) PCP-2000-H pressure control panel; (3) PVC-25-HC confining pressure / volume computer servo controller; (4) PVC-200-HP back pressure / pore-pressure volume controller; (5) SCON-2000 digital servo controller and acquisition system; (6) HPS-10-5 variant constant pressure pump; (7) WIN -CATS-ADV control and post-processing software. The entire testing process is controlled by integrated computer software named GDSLAB that can collect and deal with the data to emulate a partial stress and cyclic stress traffic load.

The soft soil sample was taken from Shenzhen Baoan Airport, which is in offshore sea level 12m~15m. It contains mica, quartz and shell debris. The moisture content and compressibility are high and the permeability is poor. Basic physical and mechanical properties of samples are shown in Table 1.

Table 1. *Physical and mechanical properties of soft clay.*

ρ/g/cm^3	w/%	e	I_L	I_P	d_s
1.63	63.4--65.2	1.57	0.98	27.9	2.71

2.2. Testing Program

Isotropic and anisotropic consolidation are took in this experiment. The frequency are 0.1Hz, 0.25Hz, 0.5Hz and 1 Hz. The cyclic stress are 20 kPa, 50 kPa and 65kPa. The test program is shown as Table 2. ε_d =15% is the failure standard. If the soil sample did not fail, the sample would be shear by 0.01%/min until broken. In order to simulate actual operation, firstly, the drain valve is going to be closed before dynamic loading. And then the initial static deviator stress is loaded, which is half of dynamic stress.

Table 2. *Un-drained cyclic triaxial test parameters.*

Test Number	Intermediate Principal Stress σ_2 /kPa	Frequency *f*/Hz	Dynamci Stress σ_d /kPa
C-1	200	0.1	20
C-2	200	0.5	20
C-3	200	1	20
C-4	200	0.1	50
C-5	200	0.5	50
C-6	200	1	50
C-7	200	0.1	65
C-8	200	0.5	65
C-9	200	1	65
C-10	225	0.1	20
C-11	225	0.5	20
C-12	225	1	20
C-13	225	0.1	50
C-14	225	0.5	50
C-15	225	1	50
C-16	225	0.1	65
C-17	225	0.5	65
C-18	225	1	65
C-19	250	0.1	20
C-20	250	0.5	20
C-21	250	1	20
C-22	250	0.1	50
C-23	250	0.5	50
C-24	250	1	50
C-25	250	0.1	65
C-26	250	0.5	65
C-27	250	1	65
C-28	200	0.25	20
C-29	200	0.25	50
C-30	250	0.25	50

3. Test Result and Analysis

3.1. The Law of Super Pore Water Pressure Under Different Frequencies

The curves of the relationship between super pore water pressure u and cyclic number N under different frequency of the same dynamic stress and isotropic consolidation $\sigma_1'=\sigma_2'=\sigma_3'$=200kPa are shown in Figure 1, 2, 3. The curves of the relationship between super pore water pressure u and cyclic number N under different frequency of the same dynamic stress and anisotropic consolidation $\sigma_1'=\sigma_2'$=225kPa, σ_3'=200kPa are shown in Figure 4, 5, 6. The curves of the relationship between super pore water pressure u and cyclic number N under different frequency of the same dynamic stress and anisotropic consolidation $\sigma_1'=\sigma_2'$=250kPa, σ_3'=200kPa are shown in Figure 7, 8, 9.

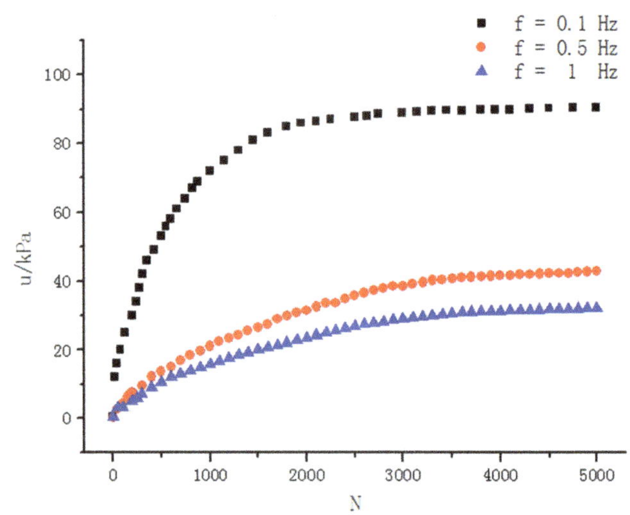

Effective stress $\sigma_1'=\sigma_2'=\sigma_3'$=200kPa, dynamic stress σ_d=20kPa

Figure 1. *Super Pore Water Pressure u against Cyclic Number N.*

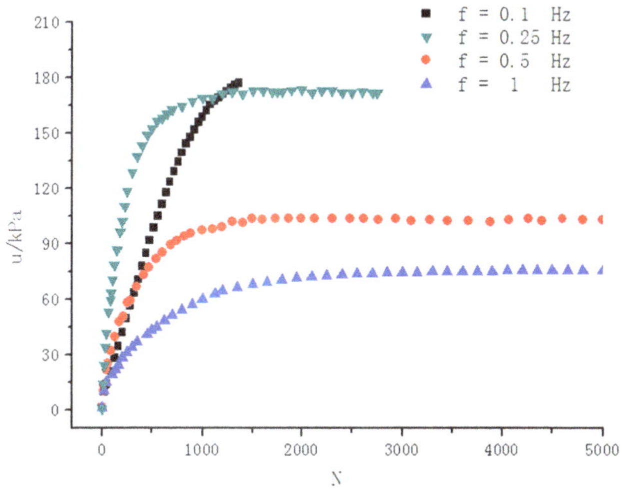

Effective stress $\sigma_1'=\sigma_2'=\sigma_3'=200kPa$, dynamic stress $\sigma_d=50kPa$

Figure 2. *Super Pore Water Pressure u against Cyclic Number N.*

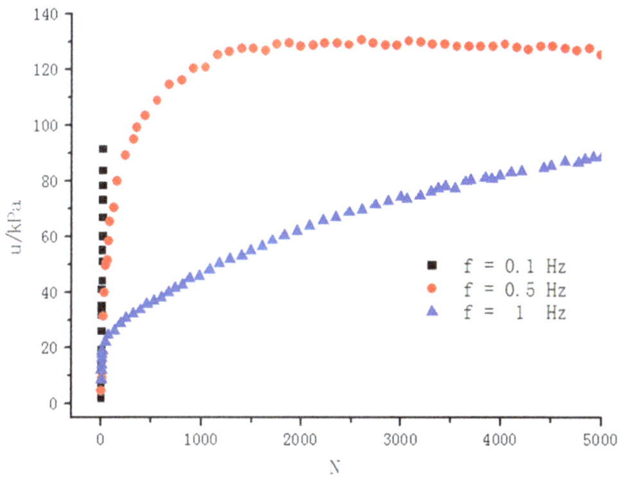

Effective stress $\sigma_1'=\sigma_2'=\sigma_3'=200kPa$, dynamic stress $\sigma_d=65kPa$

Figure 3. *Super Pore Water Pressure u against Cyclic Number N.*

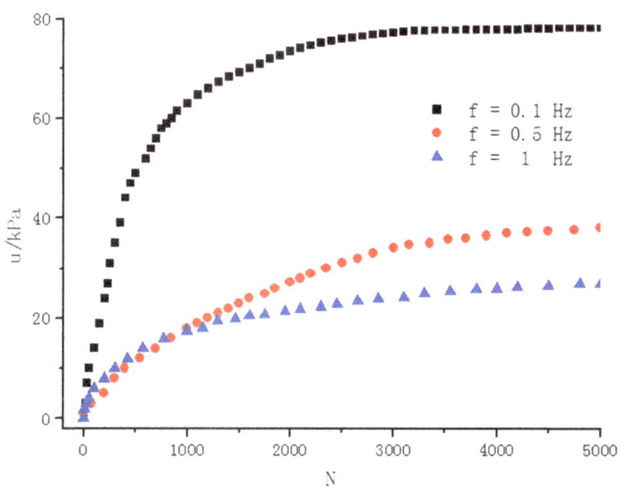

Effective stress $\sigma_1'=\sigma_2'=225kPa$, $\sigma_3'=200kpa$, dynamic stress $\sigma_d=20kPa$

Figure 4. *Super Pore Water Pressure u against Cyclic Number N.*

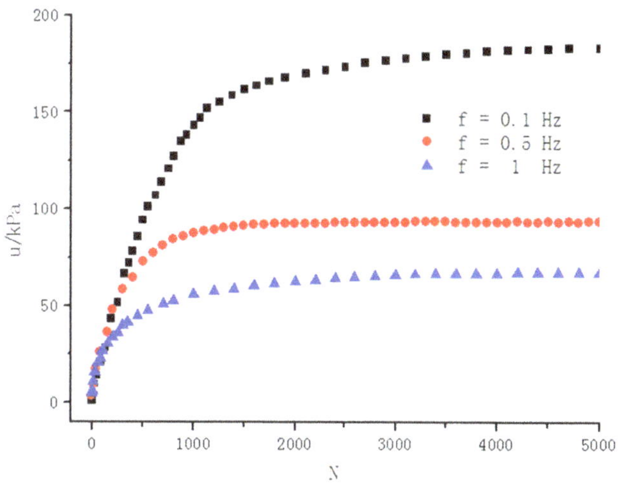

Effective stress $\sigma_1'=\sigma_2'=225kPa$, $\sigma_3'=200kpa$, dynamic stress $\sigma_d=50kPa$

Figure 5. *Super Pore Water Pressure u against Cyclic Number N.*

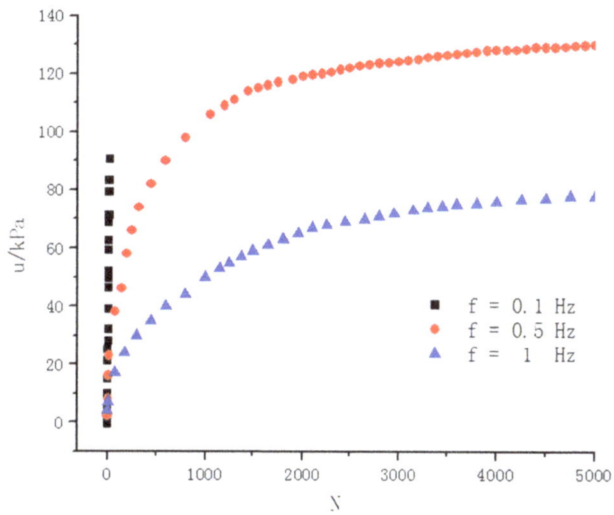

Effective stress $\sigma_1'=\sigma_2'=225kPa$, $\sigma_3'=200kpa$, dynamic stress $\sigma_d=65kPa$

Figure 6. *Super Pore Water Pressure u against Cyclic Number N.*

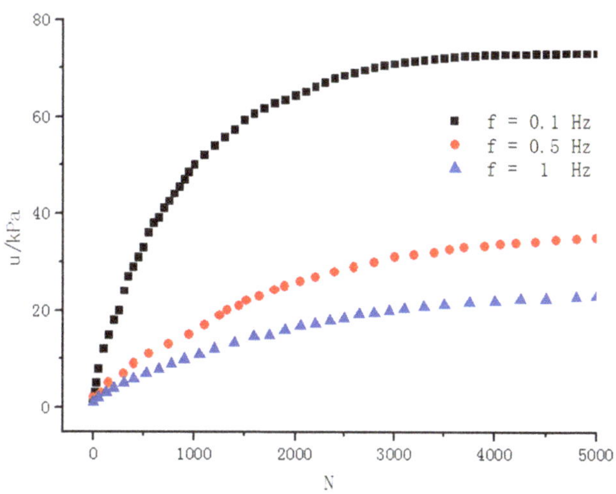

Effective stress $\sigma_1'=\sigma_2'=250kPa$, $\sigma_3'=200kpa$, dynamic stress $\sigma_d=20kPa$

Figure 7. *Super Pore Water Pressure u against Cyclic Number N.*

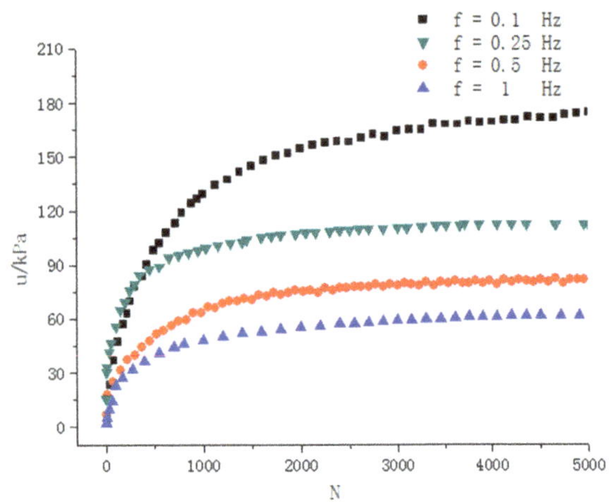

Effective stress $\sigma_{1,2}'$=250kPa, σ_3'=200kpa, dynamic stress σ_d =50kPa

Figure 8. *Super Pore Water Pressure u against Cyclic Number N.*

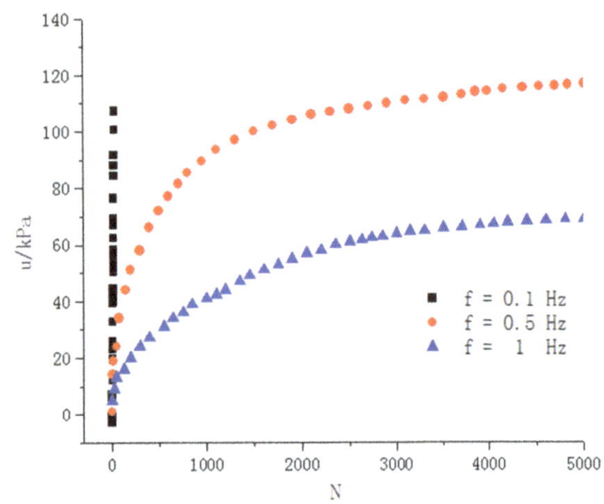

Effective stress $\sigma_1'=\sigma_2'$=250kPa, σ_3'=200kpa, dynamic stress σ_d =65kPa

Figure 9. *Super Pore Water Pressure u against Cyclic Number N.*

Since this soil samples are taken from the offshore plane following 13m~16m which lead to that the moisture content is higher than 63% generally, some of them are as high as 68%, which result in that the dynamic property is poorer than common clay. The effective stress of the sample was still lower than the similar saturated clay under the same condition even after 96 hours solidifying. Therefore the majority of energy is undertaken by pore water of the soil body during the initial period loading. No matter the growth or the final value of pore-water pressure are, this soil sample is faster and higher than the common saturated clay, which is shown in Figure 1~9.

Figure 1 shows the change of the super pore water pressure of the soil sample which is consolidated after 200kPa and then loaded by different frequencies and dynamic stress of 20kPa. Taking the 0.1Hz for example, in the initial loading period (the cyclic numbers are about 2000), the energy is mainly undertaken by the soil body and the pore water, therefore the pore water pressure rapidly grows and reaches as high as 77% of the limiting super pore water pressure. Along with the soil body compacting and the particles between the pore reducing gradually, the main pore water is working and the pore water pressure's growing rate slows down obviously. And then the pore water pressure gradually becomes stable and stay in a stage which it almost does not grow. That is called the limiting value of the pore-water pressure, which is the 45% of the effective confining pressure. Generally, the growth of the super pore water pressure can be summarized as three stages---rapid growth, slow growth and gradual staying, which is similar to the study result of the saturated soft soil around tunnel under cyclic loading by Tang Yiqun [13].

The comparison of super pore pressure water of 50kPa under 0.1Hz, 0.25Hz, 0.5Hz and 1Hz is shown in Figure 2. Comparing to the 20kPa, the final pore water pressure of 50kPa is higher than the former. That is because the higher the cyclic stress is, the higher energy impacting to the soil, which leads to that the higher energy can be absorbed by pore water under the same condition. Comparing to the 0.1Hz and 0.25Hz, although the final pore water pressure of 0.25Hz is closed to 0.1Hz, which presents that the influence of frequency on the final pore water pressure is weaken when the frequency is low enough, the increasing rate of 0.25Hz is still larger than 0.1Hz, which is closed to the result of Yasuhara [4], a study showing that the higher the frequency is, the higher the increasing rate of pore water pressure become. However, when the frequency is larger than 0.25Hz, the growth of the pore water is closed to the viewpoint of most scholars, like Wang Jun and Zhou Nianqing [11, 12], which suggested that the lower the frequency is, the higher the increasing rate of pore water pressure and the higher final pore water pressure become. This results from that lower frequency means the more time of the load working on soil body. The more energy on the soil are, the higher the final pore water pressure become since the pore water pressure become accumulative gradually and can not dissipated quickly under the same loading frequency. Therefore under the low frequency cyclic loading, the increasing rate of the pore water pressure of saturated muck soil appears periodical. During the frequency of 0.1Hz~0.25Hz, the relationship between frequency and the pore water pressure is related, namely that the higher the loading frequency is, the higher the increasing rate of pore water pressure become. However, between the frequency of 0.25Hz~1Hz, the relationship between frequency and the pore water pressure is negative, namely the higher the frequency is, the lower the increasing rate of the pore water pressure become.

Figure 3, 6, 9 shows the change of the pore water pressure of cyclic stress 65kPa under different consolidation stresses. The increasing type of the pore water pressure is the rapidly rising under 0.1Hz, which presents that the pore water pressure keep an high increasing rate until the soil to be broken from the beginning of the load to the end. According to the information above, the rapidly rising type of pore water pressure mostly happens in low frequency and high cyclic stress. The type of stable mostly happens in high frequency and low cyclic stress. And the type of gradually growing is the transitional stage between type of rapidly rising and stable.

Figure 8 shows that how the pore water pressure grow during different frequency, like 0.1Hz, 0.25Hz, 0.5Hz and 1Hz, which is different to the result above. The final pore water pressure do not increase largely as the frequency, but grow slowly after 0.5Hz. That suggests that there is an threshold of frequency between 0.1Hz~1Hz, leading to the type changing of how the pore water pressure increase. The increasing rate of pore water pressure slow down when the experimental frequency is higher than the threshold.

3.2. The Law of Super Pore Water Pressure Under Different Dynamic Stress

The curves of the relationship between super pore water pressure u and cyclic number N under different dynamic stress of the same frequency and isotropic consolidation $\sigma_1'=\sigma_2'=\sigma_3'=200$kPa are shown in Figure 10, 11, 12. The curves of the relationship between super pore water pressure u and cyclic number N under different dynamic stress of the same frequency and anisotropic consolidation $\sigma_1'=\sigma_2'=225$kPa, $\sigma_3'=200$kPa is shown in Figure 13, 14, 15. The curves of the relationship between super pore water pressure u and cyclic number N under different dynamic stress of the same frequency and anisotropic consolidation $\sigma_1'=\sigma_2'=250$kPa, $\sigma_3'=200$kPa are shown in Figure 16, 17, 18.

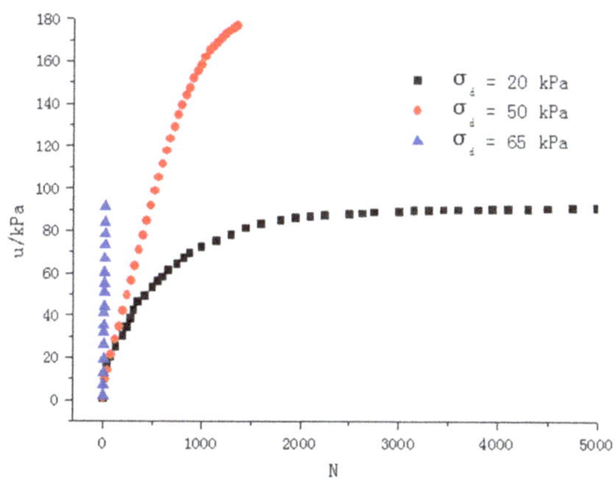

Effective stress $\sigma_1'=\sigma_2'=\sigma_3'=200$kPa, loading frequency $f=0.1$ Hz

Figure 10. *Super Pore Water Pressure u against Cyclic Number N.*

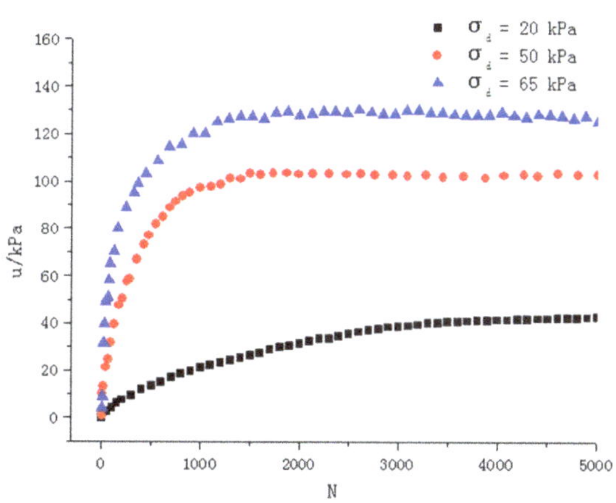

Effective stress $\sigma_1'=\sigma_2'=\sigma_3'=200$kPa, loading frequency $f=0.5$ Hz

Figure 11. *Super Pore Water Pressure u against Cyclic Number N.*

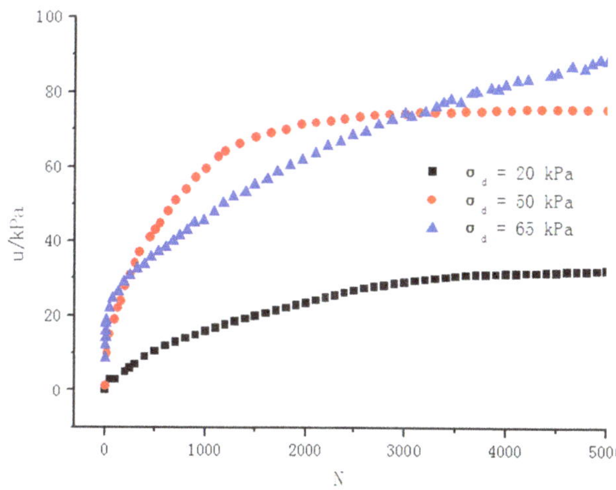

Effective stress $\sigma_1'=\sigma_2'=\sigma_3'=200$kPa, loading frequency $f=1$ Hz

Figure 12. *Super Pore Water Pressure u against Cyclic Number N.*

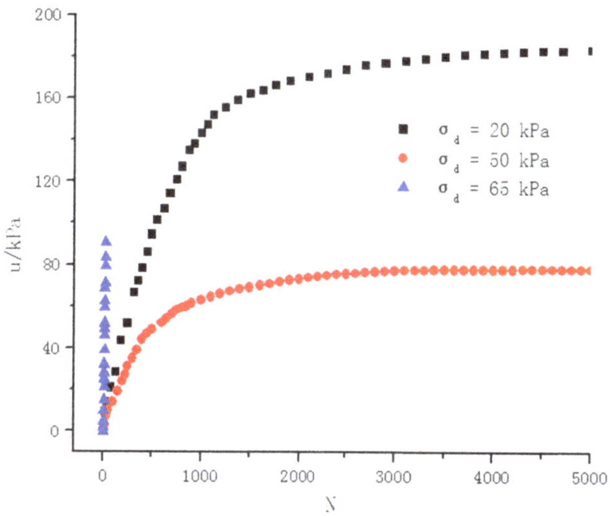

Effective stress $\sigma_1'=\sigma_2'=225$kPa, $\sigma_3'=200$kPa, loading frequency $f=0.1$ Hz

Figure 13. *Super Pore Water Pressure u against Cyclic Number N.*

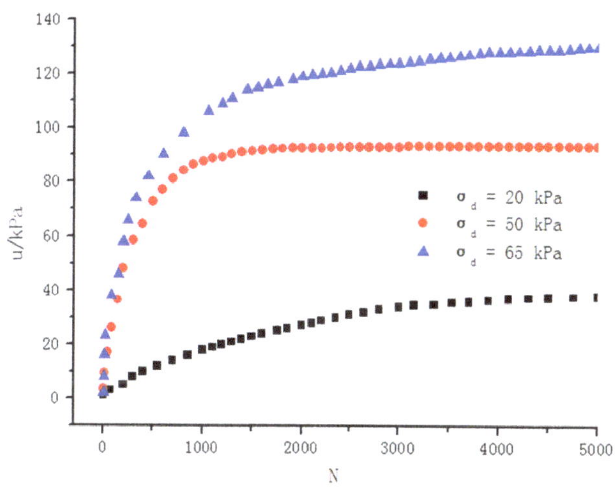

Effective stress $\sigma_1'=\sigma_2'=225$kPa, $\sigma_3'=200$kPa, loading frequency $f=0.5$ Hz

Figure 14. *Super Pore Water Pressure u against Cyclic Number N.*

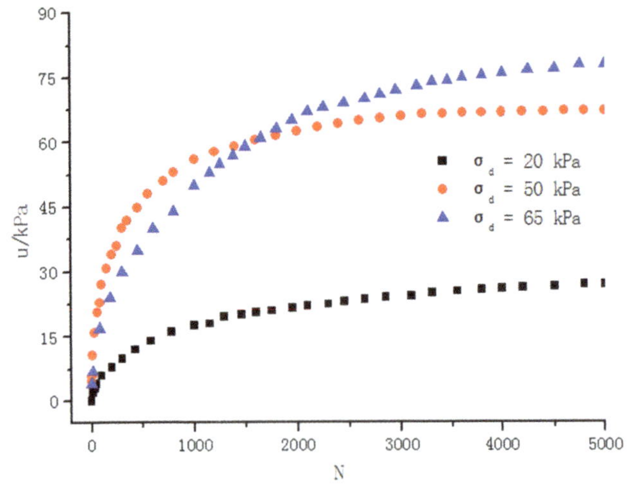

Effective stress $\sigma_1'=\sigma_2'=225kPa$, $\sigma_3'=200kPa$, loading frequency $f=1$ Hz

Figure 15. *Super Pore Water Pressure u against Cyclic Number N.*

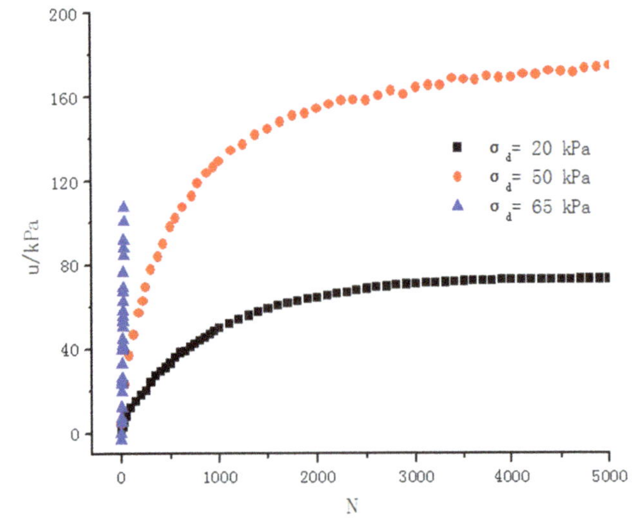

Effective stress $\sigma_1'=\sigma_2'=250kPa$, $\sigma_3'=200kPa$, loading frequency $f=0.1$ Hz

Figure 16. *Super Pore Water Pressure u against Cyclic Number N.*

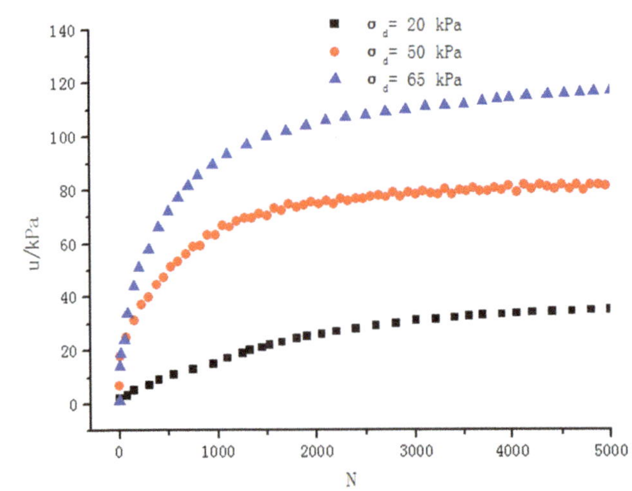

Effective stress $\sigma_1'=\sigma_2'=250kPa$, $\sigma_3'=200kPa$, loading frequency $f=0.5$ Hz

Figure 17. *Super Pore Water Pressure u against Cyclic Number N.*

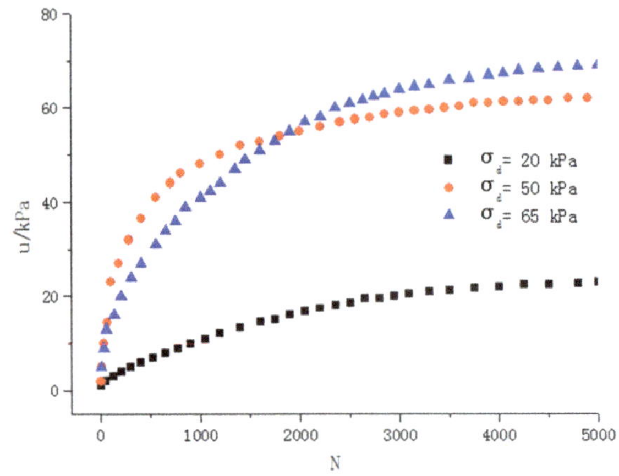

Effective stress $\sigma_1'=\sigma_2'=250kPa$, $\sigma_3'=200kPa$, loading frequency $f=1$ Hz

Figure 18. *Super Pore Water Pressure u against Cyclic Number N.*

The curve of the relationship between super pore water pressure u and cyclic number N of different cyclic stress under 0.1Hz and the isotropic consolidation is shown in Figure 10, which presents that the higher cyclic stress is, the higher increasing rate of the pore water pressure become. That is because higher cyclic stress presents more energy working on the soil body. Since the construction of saturated muck soil is poorer than common soil clay, most energy is maintained by the soil pore, which leads to the pore water increasing rapidly. In addition, considering the failure standard is 15%, the final pore water pressure of 65kPa is lower than 50kPa, which means the final pore water pressure is not proportional to the cyclic stress, since there is not enough time for the pore water pressure to increase when the energy loading on the soil body is too great. So the experimental cyclic stress should be less than 50kPa if requiring to study the development of pore water pressure under 0.1Hz according the experimental result.

Even most studies [1~6] show that the higher the cyclic stress are, the quicker the increasing rate of pore water pressure develop, but this is based on a limited cyclic numbers and cyclic stress. Figure 12, 15, 18 show that the instant pore water pressure of 50kPa is higher than 65kPa under the cyclic numbers of 0~2000, which presents that the development of pore water pressure do not follow the rule that the higher the cyclic stress is, the higher the increasing rate of pore water pressure become since the higher cyclic stress are, the quicker for the balance between effective stress and pore water pressure break. However, when the cyclic numbers are larger than 5000, the final pore water pressure still increase as the cyclic stress increase since the more energy of the pore absorb, the higher the final pore water pressure increase.

3.3. The Law of Super Pore Water Pressure Under Different Dynamic Stress

The curves of the relationship between super pore water pressure u and cyclic number N of different consolidation, the same dynamic stress and frequency are shown in Figure 19, 20, 21. The curves of the relationship between super pore water pressure u and cyclic number N of different consolidation, the same dynamic stress and frequency are shown in Figure 22, 23, 24. The curves of the relationship between super pore water pressure u and cyclic number N of different consolidation, the same dynamic stress and frequency are shown in Figure 25, 26, 27.

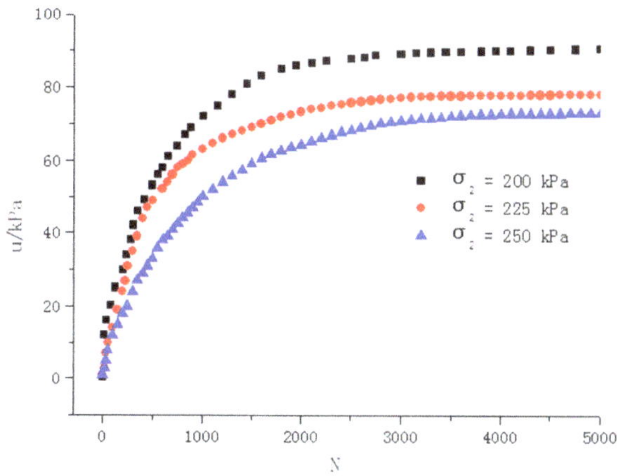

Loading frequency f = 0.1 Hz, dynamic stress σ_d = 20kPa

Figure 19. *Super Pore Water Pressure u against Cyclic Number N.*

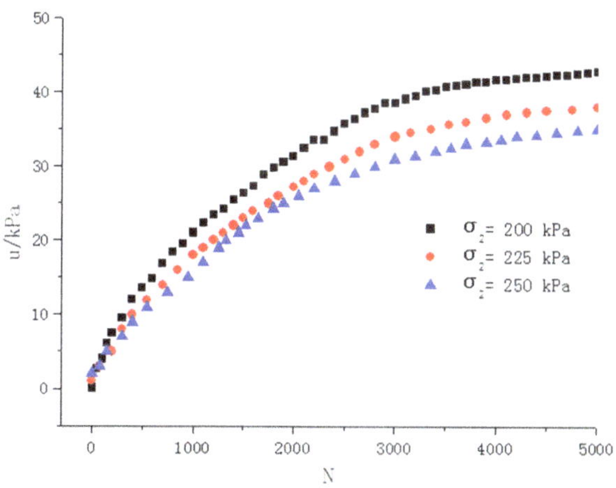

Loading frequency f = 0.5 Hz, dynamic stress σ_d = 20kPa

Figure 20. *Super Pore Water Pressure u against Cyclic Number N.*

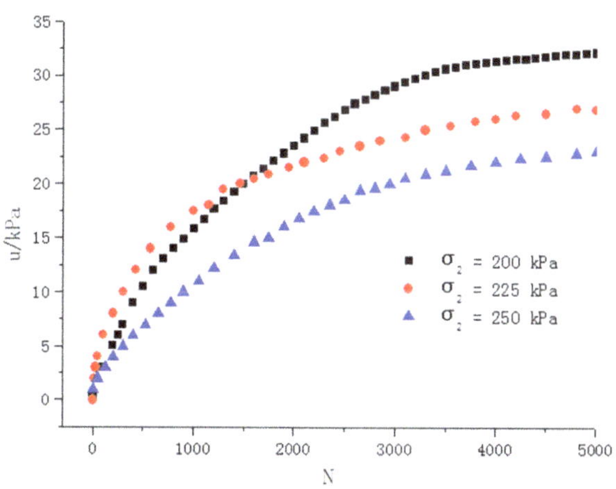

Loading frequency f = 1 Hz, dynamic stress σ_d = 20kPa

Figure 21. *Super Pore Water Pressure u against Cyclic Number N.*

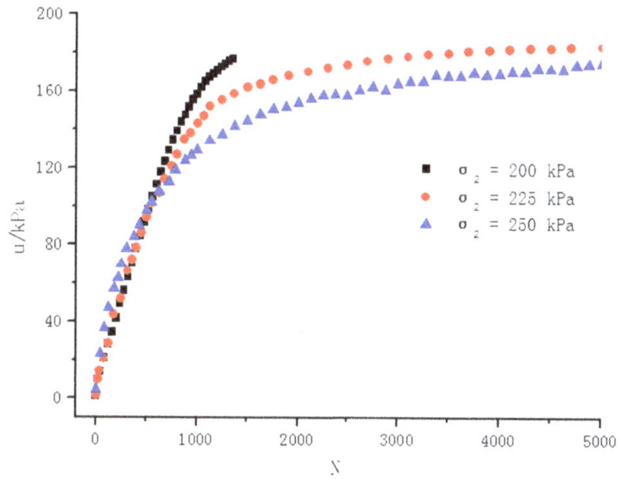

Loading frequency f = 0.1 Hz, dynamic stress σ_d = 50kPa

Figure 22. *Super Pore Water Pressure u against Cyclic Number N.*

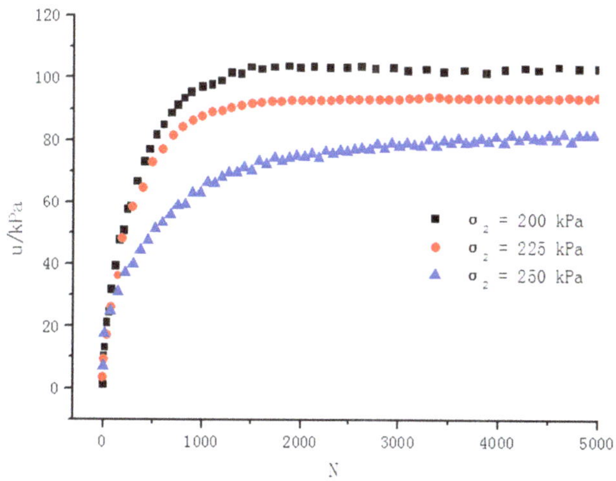

Loading frequency f = 0.5 Hz, dynamic stress σ_d = 50kPa

Figure 23. *Super Pore Water Pressure u against Cyclic Number N.*

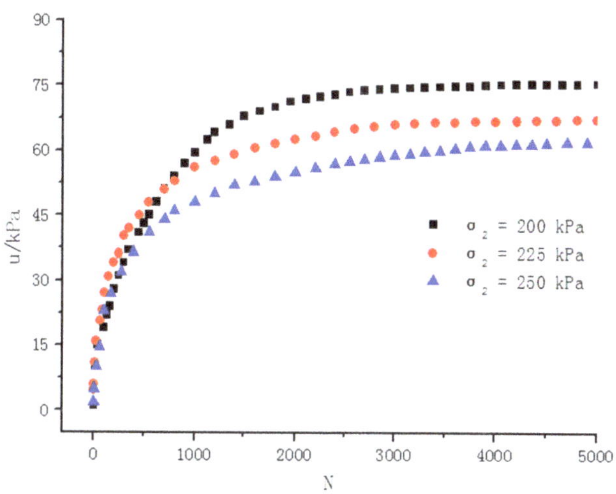

Loading frequency f = 1 Hz, dynamic stress σ_d = 50kPa

Figure 24. *Super Pore Water Pressure u against Cyclic Number N.*

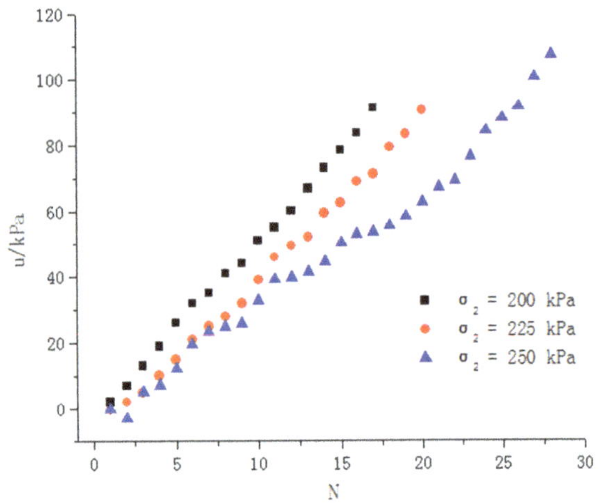

Loading frequency f = 0.1 Hz, dynamic stress σ_d = 65kPa

***Figure 25.** Super Pore Water Pressure u against Cyclic Number N.*

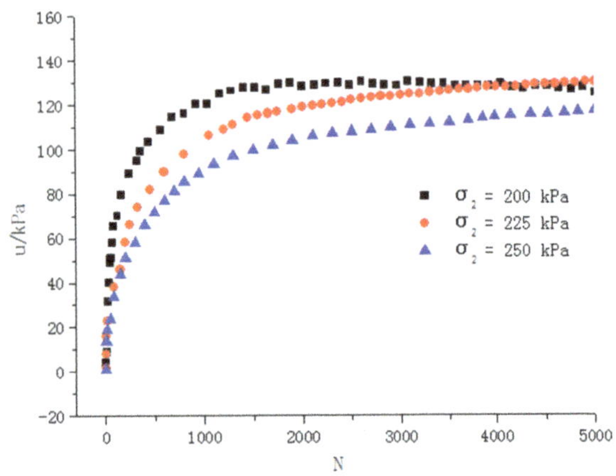

Loading frequency f = 0.5 Hz, dynamic stress σ_d = 65kPa

***Figure 26.** Super Pore Water Pressure u against Cyclic Number N.*

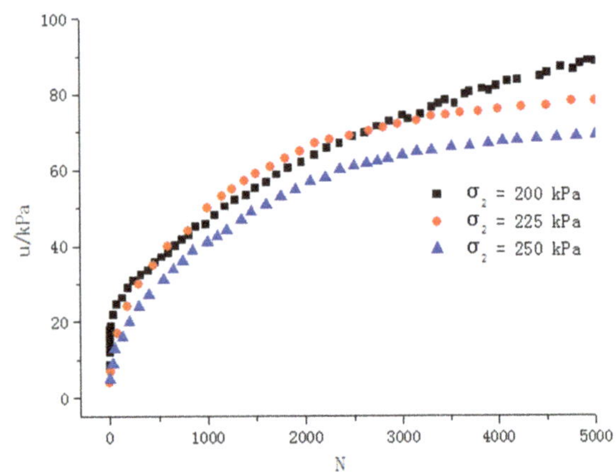

Loading frequency f =1 Hz, dynamic stress σ_d = 65kPa

***Figure 27.** Super Pore Water Pressure u against Cyclic Number N.*

The curves of the relationship between super pore water pressure u and cyclic number N of different consolidation stress under each loading frequency and 20kPa are shown in Figure 19, 20, 21. Those figures present that anisotropic consolidation lead to the different influence, comparing to loading frequency and cyclic stress, for the construction of soil. For example, since the reconstruction of soil particle leads to the increasing of the effective stress and reconstruction of drainage passage, in the equal dynamic situation, more energy can be undertaken by the construction of soil, which leads to the less energy of the pore water absorbing and the slower of the pore water pressure increasing speed. Moreover, the higher the frequency is, the more obvious the effect of anisotropy present. The comparison of different consolidation stresses under 0.1Hz are shown in Figure 19, which presents that the final pore water pressure of effective stress 250kPa is as high as the 78% of 200kPa, so that of 1Hz 69%. The dates above shows that the higher the frequency under the same consolidation are, the stronger the anisotropy become.

Moreover, Figure 22, 26 shows that the anisotropy also influence the cyclic numbers of soil developing to failure. For example, the development of super pore water pressure of 65kPa under 0.5Hz is shown in Figure 26. In the isotropic consolidation 200kPa, the pore water pressure increase rapidly when the cyclic numbers N are about 3100, which suggests that the construction of soil change. However, as the intermediate principal stress increase, it can impede, even prevent the soil broken.

According to Figure 1~27, under the traffic load of low frequency, the biggest factor affecting pore water pressure is cyclic stress, followed by loading frequency and consolidation stress. This is because although the consolidation stress can change the arrangement of soil particles, promote pore water drainage and improve soil effective stress, many structural changes are fundamentally impossible due to the low structure of the muck soil compared to clay. While at low frequency 0.1Hz ~ 0.5Hz, the time of pore water absorbing the energy from cyclic loading is similar, which weakens the effect of loading frequency. In contrast, from previous studies, cyclic stress is the key factor of pore water pressure rising, which is due to the pore water pressure absorbing energy. Therefore, the cyclic stress is the most critical factor for the growth of the pore water pressure.

4. A New Pore Pressure Model and Its Parameters

Since the development of pore pressure of muck soil under traffic loading is complicated, it is hard to build a comprehensive pore pressure model. In order to practicality, the convenience of variables should be considered firstly when choosing the model type. Yang Ping [14] suggested that the pore pressure model type of hydraulic is different under low and high frequency. Polynomial is suitable for low frequency while power exponent is suitable for high frequency. Here is the pore pressure model for muck soil in Pear River Delta, which is shown as fellow:

$$u^* = A + B\ln N + C(\ln N)^2 + D(\ln N)^3$$

u^* is the normalized pore pressure. N is the cyclic number. A, B, C and D are the experimental parameters.

Table 3. Parameters of model.

Cyclic Stress/kPa	σ_2/kPa	f/Hz	A	B	C	D
20	200	0.1	0.01416	-0.07638	0.02934	-0.00164
20	200	0.5	0.00182	-0.01074	0.00152	3.34E-04
20	200	1	0.0022	-0.00741	9.01E-04	2.69E-04
50	200	0.1	0.00786	0.09421	-0.05073	0.00769
50	200	0.5	0.03074	-0.122	0.05066	-0.00347
50	200	1	0.01615	-0.06359	0.02424	-0.00135
65	200	0.1	0.00617	0.07474	-0.04341	0.02525
65	200	0.5	-0.00699	0.04899	-0.00406	0.00131
65	200	1	0.03372	0.054	-0.01595	0.0018
20	225	0.1	0.01709	-0.10444	0.03653	-0.00222
20	225	0.5	0.00559	-0.00889	3.82E-04	3.96E-04
20	225	1	0.00182	-0.00489	2.51E-03	-6.24E-06
50	225	0.1	0.03556	-0.2321	0.07539	-0.00414
50	225	0.5	0.03466	-0.11794	0.04814	-0.00332
50	225	1	0.03135	-0.03177	0.01653	-9.96E-04
65	225	0.1	-0.00493	0.04077	-0.03315	0.02305
65	225	0.5	0.03523	-0.05824	0.03082	-0.0018
65	225	1	0.02541	-0.02273	8.56E-03	-7.23E-05
20	250	0.1	0.01602	-0.06222	0.01953	-8.32E-04
20	250	0.5	0.01055	-0.0107	9.37E-04	3.19E-04
20	250	1	5.47E-03	-1.48E-04	-0.00125	3.37E-04
50	250	0.1	0.0287	-0.17691	0.06316	-0.0036
50	250	0.5	0.05116	-0.05044	0.02104	-0.00118
50	250	1	0.01247	-0.02909	0.01477	-8.43E-04
65	250	0.1	-0.01821	0.08975	-0.06679	0.02575
65	250	0.5	0.03015	-0.03727	0.02153	-0.0011
65	250	1	0.02754	-0.01486	0.00428	2.48E-04

5. Conclusion

The author, firstly, summarize the previous studies on pore water pressure under dynamic stress and take a dynamic test of saturated muck soil, which is carried out by GCTS SPAX-2000 (modified) static and dynamic true triaxial testing system. Here are the conclusions:

(1) In the same cyclic stress and consolidation, the final pore water pressure decreases as the increasing frequency.

(2) In the same cyclic stress and consolidation, the influence of loading frequency on the increasing rate of pore water pressure is periodical. In the 0.1Hz~0.25Hz, the increasing rate of pore water pressure is raised by the frequency increasing. However, in the 0.25Hz~1Hz, the increasing rate of pore water pressure decrease as the frequency increase.

(3) In the same loading frequency and consolidation, the higher cyclic stress is, the more quickly the pore water pressure develop, as well as the accumulative pore water pressure.

(4) In the same cyclic stress and loading frequency, isotropic consolidation can increase the effective stress of soil and slow down the development of pore water pressure, but it is not the key factor.

(5) During the loading frequency of 0.1Hz~1Hz, the key factor of influencing the pore water pressure is cyclic stress and the following is the loading frequency, consolidation is the last one.

(6) Considering cyclic stress, loading frequency and consolidation, a model of pore water pressure based on an math model is built up:

$$u^* = A + B\ln N + C(\ln N)^2 + D(\ln N)^3$$

References

[1] Procter, David C., and Jalal H. Khaffaf. Cyclic triaxial tests on remoulded clays [J]. Journal of Geotechnical Engineering, 1984, 110 (10): 1431-1445.

[2] Zhou Jian, Gong Xiaonan. Study on strain soften in saturated soft clay under cyclic loading [J]. China Civil Engineering Journal, 2000, 33 (5): 75-78.

[3] Ansal, Atilla M., and Ayfer Erken. Undrained behavior of clay under cyclic shear stresses [J]. Journal of Geotechnical Engineering, 1989.

[4] Yasuhara Kazuya, Yamanouchi Toyotoshi, Hirao Kazutoshi. Cyclic strength and deformation of normally consolidated clay [J]. Soils and Foundations, 1982, 22 (3): 77-79.

[5] Zhang Yu, Tu Yangju, Fei Wenping. etc. Effect of vibration frequency on dynamic properties of saturated cohesive soil [J]. Rock and Soil Mechanics, 2006, 27 (5): 699-704

[6] Wei Xinjiang, Zhang Tao, Ding Zhi. etc. Experimental study of pore pressure model of soft clay with different consolidation degrees under subway loading [J]. Rock and Soil Mechanics, 2014, 35 (10): 2761-2768.

[7] Wang Jun, Yang Fang, Wu Yanping. etc Experiment study of coupling effect of initial shear stress and loading rate on pore water pressure model of soft clay [J]. Rock and Soil Mechanics, 2011 (S1): 111-117.

[8] Wang Yuandong, Tang Yiqun, Liao Shaoming, etc. Experimental of Pore Pressure of Reinforced Soft Clay Around Tunnel Under Subway Vibrational Loading [J] Journal of Jilin University (Earth Science Edition), 2011, 41 (1): 188-194.

[9] Ding Zhi, Zhang Tao, Wei Xinjiang, etc. Experimental study on effect of different drainage conditions on dynamic characteristics of soft clay under different degrees of consolidation [J]. Rock and Soil Mechanics, 2015 (5): 893-899.

[10] Zhao Chunyan, Zhou Shunhua, Zhuang Li. Cyclic Accumulative Pore Pressure Model of Soft Clay in the Shanghai Region [J] Journal of the China Railway Society. 2012, 34 (1): 77-82.

[11] Zhou Nianqing, Tang Yiqun, Wang Jianxiu, etc. Response characteristics of pore pressure in saturated soft clay to the metro vibration loading [J]. Chinese Journal of Geotechnical Engineering, 2006, 28 (12): 2149-2152.

[12] Wang Jun, Cai Yuanqiang, Li Xiaobing. Cyclic softening-pore pressure generation model for overconsolidated clay under cyclic loading [J]. Rock and Soil Mechanics, 2008, 29 (12): 3217-3222.

[13] Tang Yiqun, Zhang Xi, Zhao Shukai, etc. Model of pore water pressure development in saturated soft clay around subway tunnel under vibration load [J]. China Civil Engineering Journel, 2007, 40 (4): 82-86.

[14] Yang Ping, Tang Yiqun, Zhou Nianqing, etc. Study on Pore Water Pressure of Hydraulic Fill under Traffic Loading [J]. Chinese Journal of Underground Space and Engineering, 2008, 4 (2): 253-258.

The Impact of the Depth and Type of Soil Granulation Above the Bearing Capacity of the Ring Footing

Saeid Azmoodeh*, Nasser Arafati

Department of Civil Engineering, Tafresh University, Tafresh, Iran

Email address:
Saeid.azmoodeh66@gmail.com (S. Azmoodeh), Nasser.arafati@gmail.com (N. Arafati)

Abstract: Obtaining bearing capacity of the footing is among the factors influencing the footings design. In this paper, the study of the type and depth impact of the soil granulation above the bearing capacity of the ring footing through the assistance of *FLAC* software is conducted. The failure criterion intended for the soil is Mohr-Coulomb. The study of the bearing capacity of ring footing of ten thousand cubic meter of Mazut tank of Mashhad Cement Company on three types of soil, topsoil, clay and sand has been carried out. The results indicate that the loading capacity increases by granularity becoming coarser due to the increase of the contact area of the granulation and the increase of friction and the soil structure stability. Also the less the poorer soil depth (in terms of granularity) is, the more limited subsidence is, hence the bearing capacity increases.

Keywords: Ring Footing, Bearing Capacity, Tank, Type of the Soil, FLAC Software

1. Introduction

Obtaining bearing capacity of the footing is among the factors influencing the footings design. Nowadays, several numerical and laboratory studies have been carried out to determine the bearing capacity of shallow footings above the soil. In the recent years, numerical methods such as: finite difference method and finite elements method have been used widely to calculate the bearing capacity of different types of footings including strip, circular and rectangular, etc.[7]

Now using ring footing underneath the structures, which have axial symmetry such as: chimneys, water and oil storage tanks, towers and silos due to economic issues is rising. Therefore, the calculation technique of bearing capacity of these footings is required to design and build them [3]. Of course, so far researches have also been carried out on bearing capacity and settlement of the ring footings, which among them, the studies *Egorov* (1965) [6] and *Ohri* (1997) [4] can be noted, which have led to the new relationship suggestions in this regard. Mr. *Hatef* and *Boushehrian* (2002), considering "n" which is equal to the ratio of internal to external radius of the ring footing and through studying the bearing capacity above the sandy soil in the laboratory condition concluded that in n=0.4, the bearing capacity reaches its maximum amount [2].

By using FLAC software and with the assistance of coding, Wang and Zhao (2007) studied N_2 coefficient on bearing capacity of ring footing placed on cohesion-less soil. The results indicated that by increasing the ratio of "n", the coefficient of N   will also increase [9].

2. Definition of the Problem

The soils considered in the research with the Mohr-Coulomb failure criteria include: clay, sand and topsoil, which have been used in various depths underneath the footing. The characteristics desired for different types of soils can be observed in table 1. In the case of the existence of 2 layers of soil, 2, 5 and 10 meters thicknesses for the topsoil is considered.

Table 1. *Mechanical characteristics of soil.*

Type of soil	Cohesion (Kpa)	Angle friction (degree)	Poisson's ratio	Modulus of elasticity (Kpa)	Density (KN/m^3)
Sand	0	3	0.28	10000	17
Clay	25	0	0.35	20000	16
Topsoil	1	3	0.3	8000	16

The sample under study is ten thousand cubic meter tank of storing Mazut of Mashhad Cement Company, which the footing underneath this tank is of ring type with the external diameter of 34 and internal diameter of 30 meters, the thickness of the footing is also 1 meter.

Figure 1. *A view of the 10 thousand cubic meter Mazut tank of Mashhad Cement Company.*

When soil is under steady pressure of (q), based on the "Terzaghi relations", the bearing capacity of the footing circular shallow footing can be obtained from the following equation:

$$q_o=1.3CN_c+qN_q+0.3\gamma BN_\gamma \quad (1)$$

Where (C) equals cohesion, (q) represents the equivalent surcharge, (γ) soil density, (B) width or diameter of the footing and N_c, N_q and N_γ are bearing capacity coefficients related to the friction angle (φ).

In the form of ring footing *Ohri et al.* relation (1997) is used. To obtain the bearing capacity of ring footing, *Ohri* has offered a coefficient called "Ring efficiency coefficient (F_r)" which is equal to:

$$F_r = q_n / q_o \quad (2)$$

Where (q_n) equals the final bearing capacity of ring footing and (q_o) is bearing capacity of circular footing [4]. This coefficient, which is applicable for both types of footing with variable and fixed surfaces, can be obtained through having (n), which is the ratio of internal diameter to external diameter from figure 2, as it can be seen below.

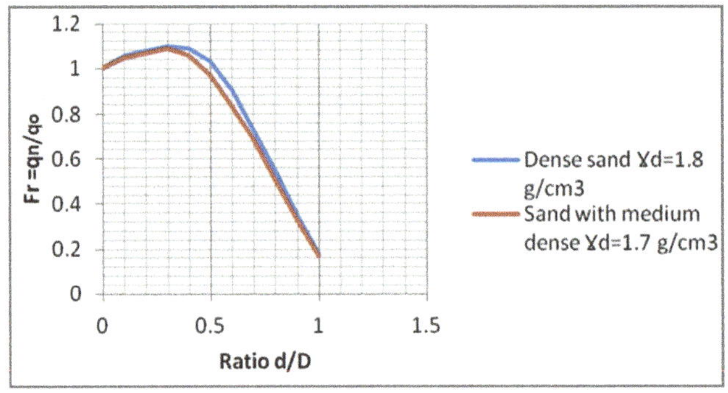

Figure 2. *Ring efficiency coefficient curve (Fr) in comparison to d/D (for variable surfaces but with identical external diameter) [4].*

In this research, also the ratio of external diameter to internal diameter of the footing is shown with (n), where (n) equals 0.88 here. Most of the researches conducted on ring footing indicates that in the case where $n \geq 0.7$, the failure underneath the footing changes from the general mode to the positional mode [5], in this case instead of using (C) and (φ), we should apply ($\acute{C}$) and ($\acute{\varphi}$), which are corrected as follows [1]:

$$\acute{C} = 2/3\ C \quad (3)$$

$$\acute{\varphi} = \tan^{-1}(2/3\ \tan\ \varphi) \quad (4)$$

It should be noted that in a state that the tank is full, the pressure resulting from the weight of Mazut inside the tank is exerted to all the surfaces of the circle, for this reason the bearing capacity is calculated first in the state of n=0 and then for n=0.88, which is equal to the real conditions of the model, until finally it is specified which soil will bear the exerted pressure from the tank. Of course this pressure with regard to the density of Mazut (γ=0.9 gr/cm^3) and the volume of 10 thousand cubic meter of the tank will be equal to 90000 KN.

3. Numerical Simulation

With regard to the external diameter and settlement distribution and tension underneath the footing, the bearing considered for this model is flexible. Modeling is done with the help of FLAC software, which is limited on the basis of Difference Method. Since the analysis under axial symmetry has the modeling ability of problems, which has an axis and the entire model studied from the rotation of a radial plate around symmetry axis can be modeled [10]. To analyze the ring footing desired, the axially symmetric condition, while the "Y" axis is the entire symmetry axis, the entire model geometric shape (including soiled bed and bearing) is modeled through rotation of a radial plate around "Y" axis. With these conditions, only the right side of the model will be studied.

In order to estimate the bearing capacity of the footing, "Cross tangents" method on the graph of load-settlement is used. It is necessary to remind that in the state of the existence of 2 or 3 soil layers, an interface level between the soil layers to create full cohesion and uniformity between two areas with different zoning should be taken into account, where this work is done in FLAC software through assigning vertical and sheer rigidity of adjacent zones (unit: tension over displacement) [10]. Figure 3 shows the zones intended for 3 soil layers through exerting the interface level conditions.

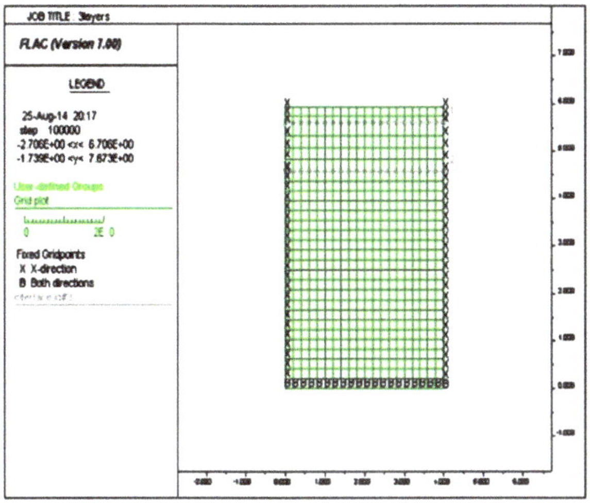

Figure 3. *Intended zone for 3 soil layers.*

4. Analysis and Results

Bearing capacity of ring footing on clay and sandy soil in modes n=0,0.88 was studied, where the load-settlement graph related to it can be seen in figure 4. As it is seen, the bearing capacity of the footing above the sandy soil is more than the period, when the soil underneath the footing is from the clay type.

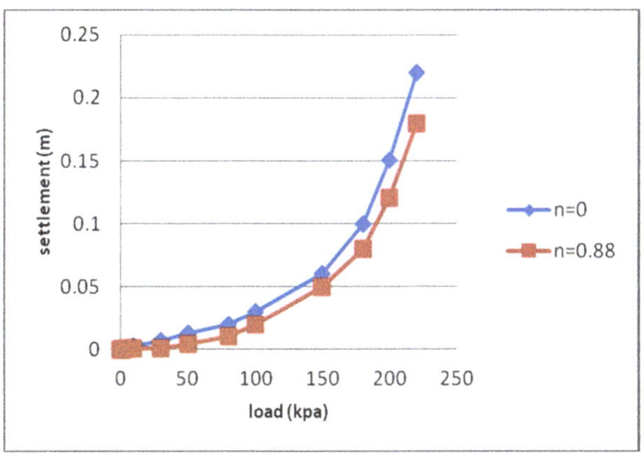

a) Sandy soil

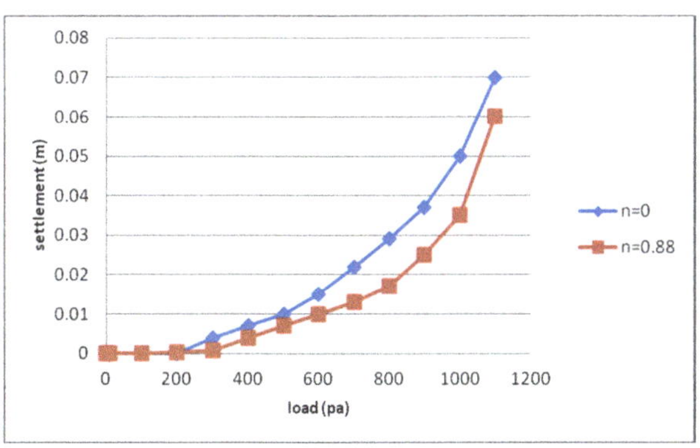

b) Clay soil

Figure 4. *The figure of load- settlement for clay and sand.*

In the mode of using 2 layers of soil, by considering different thickness for upper layers, load- settlement graph has been shown.

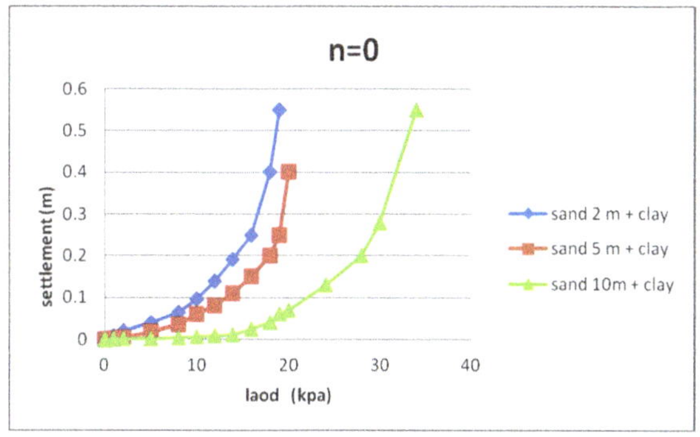

a) Sandy soil on clay, n=0

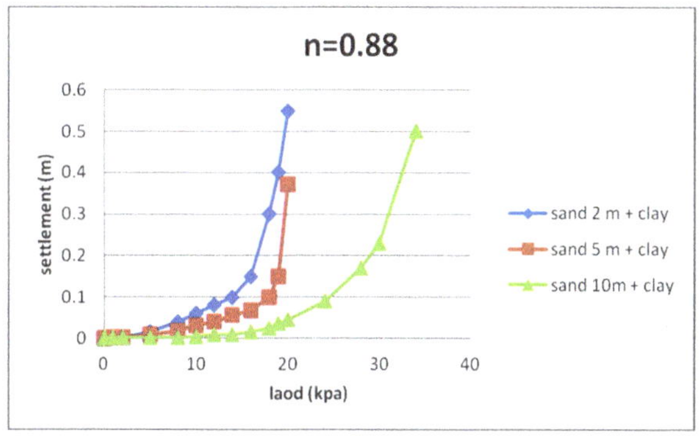

b) Sandy soil on clay, n=0.88

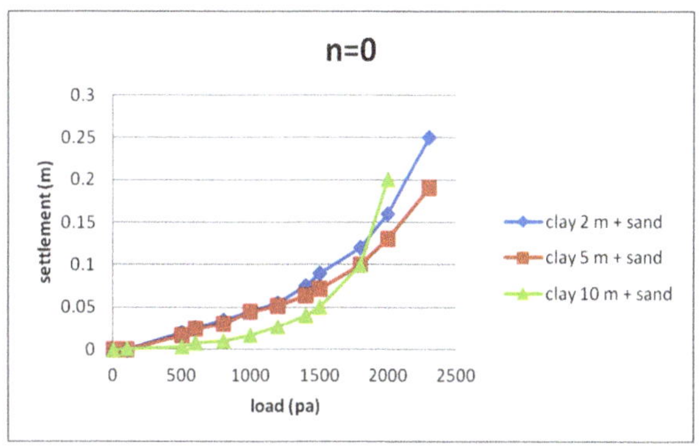

c) Clay above the sand soil, n=0

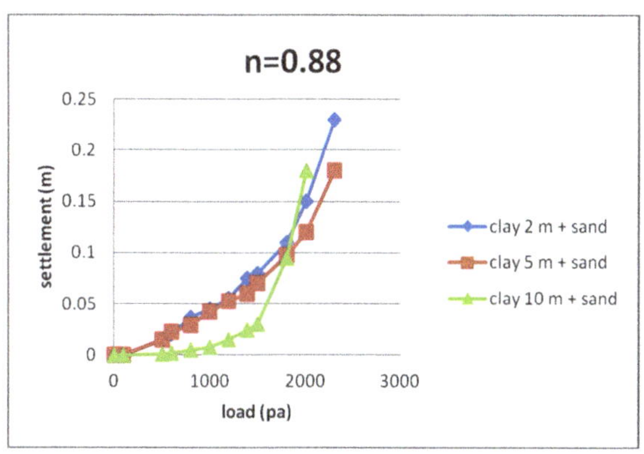

d) Clay above the sand soil, n=0.88

Figure 5. *Graph of load- settlement in the mode of existence of 2 soil layers.*

Figure 6. bearing capacity graph in terms of the depth shows the soil layer.

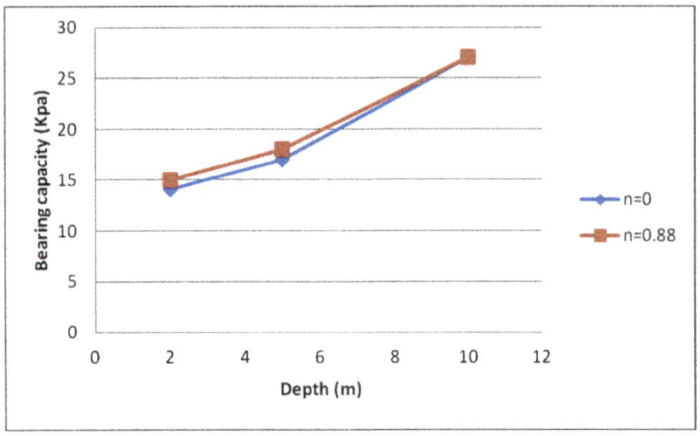

a) Upper layer of sandy soil

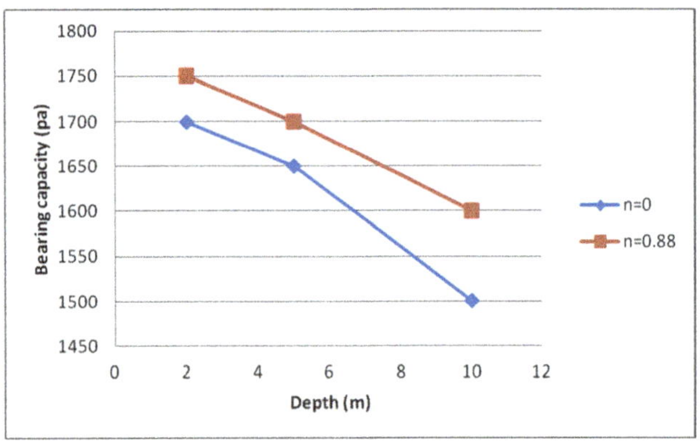

b) Upper layer of clay soil

Figure 6. *Graph of bearing capacity in terms of depth.*

From the graph, the bearing capacity in terms of depth can be seen, when the depth of the sand layer increases, the bearing capacity will also increase and this is when through the increase of the clay layer depth, the bearing capacity decreases. The range of subsidence and displacements of the soil in the mode of 2 layers of soil can be seen in figure 7.

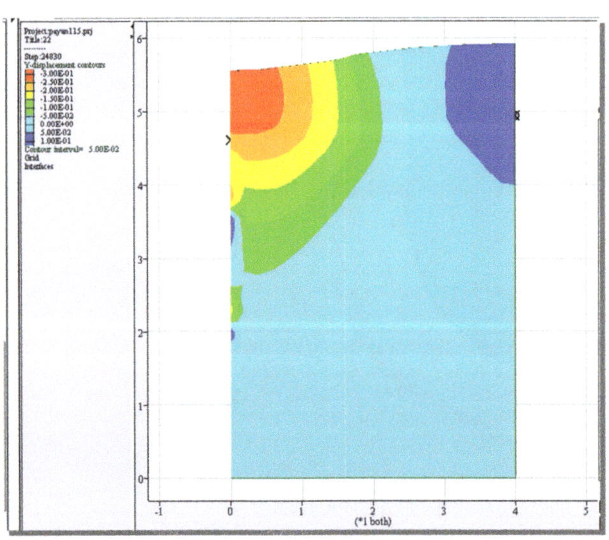

Figure 7. *Settlement in the mode of existence of 2 soil layers.*

The load- settlement graph obtained from the existence of 3 soil layers underneath the ring footing, in a mode that the Topsoil depth is 2 meters and it is in the upper layer has been given in figure 8.

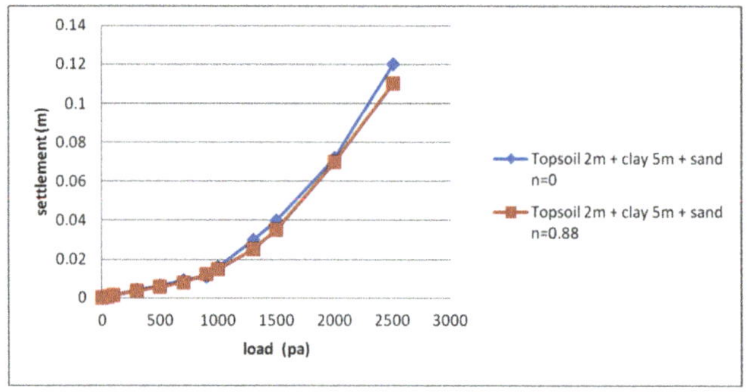

a) Topsoil on fine granulated soil

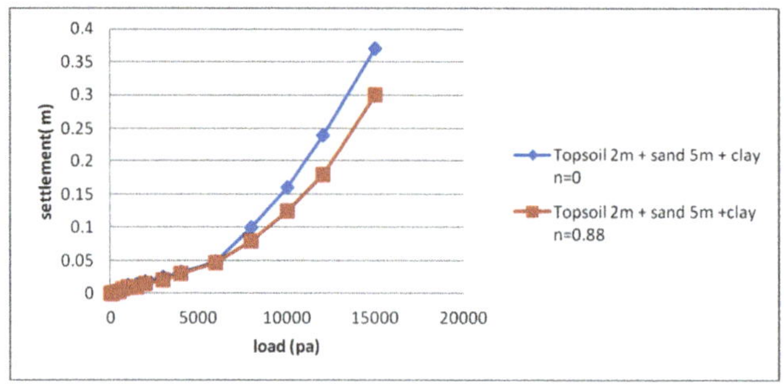

b) Topsoil on coarse granulated soil

Figure 8. *Subside in the mode of the existence of the 3 layer soil.*

The failure form of the soil taken place can also be seen in the following forms. As it can be seen in figure 9, the existence of coarse granulated soil underneath the Topsoil causes settlement being limited and ultimately, causes the increase of the bearing capacity. Also settlement in the Topsoil will afect the the clay very rapidly, when there is clay underneath it and this is when settlement above the coarse granulated sandy soil is very little.

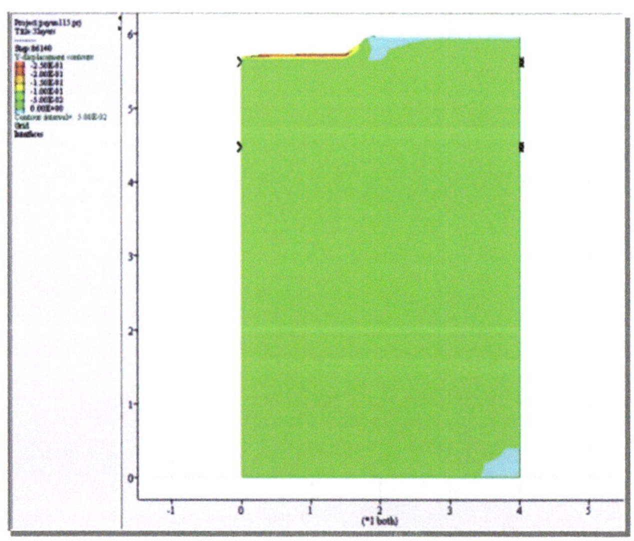

a) Topsoil on clay soil

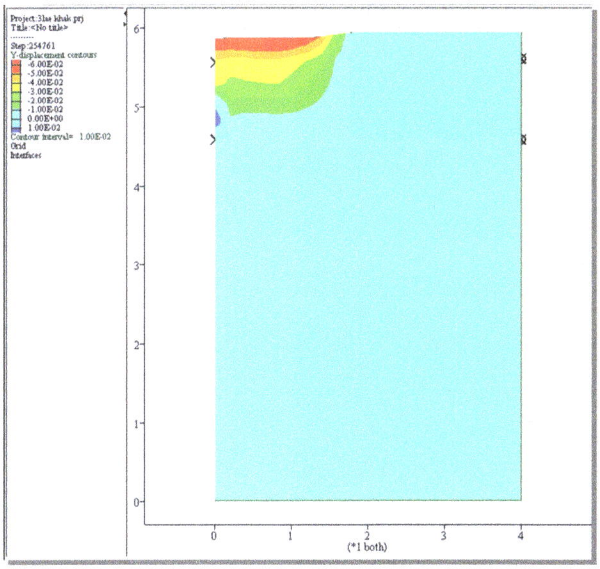

b) Topsoil on sandy soil

Figure 9. *Soil failure in the mode of exerting load.*

5. Discussion and Conclusion

Numerical simulation to estimated bearing capacity of ring footing on 1, 2 & 3 soil layers in different depths exists. topsoil, clay and sandy soil with "Elasto-plastic" behavior and based on the Mohr-Coulomb "Yield criteria" was the granules with each other and as a result of the increase of the friction and soil structure stability. About the 2 layers of soil, the bearing capacity of the footing in two modes n=0 and n=0.88 above the sandy soil is much more than the time, when the soil underneath the footing is of the clay type. Also with the increase of the clay layer depth, the subsidence will be more and the bearing capacity will decrease. symmetric conditions was calculated. The calculations show that the bearing capacity has increased by coarsening of garrulity bearing capacity with the help of FLAC software in the axially. modeled and the reason can be the increase of the contact surface of In the mode of 3 layers of soil the results indicate that with the decrease of the depth of the weaker layer (in terms of granulation) the failures will be restricted and the tolerance of the soil under the pressure will be more.

References

[1] Tahoni SH.(2011).His translation of the book Principles of Geotechnical Engineering Braja M. Das, pp.150-153.

[2] Boushehrian JH, Hataf N (2002). Expermental and numerical investigation of the bearing capacity of model circular and ring footing on reinforced sand. Geotext Geomemberanes 2003; 21: 241-56.

[3] Hataf, N, Razavi, MR, (2003) Behavior of ring footing on sand. Iranian Journal of science and Technology, Transaction B, vol. 27, pp.47-56.

[4] Ohri, M.L, Purhit, D.G.M, Dubey, M.L, (1997). Behavior of ring footing on dune sand overlaying dense sand. Pres In ternational conference of civil Engineers, Tehran, Iran. p.4-6.

[5] Fisher, K.(1957).Zur Berechnung der setzung von foundamenten in der form einer kreisformigen ringflache.Der Bauingenieur, vol. 32, No.5. German, pp.172-174.

[6] Egorov, K.E. (1965). Caclculation of bed for foundation with ring footing. Proceedings of the sixth International Conference on soil mechanical foundation of engineers. Vol.2, Montreal, pp. 41-43.

[7] KumarJ, Ghosh P. Bearing capacity factor Nc for ring footings using the method of characterishtics. Can Geotech J (2005); 40(3): 1474-84.

[8] Terzaghi K. Teoretical soil mechanics. New York: John Wiley and Sons; (1943). 1256-62.

[9] Zhao L. and Wang, J.H. (2007), Vertical bearing capacity for ring footing, Computers and Geotechnical 35 (2009) 292-304.

[10] Zamirian S. (2012), The book modeling and analysis of soil and rock structures in FLAC, Siavash mind. p.123-148.

A Computer Simulated Model for Load-Application Capacity Check of Chicken Egg

Hasan Dehghanzadeh Najm Abad[1,*], Eisa Salajegheh[2], Amir Ahmad Hedayat[1]

[1]Department of Civil Engineering, Azad University, Kerman, Iran
[2]Department of Civil Engineering, Bahonar University, Kerman, Iran

Email address:
hassan.dehghanzadeh@yahoo.com (H. D. N. Abad), eysasala@mail.uk.ac.ir (E. Salajegheh), amirahmad1356@yahoo.com (A. A. Hedayat)

Abstract: The material, form and geometric shape of egghavebeen designed in a way that nobody could ever break an egg squeezing it lengthwise. This research is based on the pattern of human palm. Two soft and concave seating bases are built in order to hold the two ends of the chicken egg. Differentchicken eggs with various geometric shapes were examined by Zwick compression strength test while they were raw, baked and empty. The maximum failing force (breaking of the egg) was 988 Newton.Static linear and nonlinear analysis and modeling of material and geometry is done by Ansys software or with Shell181 element. Early results show that chicken egg content has no effect on its load-application capacity and the reason of breaking are exceeded orbital tensile stresses. With the same thickness, eggs with length to width ratio of 1.4 have the highest load-application capacity and eggs with length to width ratio of 1.7 have the lowest load-application capacity.

Keywords: Chicken Egg, Geometric Properties, Axial Compression Strength, Linear and Nonlinear Static Analysis of Geometry and Material

1. Introduction

Chicken egg is a wonderful thing created by God which has a beautiful geometry and advanced engineered structure. With full safety and nutritional features, this structure is an appropriate place for the growth of a being. This structure has a protected outer shell (cuticle) made of lime with microscopic pores which allow carbon dioxide exchange with outside and let air into the shell. Under the chicken egg shell there are two transparent protein membranes which have defensive role against bacterial invasion. The combination of membranes and calcareous shell forms a composite structure which is the main load-bearing structure. In Figure1 the transparent protein membrane (made of human hair) and in Figure2 the Mammillary sponge layer (or knob) is shown. Hard calcite crystals are made over Mammillary knobs and these layers (or knobs)act as composites. Indeed, the transparent membrane is like a paper that is tensile resistant but has a weak performance under compression. So, in the composite element the protein membrane is utilized as the tensile resistant member and the hard calcite layer is utilized as compressive resistant member. Meanwhile, in longitudinal load application of the two ends of the chicken egg, in meridional direction we have compressive forces an in orbital direction we have both compressive and tensile forces.

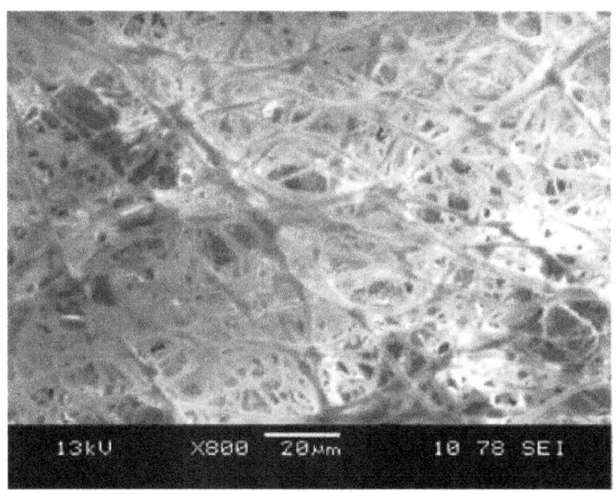

Figure 1. *Transparent protein membrane [2]*

Figure 2. Mammillary sponge layer [2]

The main food sources of the embryo in the chicken egg are yolk and albumen which are stored for 21 days. The air cell and Chalaza bands are other parts of this structure. The air cell provides oxygen and the Chalaza bands suspend the yolk in the center of the albumen. This research deals with geometric structure, structural materials and its load-bearing capacity. This research has nothing to do with the formation of the environmental, chemical properties and effective factors on these qualities. This work style was used in our previous research in ValiAsr University of Rafsanjan and was done by triaxial compression test and maximum failure load (breaking of the egg) of 750 Newton was measured. The experimental results were analyzed by Ansys linear static analysis. In this research, in order to achieve better results, the Zwick strength testing machine was used which belonged to Kerman University. Meanwhile, the results were analyzed by Ansys nonlinear static analysis.

Shell structures are one of the greatest structures because they transmit forces in numerous directions. Many natural forms are made of shells. Water and gas containers, silos, bow-shaped dams, fuselage, ship and automobile bodies are examples of shell structures. In the past because of the lack of tensile-resistant materials, brick and clay were used for construction of shell structures such as domes and arched vaults.Due to its non-expandable geometry, the egg shell has high mechanical and load-bearing capacity. This structure is symmetrical in its longitudinal axis but is asymmetric in its transversal axis. So, this structure has a narrow tip and a flat bottom.

2. Description of the Work

First, some samples of chicken egg are provided and tested longitudinally by the machine. The loading is done incrementally on a small area of the narrow tip and the flat bottom is placed on the seating base. The maximum failure load is measured for every sample. Then samples are modeled in Ansys software. The maximum load is applied and through an experimental method all stresses and reasons of failure (breaking) in this structure are analyzed. In the research of Periann and his colleagues [2] and P. Coucke and his colleagues [3], a nondestructive method was used. In the research of Ashley Lloyd and his colleagues [4] the samples were placed between two flat metal plates. In this research all samples are located in soft and flexible seating bases that are in compliance with the two ends of an egg. In this way, the transmitted load is applied superficially to both narrow tip and flat bottom of eggs. The load-application rate is 0.5mm per minute. In this research the load-application rate is not adapted to Ahmeds [5]. Nedomova [6], Ketelare [7],Voisey [8] and Buchar [9] research studies.

We provided 20 raw eggs, 20 baked eggs and 20 empty eggs. For each part a variety of white, colored, round, oval, small and big chicken eggs are selected. We baked eggs for 20 minutes in boiled water. The cracked eggs are omitted. By a syringe needle we emptied eggs through a small hole in the top of the egg. After this the hole is filled by adhesive.

For this experiment we used Zwick strength testing machine in Bahonar University of Kerman (Fig.3). This machine was manufactured in Germany and is used for load-application of a wide range of materials. This machine is controlled by an attached computer and has special options for customizing sample dimensions, load-application rate etc. the outputs are in the format of excel and diagrams.

Figure 3. Zwick testing machine

Figure 4. Turning for creating a seating base

In this research the load application is done widely on a small area of the narrow tip and a small area of the egg bottom is considered as its seating base. So, we need deep and flexible seating bases to be installed on two sides of a metal shaft then we connect shafts to the machine. The seating base is made of compressed Teflon cylindrical tube with 10cm diameter and 15cm of height (Fig.4 and Fig.5). In a casting workshop we used the biggest egg and sand to provide a metal plate with 5mm thickness and in turning workshop we cut the quadrant of its two ends to provide a device to deepen the Teflon cylinder. According to main sample, the depths of these bases are all 11mm. Then these cylinders were installed on metal shafts with 10cm diameter and 20cm of height.

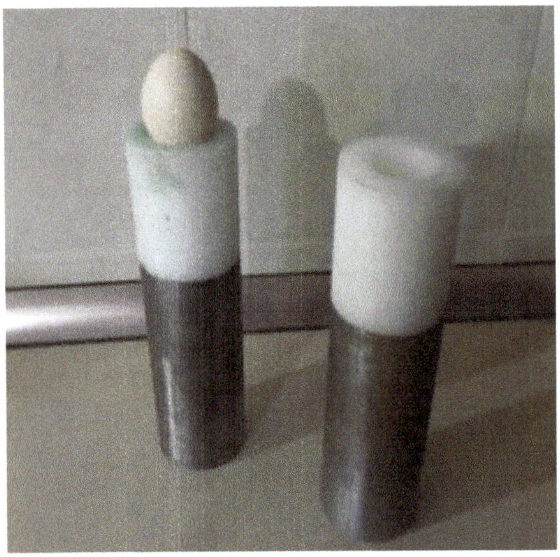

Figure 5. *Seating base*

3. Testing

Geometric properties such as length, width and shell thickness (after breaking of the egg shell) of all samples are measured precisely and then are placed in testing machine. After necessary settings, the machine starts to apply the load and the results are recorded. Many of samples broke early after the load-application because of various reasons such as slipping from the seating base, existence of cracks or other weak points that were not visible by eye, or low strength of the shell. These samples were not considered in calculations. To make the seating base a big egg was used so the base was loose for some samples. In this case we used foam sheets to fix the egg in the seating base (Fig.6). Recorded strains are caused by compression of the foam, Teflon and samples so they are not reliable for the egg structure.

Figure 6. *Placement of the sample inside the machine*

In this regard ten samples were chosen among different types of eggs which their geometric properties are mentioned in Table2. The highest load-bearing capacity is for rows number 1 and 2. In these samples the length to width ratio is 1.42 and 1.41. Also, the lowest load-bearing capacity is for row numbers 7 and 9. In these samples, with equal thickness, the length to width ratio is 1.7 and 1.68. Tthe results show that eggs with more natural and more spherical shapes are more resistant (an approval for Jacobs and his colleagues [10] and Narushin and his colleagues [11], research). In other words, longer eggs have les resistance. This matter approves Ebubekir Altuntas research [12] (in Abubekir research, SI (shape index) is width to length ratio). Meanwhile, the highest resistance occurs in the highest width (Anderson and et al.[13]).

Table 1. *Geometric properties of samples*

NO	Contents	Fac. / Home	Color	Length	Width	Thickness	Weight	Press	L / Wi
1	Empty	Home	Brown	55.4	39	0.35	61	988	1.42
2	Raw	Home	Brown	60	42.5	0.35	58	878	1.41
3	Empty	Factory	White	60	41	0.38	61	787	1.46
4	Raw	Home	Brown	56	35.5	0.36	59	576	1.58
5	cooked	Factory	White	54.5	33	0.35	57	365	1.65
6	Raw	Factory	White	53.5	33	0.35	60	364	1.62
7	Raw	Home	Brown	54.5	32	0.35	61	326	1.70
8	cooked	Home	White	53	32.5	0.33	56	249	1.63
9	cooked	Home	Brown	60.5	36	0.35	57	163	1.68
10	Empty	Factory	White	57	35.5	0.34	59	110	1.61

The above results confirm our previous researches. The geometric properties of samples are in Table 1.

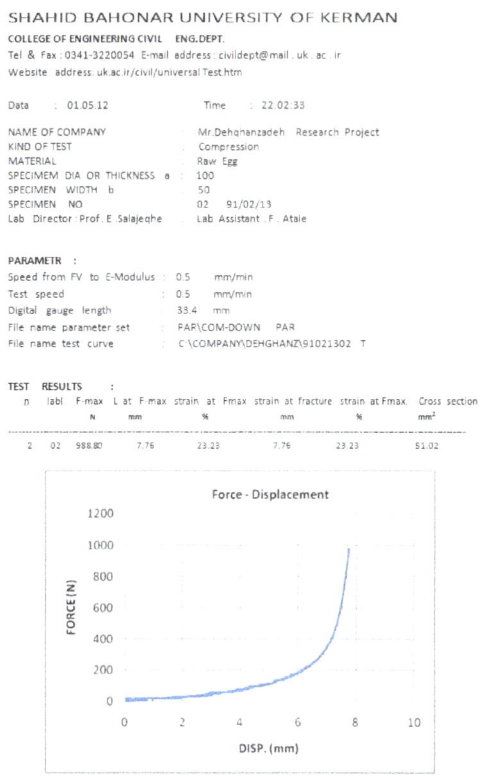
SHAHID BAHONAR UNIVERSITY OF KERMAN

COLLEGE OF ENGINEERING CIVIL ENG.DEPT.

Tel & Fax : 0341-3220054 E-mail address : civildept@mail . uk . ac . ir

Website address: uk.ac.ir/civil/universal Test.htm

Data : 01.05.12 Time : 22:02:33

NAME OF COMPANY : Mr.Dehqhanzadeh Research Project
KIND OF TEST : Compression
MATERIAL : Raw Egg
SPECIMEM DIA OR THICKNESS a : 100
SPECIMEN WIDTH b : 50
SPECIMEN NO : 02 91/02/13
Lab Director : Prof . E .Salajeqhe : Lab Assistant : F . Ataie

PARAMETR :
Speed from FV to E-Modulus : 0.5 mm/min
Test speed : 0.5 mm/min
Digital gauge length : 33.4 mm
File name parameter set : PAR\COM-DOWN PAR
File name test curve : C:\COMPANY\DEHGHANZ\91021302 T

TEST RESULTS :

n	labl	F-max N	L at F-max mm	strain at Fmax %	strain at fracture mm	strain at Fmax %	Cross section mm²
2	02	988.80	7.76	23.23	7.76	23.23	51.02

Figure 7. *The diagram of force-displacement with the capacity of 576Newton*

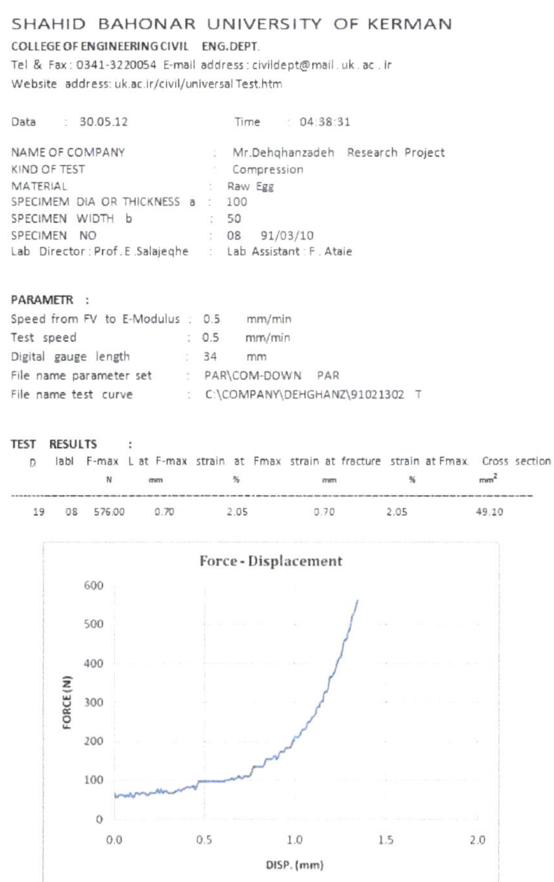
SHAHID BAHONAR UNIVERSITY OF KERMAN

COLLEGE OF ENGINEERING CIVIL ENG.DEPT.

Tel & Fax : 0341-3220054 E-mail address : civildept@mail . uk . ac . ir

Website address: uk.ac.ir/civil/universal Test.htm

Data : 30.05.12 Time : 04:38:31

NAME OF COMPANY : Mr.Dehqhanzadeh Research Project
KIND OF TEST : Compression
MATERIAL : Raw Egg
SPECIMEM DIA OR THICKNESS a : 100
SPECIMEN WIDTH b : 50
SPECIMEN NO : 08 91/03/10
Lab Director : Prof . E .Salajeqhe : Lab Assistant : F . Ataie

PARAMETR :
Speed from FV to E-Modulus : 0.5 mm/min
Test speed : 0.5 mm/min
Digital gauge length : 34 mm
File name parameter set : PAR\COM-DOWN PAR
File name test curve : C:\COMPANY\DEHGHANZ\91021302 T

TEST RESULTS :

n	labl	F-max N	L at F-max mm	strain at Fmax %	strain at fracture mm	strain at Fmax %	Cross section mm²
19	08	576.00	0.70	2.05	0.70	2.05	49.10

Figure 8. *The diagram of force-displacement with the capacity of 988Newton*

The outputs from test results of rows number 1 and 4 are illustrated in Figure 7 and Figure 8 The great displacement is caused by usage of foam and its compression. The aim of this test is to determine the load-bearing capacity of the chicken egg, so we had to put foam under the samples in order to prevent local failure and to prevent stress concentration. The strain- stress diagram of all samples have relatively uniform pattern. In calculation of stress, the extremity of the ring with lowest area was used. In this method, depending on the created seating base, the maximum applied load on a small area of the egg's narrow tip was 988Newton. Whereas in Macleod research, regardless of the load-application method, the maximum applied load was 40 Newton, in Ashley Lloyd research this figure was 200 Newton, in De Ktelaere research this figure was 37.8 Newton, in Zlatica, Vitorovic and Lukic [14] this figure was 35.3, in Pilat [15] research this figure was 10.5 Newton and in Tung [16] research the maximum applied load was 31Newton. In our previous research this figure was 750Newton.

4. Modeling of Chicken Egg in AutoCAD and Ansys Software

For the creation of a model in Ansys environment, version 13, first we scanned a sample chicken egg which its shape was more proper and more real. Then we started to create a real model in AutoCAD. The result was saved in Sat. format and was analyzed in Ansys. The chicken egg longitudinal axis is X and its transversal axis is Y.

The Z axis is a transversal axis, too. Based on the experimental pattern and the depth of the seating base on the two ends of the chicken egg, two areas were created as the load-application area and support area with 11mm of depth. The Shell181 was used as a layer with 0.35 mm of thickness, without considering the impact of composite. This element is appropriate for the analysis of thick and thin Shell structures. It is a 4-noded element with 6 degrees of freedom at each node (three translational and three rotational) which is appropriate for linear analysis of great rotations and nonlinear analysis of strains.

4.1. Elastic Modulus of Chicken Egg

According to the Coak and colleagues research, the elastic modulus was 31000 Mega Pascal and according to Tang research, by using Reissner method [17], the elastic modulus of a chicken egg was 46000 Mega Pascal. This Figure was 55000 Mega Pascal in Meloyd research and was 30000 mPa in Bain [18] research. In this research, the elastic modulus of a chicken egg was considered 47000 Mega Pascal and the Poisson's ratio was 0.25.

4.2. Various Analysis of a Chicken Egg in Ansys Software

The properties of samples including length, width and thickness of the shell in mm, the amount of transmitted load in Mega Pascal, load-application area in square meters and the length to width ratio is based on Table1. The seating base depth is 11 mm, so in Ansys, the load-application area of the two ends were considered to the depth of 11mm into the model and were bounded to translation and rotation (all dofs=0)

4.3. Linear Static Analysis

According to rows number 1, 2, 7 and 9 in Ansys Software with elastic modulus of 47000 Mega Pascal and 0.25 Poisson's ratio, we considered 10 steps of load applying. According to experimental loading, the maximum load was one thousand Newton. In this analysis all steps are loaded. The unit of displacement is mm and stress unit is Mega Pascal. The Sx stress represents meridional stresses along longitudinal axis of the egg and Sy represents orbital stresses along transversal axis. The negative numbers represent compressive stress and positive figures represent tensile stress. Svon represents Von Mises stress. The center index is related to average stress in the middle of the sample. The Table 2shows the results of linear static analysis and Figure9 and Figure 10 illustrates Sy and Sx stresses in samples of row 1.

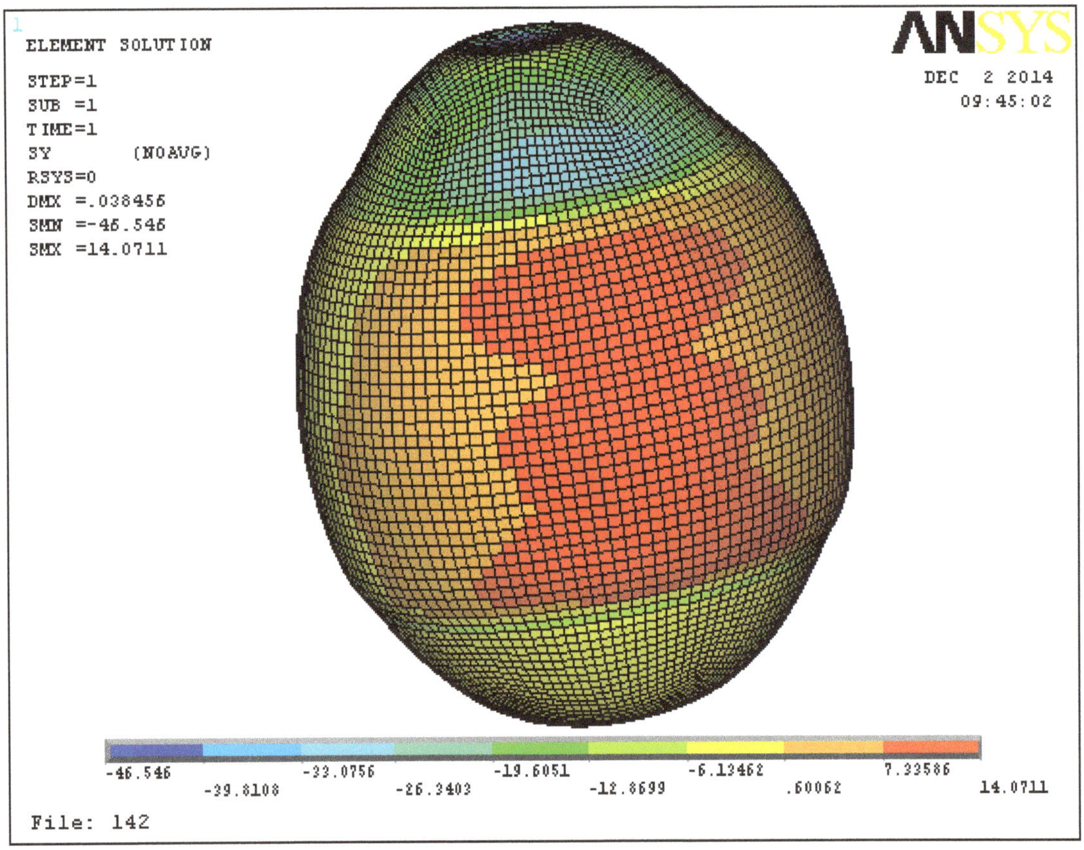

Figure 9. *Sy stress for sample1*

Table 2. *Results of linear static analysis*

Analyse	No	Length	Width	Thickness	L / Wi	Press	Area	Def.	Sx)min	Sx)max	Sx)center	Sy)min	Sy)max	Sy)center	S)von	S)von0center	step
L_S	1	55.4	39	0.35	1.42	0.824	1213.7	0.038	-33.86	4.76	-14	-46.55	14.071	10	45.2	22	10
L_S	2	60	42.5	0.35	1.41	0.617	1621.4	0.1041	-31.99	6.6	-12	-91.34	15.06	9	89	14	10
L_S	9	60.5	36	0.35	1.68	0.744	1343.2	0.098	-31.47	2.99	-14	-98.65	21.54	14	96.2	11	10
L_S	7	54.5	32	0.35	1.7	0.895	1116.9	0.06	-32.62	1.39	-15	-80.15	11.28	6	78.98	30	10

4.4. Nonlinear Static Analysis of Geometry and Materials

For Nonlinear Static Analysis of Geometry and Materials, after creating a model based on available samples, linear profile of materials including elastic modulus and Poisson's ratio are respectively 47000Mega Pascal and 0.25 and nonlinear profiles of materials according to relation number "(1-4-8)" are yield stress of 36 Mega Pascal and Tangent modulus of 1 Mega Pascal. For calculating the yield stress, the maximum bearing load derived from experimental results is divided by the cross section of the seating base bottom (circle perimeter* thickness).

$$\sigma = -1000/(3.14 * 25 * 0.35) = -36.38\text{MPa} \qquad (1\text{-}4\text{-}8)$$

The tangent modulus of the chicken egg is considered one because its Tangent modulus is not available and is very brittle. We considered 10 steps of load applying. According to experimental loading, the maximum load was one thousand Newton. The Table3represents nonlinear static analysis results and Figure 11 and Figure 12 represent

Svonand Sx stresses of the samples of the row number 7.

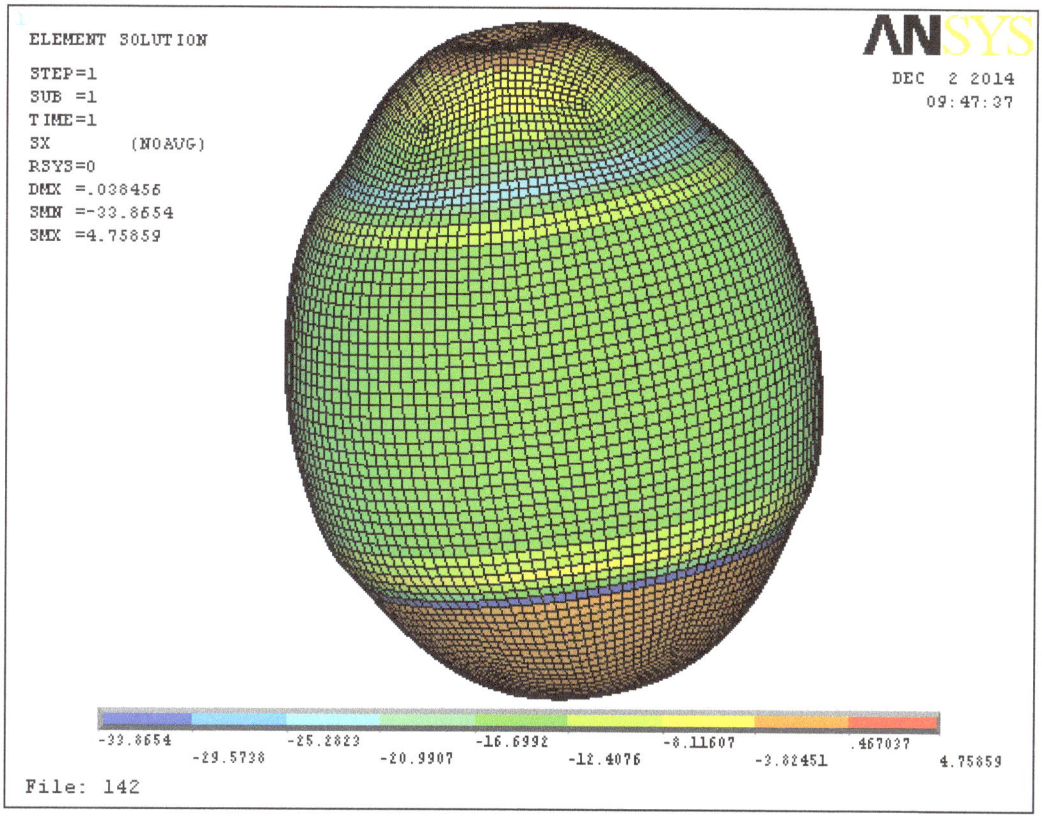

Figure 10. *Sx stress for sample1*

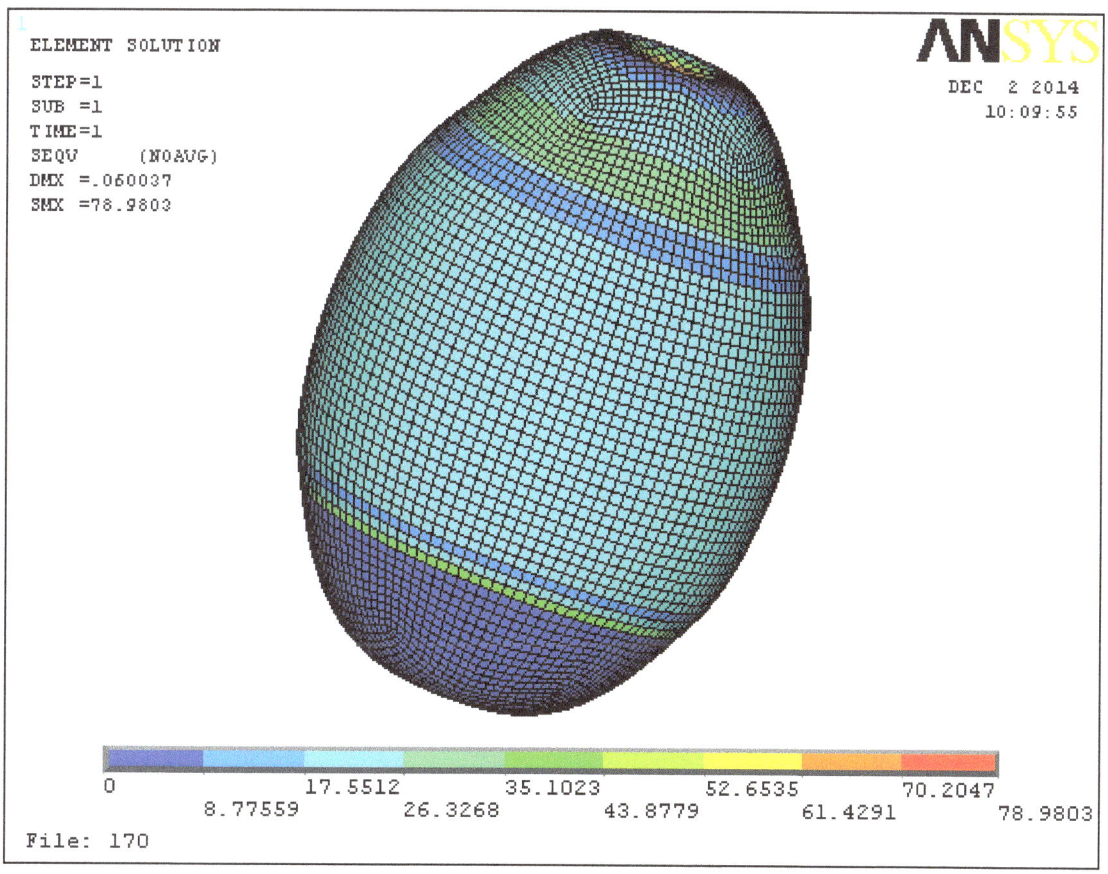

Figure 11. *Svon stress for sample 7*

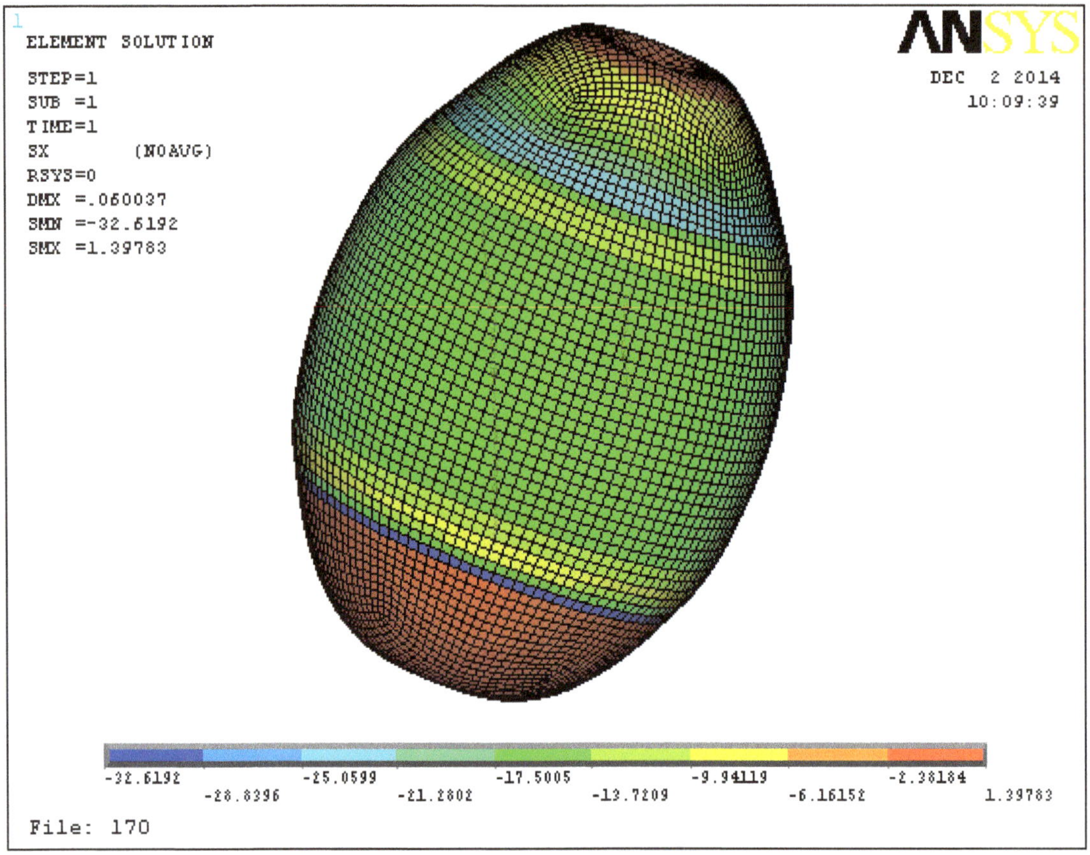

Figure 12. Sx stress for sample 7

Table 3. Nonlinear static analysis results

Analyse	No	Length	Width	Thickness	L / Wi	Press	Area	Def.	Sx)min	Sx)max	Sx)center	Sy)min	Sy)max	Sy)center	S)von	S)von0center	step
NL_S	1	55.4	39	0.35	1.42	0.824	1213.7	0.041	-30.61	4.99	-12	-38.06	14.37	11	36	20	10
NL_S	2	60	42.5	0.35	1.41	0.617	1621.4	0.13	-19.24	4.02	-7	-41.5	30.56	-2	36	10	6
NL_S	9	60.5	36	0.35	1.68	0.744	1343.2	0.145	-18.94	2.81	-7	-40.45	41.11	1	36.02	10	6
NL_S	7	54.5	32	0.35	1.7	0.895	1116.9	0.066	-26.16	1.65	-12	-40.7	17.43	13	36	10	8

4.5. Displacement of Load-Application Point and the Lean Area

When changing the support area in an experimental model, the results of load application are very weak and samples fail under very small loads. So, we changed the lean and load-application location in Ansys model. We applied load on the flat tip of the egg and moved the lean area to the narrow tip. Then we realized that with 1000Newtons of load, Sx stresses stay the same by Sy stresses sharply increase. These results are shown in Table 4and ros22 and 99 are new models where the load and the lean location have changed.

Table 4. Results of moving the leant area in Ansys model

Analyse	No	Length	Width	Thickness	L / Wi	Press	Area	Def.	Sx)min	Sx)max	Sx)center	Sy)min	Sy)max	Sy)center	S)von	S)von0center	step
L_S	2	60	42.5	0.35	1.4	0.6	1621.39	0.104	-31.99	6.6	-12	-91.34	15.06	9	89	14	10
L_S	22	60	42.5	0.35	1.4	0.5	1874.13	0.199	-25.8	1.32	-13	-147	58.5	7	147	20	10
L_S	9	60.5	36	0.35	1.68	0.74	1343.24	0.098	-31.47	2.99	-14	-98.65	21.54	14	96.2	11	10
L_S	99	60.5	36	0.35	1.68	0.65	1538	0.18	-26	2.45	-13	-155	67.85	-0.4	155	20	10

5. Results

Applying load from the narrow tip and placing the lean area on the flat bottom of the chicken egg cause orthogonal tensile stresses which are orbital tensile forces (Sy) of the model. Because of egg brittle calcareous shell which has low tensile resistance, the shell fails and cracks. But it can withstand compressive forces because of its two-sided vaulted shape. To prove this matter, we press a cracked sample along its longitudinal axis. As it is shown in the Figure 13, the vertical cracks are opened in an orbital pattern. In experimental samples no buckling is observed in the egg shell (no deformation). Also, in Ansys model the bending stresses are negligible and this shows that in such structures buckling is very weak.

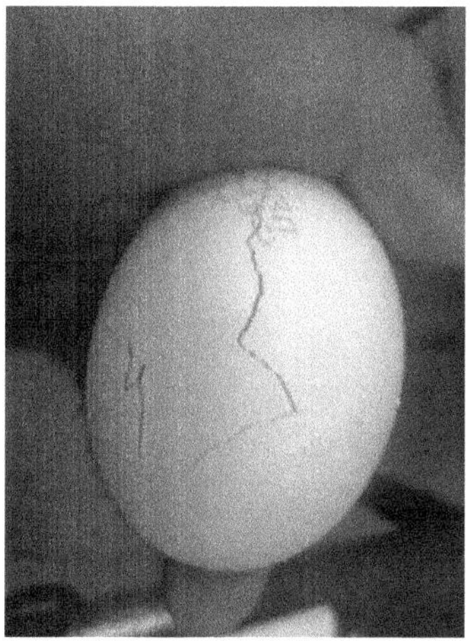

Figure 13. cracks opening

6. Discussion and Conclusions

The above results show that all meridional compressive stresses (Sx) are compressive, but due to stress concentration the lean areas are exceptions. But orbital stresses (Sy) are compressive in some points and in adjacent points to the lean areas stresses are tensile. These tensile stresses cause crack along the longitudinal axis of the chicken egg and breaks the egg from two sides.

The lean status changing from All Dof=0 to Ux,y,z=0 has no significant effect on Ansys results. Maximum tensile stresses are formed on the flat side of the egg and are transmitted to other parts. Maximum compressive stresses of Sx and Sy are formed by concentration of stress in the lean or in load-application surfaces.

According to relation 1-4-8, if we consider -36Mega Pascal for the allowable stress, we will realize that all compressive stresses of Sx are in allowable limits and Sy stresses exceed allowable limit.

The results show that chicken egg content has no important role in load-bearing capacity and it's geometric shape of the eggshell which has the main role. In above conditions, the egg is capable of withstanding 988 Newton of load. Meanwhile, we can improve egg bearing ability by providing soft, flexible and deep seating bases for the narrow tip and flat bottom of the egg because these bases prevent brittle shell from failure (break).

The geometric form and shape of an egg have the main role in its load-bearing capacity. This form changes the load into orbital forces while there is a little amount of bending. The calcareous eggshell shows a proper behavior under compressive loads. But it has low tensile resistance and this causes tensile cracks at the highest part of the egg in contact to lean area. Creation of these cracks and longitudinal expansion of them to the other side is the reason of egg failure (break).

We use a needle to empty the egg content. If we pimp air inside the eggshell it will explode loudly. This shows that the eggshell can withstand no pressure from inside. At the end of growth period this feature lets the chicken to easily come out of the egg.

Experimental results show that in chicken eggs with the same shell thickness, eggs with length to width ratio of 1.5 have the highest load-bearing capacity and eggs with length to width ratio of 1.7 have the lowest load-bearing capacity. But in software models, maximum compressive stresses usually occur in the lean area and only the maximum orbital tensile stress occurs outside and near the lean area. For various models no same (equal) procedure was observed.

The total applied load changes into meridional and orbital forces and moments are little. The meridional forces are mostly compressive with an exception in the seating bases which are caused by stress concentration. But orbital forces are compressive in some parts and in the bottom of the lean area are tensile. These tensile forces cause orbital tensile stresses and create cracks in this part. The longitudinal expansion of these cracks causes the egg to fail (break).

The results show that if we build structures based on the shape of chicken egg we can build roofs with thin thickness and large spans. Even if we consider orbital and meridional reactions in parts of this structure, we still can build it.

If we use tensile resistant elements in an orbital or angled way in order to prevent tensile forces that cause cracks in the lean area, we can achieve higher load-bearing capacity. This shows the role of composite layers. In an eggshell, if the calcareous crystals are safely connected to the internal membranes, we will achieve higher load-bearing capacity. So, the protein membranes have two roles, a barrier for bacterial invasions and a structural role as a composite structure. The later can be studied as "The Analysis of structural resistance of a chicken egg with composite shell" in another research.

References

[1] M. Farkhooii, T. Sigaroodi and F. Niknafs," a complete guide to poultry," Iran institute of aviculture,1999.http://www.mardoman.net/?id=596&pt=full&sn=articles

[2] C. Perianu a, B. De Ketelaere a, B. Pluymers b, W. Desmet b, J. De Baerdemaeker a, E. Decuypere aa Division MeBioS, KU Leuven, Dept Biosystems, KasteelparkArenberg 30, B-3001 Heverlee, Belgium Department of Mechanical Engineering, Division PMA, KU Leuven, Celestijnenlaan 300B, B-3001 Heverlee, Belgium. biosystems engineering 106 (2010) 79e85. journal homepage: www.elsevier.com/locate/issn/15375110

[3] P. Coucke, G. Jacobs and ets, "comparative analysis of the Static and Dynamic Mechanical Eggshell Behaviour of a Chicken Egg", In: Proceeding of international Conference on Noise and Vibration Engineering, Long Beach, CA, USA, (1998).

[4] Ashley Lloyd, Tara Vaziri, Sam Yang, "A composite of crystal inelayers and membrane proteins: A mechanical, structural, and compositional investigation of eggshells", April 18, 2011.

[5] Ahmed, A. M. H., Rodriguez-Navarro, A. B., Vidal, M. L., Gautron, J., Garcia-Ruiz, J. M., &Nys, Y. "Changes in eggshell mechanical properties, crystallographic texture and in matrix proteins induced by moult in hens", British Poultry Science,(2005),pp 46, 268–279.

[6] S. Nedomova, J. Trnka, P. Dvorakova, "Hen`s eggshell strength under impact loadind", Mendel university agriculture and foresty, Department of food technology, (2009) 350-357

[7] De Ketelaere, B. Govaerts, T. Couke, P.Dewil, E. Visseher, T. Decuypere, "Measuring the eggshell strength of 6 different strains of laying hens: Techniques and comparison", British Poultry Science, (2002). 43, 238–244.

[8] Voisey, P. W., & Hunt, J. R.. "Effect of compression speed on the behaviour of eggshells". Journal of Agricultural Engineering Research,(1969), 14, 40–46.

[9] Buchar, J. Simeonovova, "On the identification of the eggshell elastic properties under quasistatic compression", In: Proceeding of 19th CAD- FEM users meetig, Berlin, Germany, (2001).

[10] Jacob, J. P. Milles, R. D.& Mather, F. B.."Egg quality", University of Florida extension, Institute of food and agricultural science,(2000), p. 11.

[11] Narushin, V.G., van Kepmen, T. A., Wineland, M. T., & Christensen, " Comparing infrared spectroscopy and egg size measurements for predicting eggshell quality", Biosystem Engineering, (2004).

[12] Ebubekir Altuntas, Ahmet Sekerog, "Effect of egg shape index on mechanical properties of chicken eggs", CANADIAN Received 28 May 2007; received in revised form 22 August 2007; accepted 30 August 2007 Available online 24 October 2007.

[13] Anderson, K. E., Tharrington, J. B., Curtis, P. A., & Jones, F. T., "Shell characteristics of eggs from historic strains of single comb white leghorn chickens and relationship of egg shape to shell strength. International", Journal of Poultry Science, (2004),3, 17–19.

[14] Zlatica, P., Vitorovic, D., Lukic, M., & Spasojevic, I. "Improving egg shell quality by replacement of pulverised limestone by granular limestone in the hen diet",ActaVeterinaria (Beograd),(2003), 53, 35–40.

[15] Polat, R., Tarhan, S., C¸ etin, M., & Atay, U.,"Mechanical behavior under compression loading and some physical parameters of Japanese quail (Coturnixcoturnix japonica) eggs", Czech Journal of Animal Science,(2007),52, 50–56.

[16] Tung, M. A., Staley, L. M., Richards, J. F., "Estimation Of Young`s Modulus And Failurs Stresses In The Hen`s Egg Shell", Canadian Agriculcture Engineering, Vol. 11, No. 1, MAY 1969.

[17] Reissner, E., "Stresses and Small Displacements of Shallow Spherical Shells", II. J. Math, 1947.

[18] M.M. Bain, "Eggshell strength: a mechanical/ultrastrucutral evaluation" Ph. D. Thesis, University of Glasgow, Scotland(1992).

25

Evaluation of Ten Thousand Cubic Meters Mazut Tank Behavior Against Explosive Charges

Saeid Azmoodeh[*], **Nasser Arafati**

Department of Civil Engineering, University of Tafresh, Tafresh, Iran

Email address:

Saeid.azmoodeh66@gmail.com (S. Azmoodeh), Nasser.arafati@gmail.com (N. Arafati)

Abstract: Considering the terrorist threats and blasts, which may occur to destroy a structure, the effect of explosive on a 10 thousand cubic meter Mazut tank of Mashhad Cement Company is studied. This article studies the behavior of 10 thousand cubic meter tank of storing Mazut with the help of ABQUS^{3D} software. The explosive considered is TNT, which is studied in distances of 15, 25, 50, 100, 200 meters to the structure and with various thicknesses of the tank wall. Finally, the results indicate that by the increase of the wall thickness, the blast effect will mitigate. Also due to the surface wave generated, by the reduction of the blast distance to the structure, the effectiveness of the blast wave will be greater and ultimately, in a distance less than 15 meters, the tank will collapse.

Keywords: Tank, Explosive Loading, ABAQUS Software, Performance Evaluation

---

1. Introduction

Considering the terrorist threats and blasts, which may occur to destroy a structure, the effect of explosive on a 10 thousand cubic meter Mazut tank of Mashhad Cement Company is studied.

According to the definition of chapter 21of Iran National structure regulations (Passive defense), explosion is a reaction, in which the burning rate of materials is done with a speed much more than the speed of sound, resulting in the generation of a very high pressure and temperature gradient and the shock wave is generated immediately, which will spread with a high speed [4]. In terms of the classification, the explosion type can be natural, physical, nuclear and chemical. In terms of physical, the explosions caused by the sudden failure of the tanks under pressure or volcanic eruption can be referred. In nuclear explosions it is based on the core separation and the Neutrons and Protons distribution loading, whereas in chemical explosion, the rapid oxidization of elements such as Carbon and Hydrogen can be noted. The explosives can be in the form of solid, liquid and gas, but the explosion of solids has usually the greatest effect in terms of the force released [1]. The explosives can also be classified in terms of the primary and secondary forces, which are exerted on the structures or they can be classified due to the activation of these materials through using condiments such as spark, flame or strike. The secondary stage of an explosion is in fact, the wave generated from that explosion, which can cause a lot of damages to the environment. Among very common explosives, TNT (Tri Nitro Toluene) can be referred.

The explosive, TNT can generate a pressure equal to 300 Kb and a temperature of about 3000 to 4000 Celsius degree. After the explosion the expanded hot gas and a layer of compressed air generate (explosion wave). This wave raises the amount of environment atmospheric pressure at about the gas pressure and the pressure behind the wave will be lower and a relative vacuum due to air intake behind it is generated. This cycle is completed in the next round and it generates the explosive shocks. The pressure exerted by the explosion shocks mitigate with the distance increasing from the explosion source. The pressure graph in terms of distance from the blast site is shown in figure 1 [1].

Between the years 1928 to 1945 a series of tests were carried out by the headquarters of the Japanese army on the striking blast to a three-storey building with a change in the thickness of the slabs. The result showed that by the increase of the slab thickness in the ceiling of the roof, the intrusion of the bomb inside the building is prevented [6].

Bing Li & Toe Chien (2008)[5], analyzed two reinforced

concrete frames with Finite Elements Method (FEM), which one lacked a wall and the other one was with a wall and compared their dynamic response due to the effect of the distant blast loading with each other. *Ehsan Mohtashami & Ahmad Shooshtari* (2010) [2] studied the steel structure behavior as a result of the blast occurring nearby. The existing results indicate that with the increase of the column angles, the improved safety measures and the structure displacement are also reduced. By studying the behavior of concrete constructions against the blast loads, Mohammad *Bozorgvar & Ahmad Shooshtari* (2011)[3] also concluded that if the building is designed and constructed on the basis of stand-off seismic standards, it must be precisely revised for the stand-off against the blast loads.

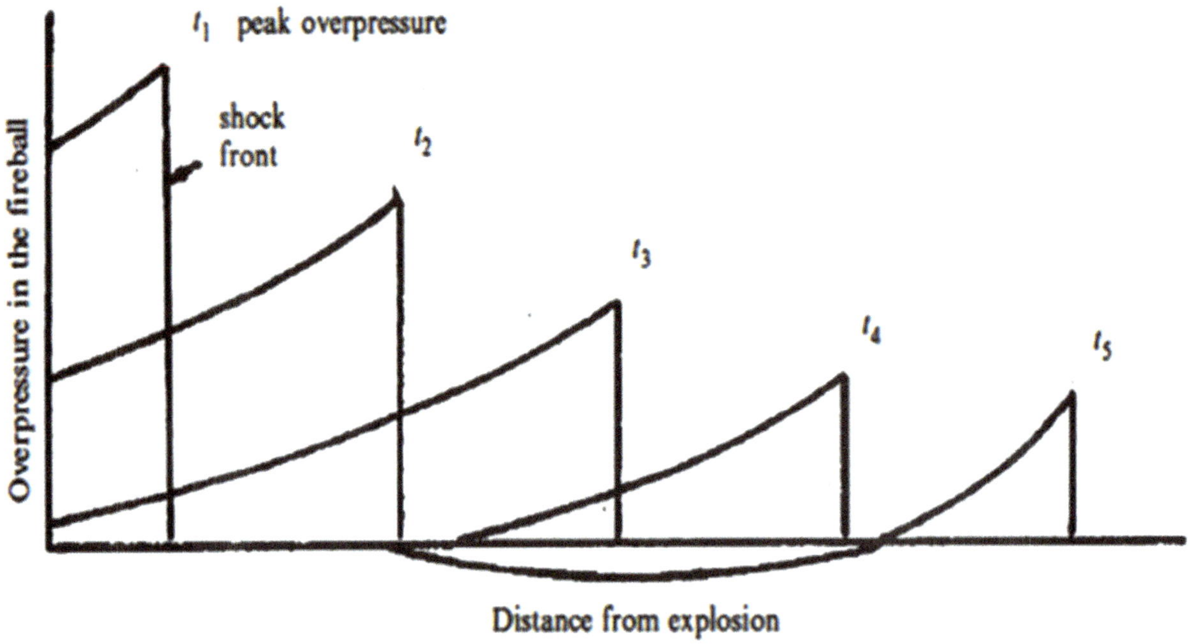

Figure 1. *Blast wave [8].*

2. The Structure Under Study

As it is shown in figure 2, the structure considered is the 10 thousand cubic meter tank of Mashhad Cement Company, which its foundation is ring shape with an interior diameter of 30 meters and exterior diameter of 34 meters. The height of the tank is 10 meters and the thickness of the foundation is 1 meter.

Figure 2. *A view of the 10 thousand cubic meter tank of Mashhad Cement Company.*

The thickness of the wall, ceiling and floor of the steel tank in 5, 10 and 15 centimeter forms were studied. Also the weight of TNT caused by the blast was considered 10 tons and the distances of the blast occurrence were assumed 15, 25, 50, 100 and 200 meters.

3. Explosive Loading

Considering that the blast intended occurs on the surface, the waves at the blast surface because of the reflection from the ground are retrofitted and as in this mode the explosives are so close to or in contact with the ground surface, therefore, the blast waves are released and spread in a semi-spherical form.

(Z) the scaled distance, is defined as follows:

$$Z = R/(W)^{1/3} \tag{1}$$

In this equation (W) is TNT equivalent weight in terms of kilo grams for the explosives intended and (R) is the distance of the materials to the target (Stand-off) in terms of meter. This parameter was first stated independently by B. Hopkinson in 1915 and then by *C. Cranz* in 1926 [6].

The primary pressure is generated at a distance from the blast centroid is much more than the atmospheric pressure. This pressure generated is called “Over pressure”, which is shown with (P_s) and (P_{so}) is the maximum value. The period of time, which (P_{so}) is added to the pressure of the environment is very sudden, as it can be considered as instantaneous.

As it is seen in figure 5, the pressure wave expands radially from the blast centroid with a reducing speed, which this speed is always more than the speed of the sound in the environment. The pressure changes generated in every point, which is at a distance from the blast centroid is as shown in figure 5.

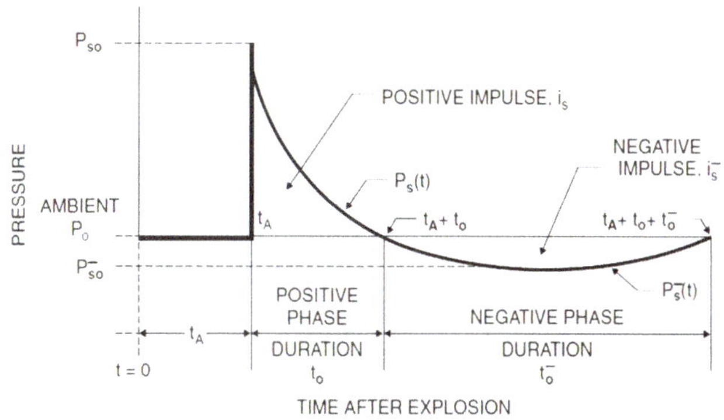

Figure 3. *The pressure graph of blast wave of the time [9].*

In a moment such as "Arrival Time" (t_A), which is the time of the wave ascending, it will reach the point intended and the amount of pressure will increase instantaneously to the amount of (P_{so}) and then it will decrease to reach the environment pressure. This decreasing process is done in duration of (t_o). This duration from (t_A) to (t_o) is called "Positive phase". After that the negative phase will start and the pressure will reach below atmosphere. The duration of negative phase is shown with (t_o^-), which is usually more than positive phase. however, due to the slight negative phase; it can be disregarded in designing and analyzing.

To mitigate the over pressure with the time the following exponential equation is suggested:

$$P_{(s)} / P_{so} = [1- (t-t_A/t_o)\ ]e^{-(t-tA)/to} \qquad (2)$$

To calculate (P_{so}) an equation by *Kinney & Graham* (1985) has been suggested, which is as the following form of equation [7]:

$$P_{so}/P_a=\{808[1+(z/4.5)2]\}/\{\sqrt{1}+(z/0.048)2.\sqrt{1}+(z/1.32)2.\sqrt{1}+(z/0.32)2\} \qquad (3)$$

(P_{so}) is the peak over pressure and (P_a) is equivalent to the atmospheric pressure in terms of (kg/cm^2), which is usually considered equivalent to 100 KN per cubic meter. The cases which have been stated so far are in relation to the pressure increase in one point in the open air, however, if at this point there is a level or a rigid structure, the over pressure maximum amount will not be (P_{so}) anymore [7], when the pressure primary wave hits a rigid surface, it is reflected from the surface, therefore, the waves are compressed on the surface momentarily, so the peak pressure generated on the surface is definitely more than the amount of the pressure, which is generated in the absence of a liquid existing at that point. The problem can be expressed in this way that when a structure is located in the direction of a blast wave, the interaction between the air moving and the body of the structure results in reducing the waves velocity and the kinetic energy of the particles in the air suddenly reaches zero, thus an increasing loading will be exerted to the structure. The striking pressure has been presented with (P_r), which its amount is subject to the amount of (P_{so}) at the same point [6].

The amount of (P_r) is obtained from the following equation:

$$P_r = 2P_{so}\ (4P_{so}+7P_a)/(P_{so}+7P_a) \qquad (4)$$

This equation was obtained by *Rankine* (1870)[11], assuming that the air behaves like an ideal gas. In this equation (P_r) in comparison to the pressure generated on the surface of the structure is more at a point, where the connecting line of that point to the centroid of the blast is perpendicular to the surface of the structure.

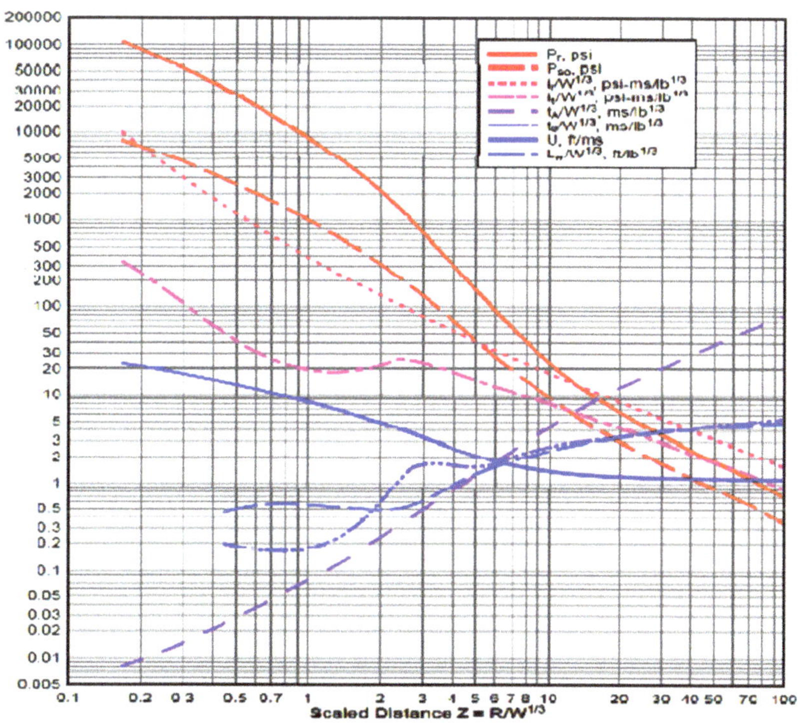

Figure 4. *Shock wave parameters of the positive phase for spherical surface blasts [10].*

The duration of the positive phase (t_o) is a function of (P_{so}) and also the amount of the total energy released in the blast. The equation to determine (t_o) was presented by *Kinney & Graham* (1985)[11] as follows:

$$t/^{3}\sqrt{w}=\{980[1+(z/0.45)10]\}/\{[1+(z/0.02)3].[1+(z/0.75)6].[1+(z/6.9)2]-1\} \quad (5)$$

Where (W) is the weight of the explosive and (Z) is equal to the distance scaled, while the positive phase duration is obtained in terms of Millisecond.

As well as the theoretical equations, which have been stated so far to calculate the blast load parameters, some graphs have also been provided based on the practical tests in this regard. In figure 4, the graph of different parameters have been shown in the form of a function to the distance scaled of (Z) for the positive phase. These curves are related to the surface blast [3].

Figure 5, The loading history considered is seen in a mode, which the distance of the blast from the structure is 100 meters.

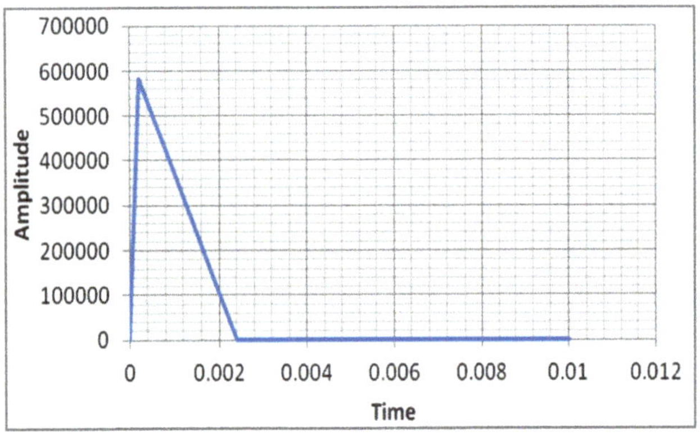

Figure 5. The history of explosive load.

4. Analysis of the Structure Against the Explosive Loading

To study the behavior of the structure under explosive loading, a 3D form and *Abaqus V.6* software are used. In this research, considering the foundation of this structure, which is from concrete, the pressure resistance intended for it is assumed 35 Mega Pascal, the Poisson ratio 0.2 and elasticity modulus 20 and 59 Giga Pascal and the relative density 2400 kg per cubic meter. The concrete property in the plastic area has been presented in tables 1 and 2.

Table 1. The values of stress, strain and degradation of plastic concrete in tensile mode.

Parameter degradation in stretch	Strain cracking	Tensile strength (Mpa)
0.00	0.000000	5.3
0.25	0.000176	5.81
0.99	0.001539	0.58

Table 2. The values of stress, strain and degradation of plastic concrete in pressure.

Parameter degradation in pressure	Non- elastic strain	Compressive strength(Mpa)
0.000	0.00000	17.5
0.112	0.00038	25.7
0.429	0.00189	34.9
0.466	0.00218	35
0.701	0.00456	28

To make a model of the tank, which is made of steel, in the area of the elasticity, the elastic modulus is assumed 210 Giga Pascal, Poisson coefficient 0.3 and relative density 7800 kg per cubic meter and the properties of steel in the plastic area for blast analysis can also be seen in table 3.

Table 3. The values of steel actual stress and strain.

Plastic strain	Real tension (Mpa)
0.000	300
0.025	350
0.100	375
0.200	394
0.350	400

5. The Structure Response to Explosive Loading

In figure 6, Mazut 10 thousand cubic meter tank model made can be seen, the ceiling, floor and wall sections, which are of the same type, have been considered to be integrated.

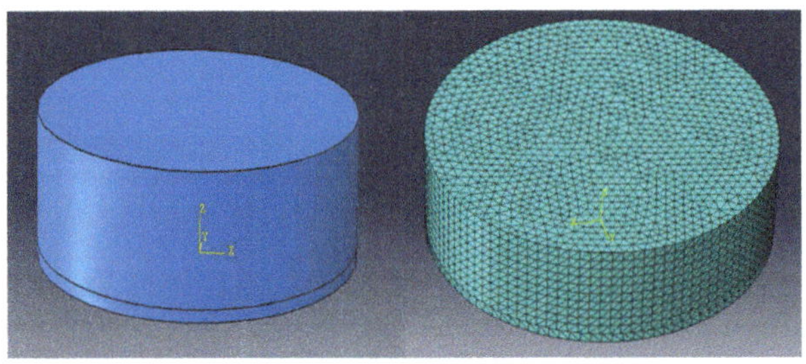

Figure 6. Tank model made in Abaqus software.

Figure 7, The graph related to displacement shows the ratio of the time for various distances of blast to the structure.

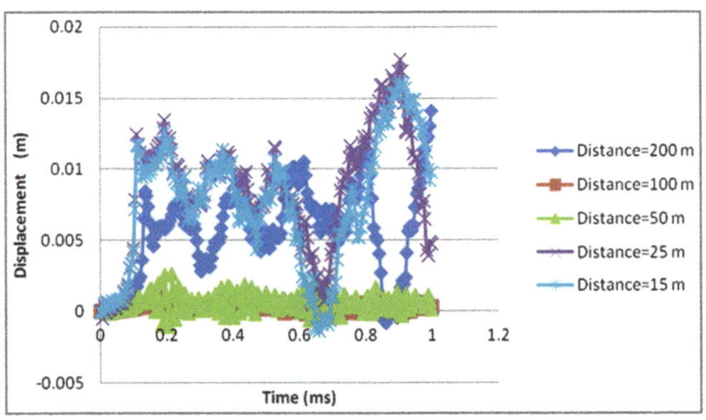

a)Thickness of 5 centimeters

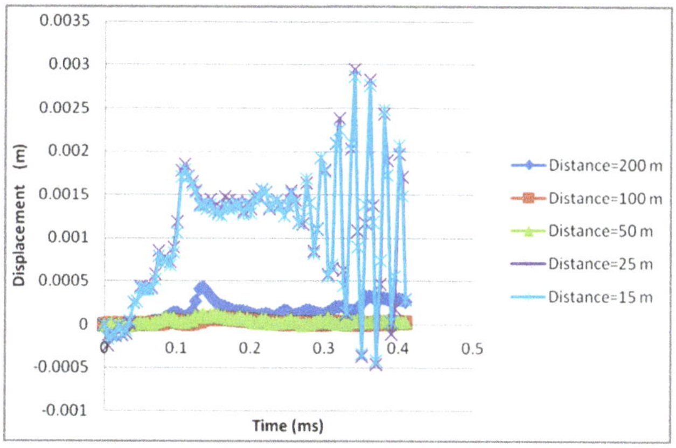

b)Thickness of 10 centimeters

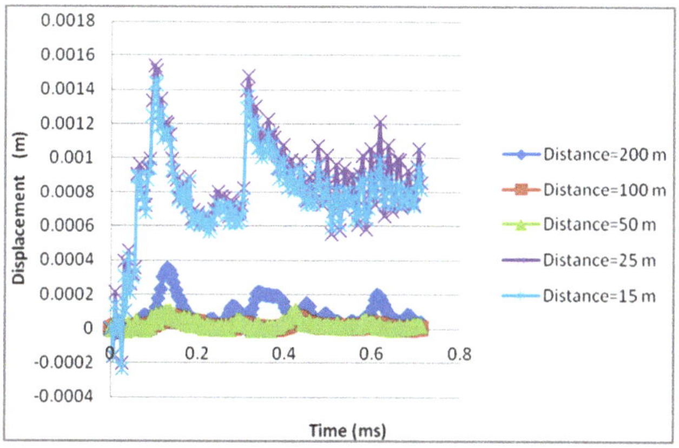

c)Thickness of 15 centimeters

Figure 7. *Graph of displacement to the time for various distances of the blast site from the structure.*

Figure 7 shows that with the increase of the blast distance to the structure, the effect of the blast wave will be mitigated. Also in low thicknesses, the displacements are more than the mode, which the thickness of the tank steel wall is more.

In figure 8, a view of the structure deformation in the vertical direction, while the thickness of the tank is 5 centimeters and the distance of the tank is 25 meters from the from the blast site, has been shown.

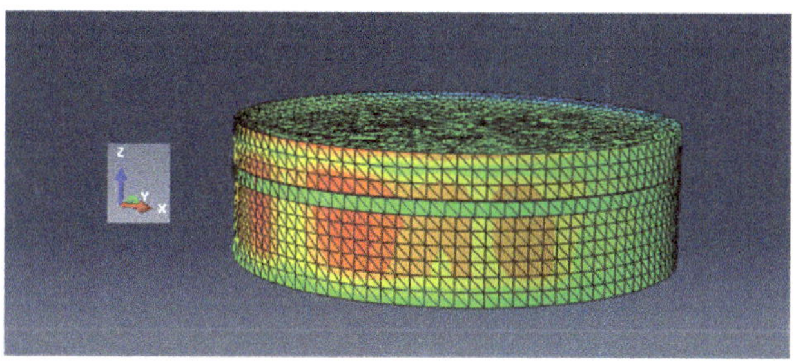

a)A general overview of the model

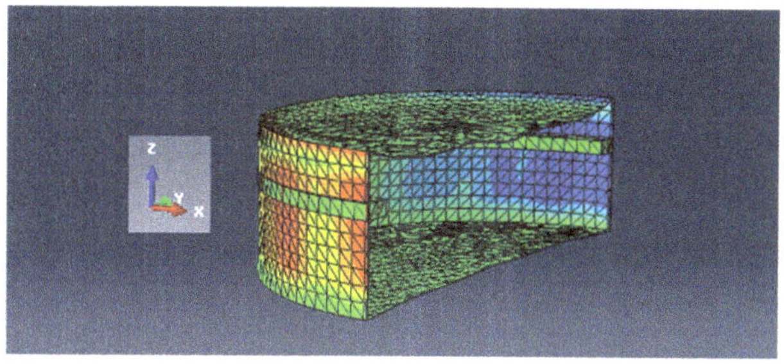

b)A ross-sectional overview of the model

Figure 8. *A deformation overview of the model under explosive load.*

The displacement graph in terms of time for the thicknesses of 5, 10 and 15 centimeters of the tank in a mode, which the distance of the explosive site from the structure is 15 and 100 meters, which has been presented in figure 9, it can be realized that the more the thickness of the tank wall becomes, the less the blast effect will be and the thickness also plays an important role in the displacements.

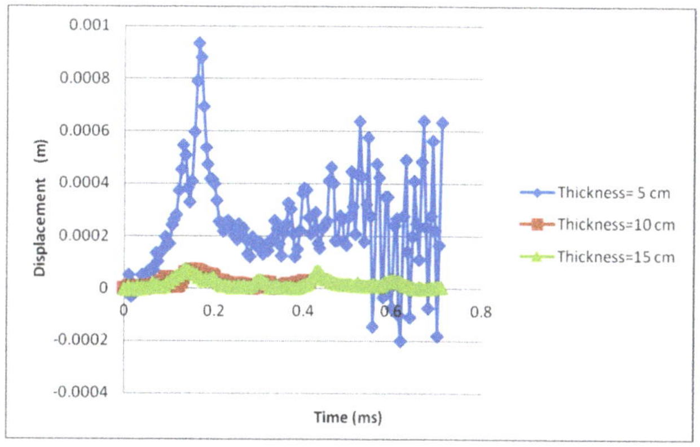

a)Distance of the explosion from the structure, 100 meters

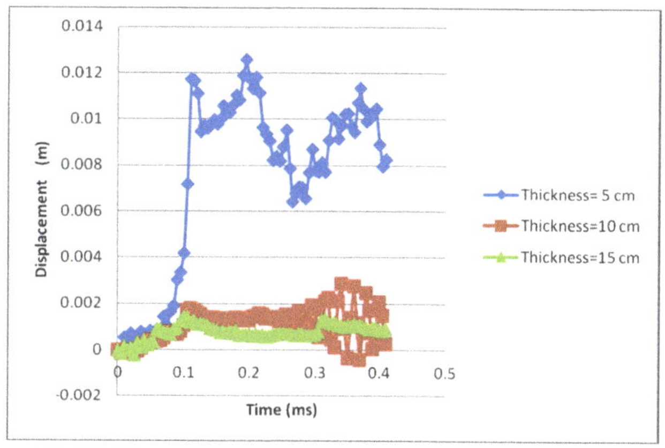

b)Distance of the explosion from the structure, 15 meters

Figure 9. *Displacement in terms of time for 15 and 100 meters distances of the explosion.*

Energy graph for the whole structure can also be seen in figure 10.

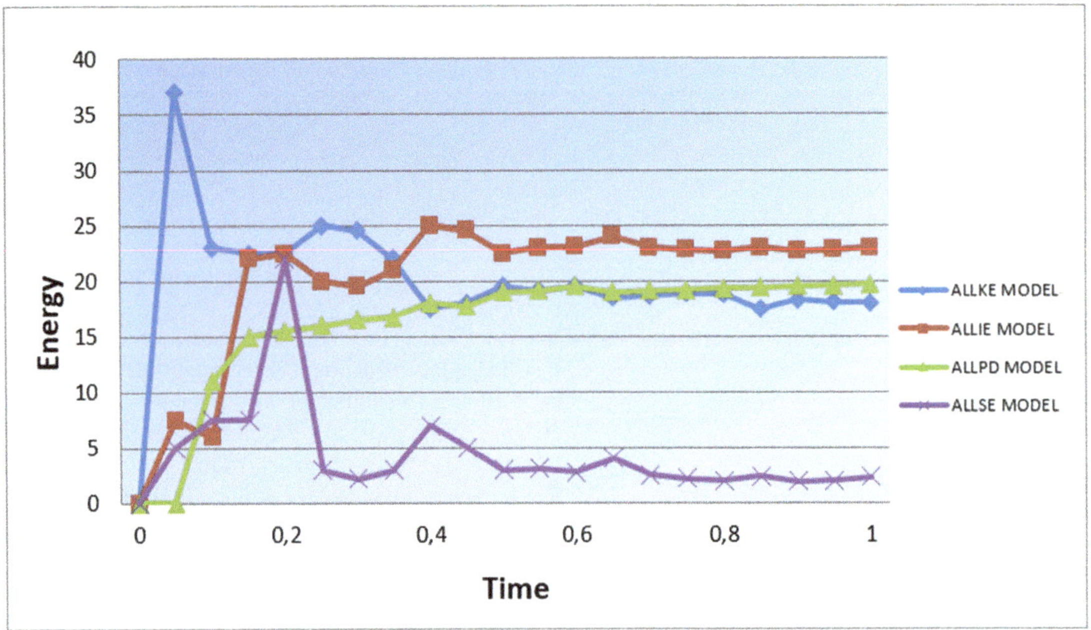

Figure 10. *The graph of energy of the structure.*

6. Discussion and Conclusion

The 10 thousand cubic meter tank centroid made is with the diameter of 34 centimeters and the height of 10 meters, while its foundation is ring shape with an interior dimension of 30 meters, which can be seen with the help of *Abaqus 3D* software. The structure was analyzed in the mode of explosive loading. The type of explosive has been assumed TNT and its weight has been assumed 10 tons. The blast site distances intended from the structure, 15, 25, 50, 100 and 200 meters for the thicknesses of 5, 10 and 15 centimeters of the tank were exerted in the calculations. The conclusions indicated that with the increase of the blast distance from the tank, the intensity of the wave obtained from the blast is strongly reduced, This is while the thickness considered for the tank plays a significant role in the sustainability and mitigation of the displacements and with the increase of the thickness of the steel tank, the safety coefficient of the structure against blast loads will increase.

References

[1] The most comprehensive reference book used Abaqus. Sohail Srvshnya and Nima Paradise. [2008].pp, 124-150.

[2] E. Mohtashami, Sinai, Sinai, Ahmed Shoshtari. "The assessment of the behavior of steel frames against explosion charges" Fifth National Congress of Civil Engineering, University of [2010].p, 45, 58, 67.

[3] Mahmoud Imam, Ahmad Shoshtari. "Blast effects on reinforced concrete buildings earthquake resistant," the Sixth National Congress of Civil Engineering. Semnan University in [2010].p34, 86,111.

[4] National Building Regulations, "Section 21: Civil Defense"[2008].pp3-6.

[5] Bing, L., Tso Chien, P. and Anan., N.,[2008]"A case study of the effect of cladding panels on the response of reinforced concrete frames subjected to distant blast loading". Elsevier, Singapore. pp, 130-142.

[6] Hesamoddin forearm Yousefi, Shariatmadari. "Check flexible filler walls and reinforced concrete structures under load in response to lateral stiffness remote detonation" thesis, Ferdowsi University. [2011].p, 75,120.

[7] Kinney, G.F. and Graham, K.J. [1985], Explosive Shocks in Air, Springer-Verlag, New York, 2nd edn.

[8] M.Y.H. Bangash "Shock, Impact and Explosion" Structural Analysis and Design. [2008] page 430.

[9] Donald O. Dusenberry "Handbook for Blast Resistant Design of Buildings" [2010] page176.

[10] A. Remennikov. "The state of the art of explosive loads characterisation" [2007].p12.

[11] Emily L. Guzas. and Christopher J. Earls "Air blast load generation for simulating structural response"[2010].pp433-437.

26

The study of different geometrical parameter effects for single grid layers of space structure on performance level for vertical load of earthquake

Hamid Shahrabadi

Department of Civil Engineering, Chabahar Maritime University, Chabahar, Iran

Email address:
h.shahrabadi@cmu.ac.ir

Abstract: Acceptance and popularity of space structure among architectures and designers is due to reasons such as: great spans covering, magnificent architecture, extremely low weight, the ease of building process, speed in installation, quality of response to earthquake, etc. Considering that usually large population live in space structured buildings, resistance and safety of the structures against destructive factors such as earthquake is very important. In this paper, we examine the effects of different geometrical parameters such as: type and shape of support on performance level of single grid layer of space structures and their performance at different risk stages are examined based on the regulations stated in Iranian Standard for Retrofitting of Existing Buildings. Since there is no acceptance criteria in international codes defined for determination of performance level (among Iranian code) in space structures, these criteria are suggested primarily. Finally performance points and levels of these structures are obtained by illustrating the capacity curves. Nonlinear static analyses (pushover) are carried out on the selected models using SAP and ANSYS software. Performance levels and capacity spectrums are obtained by FEMA and ATC40 bylaw.

Keywords: Single Grid Layer Space Structure, Performance Level, Capacity Spectrum, Nonlinear Static Analysis, Pushover Analysis

1. Introduction

Space structures have three-dimensional behavior and cannot be analyzed in plane state. These structures are very regular in geometry and are repeated by a specific pattern next to each other. Steel is the most frequent material in space structures, but aluminum; and wood could also be used.

Space structures have high degree of indeterminacy. The configuration and connections type prevent total failure in the system due to elimination of one element of the structure. Thus, if some elements in a grid lose their functionality due to the applied loads, other elements can corporate in tolerating the loads and can distribute the forces tolerated by the lost elements. Thus, in these structures, extreme damages do not cause sudden collapses. In recent years the use of these structures is increasingly expanding because of improvements in civil and architecture engineering, the software, welding, casting and forging industry, advanced machinery for product and assembly of prefabricated units. Different types of space structures are:

Plane Grids: The construction of these structures is 2-sided, 3-sided, 4-sided, single layer, double layer, triple layer or multi layer. Single layer plane grids are usually used for roofing small spans. For bigger spans double, triple or multi layer grids are used. Connections in single layer plane grids are rigid for stability of the structures. In double, triple or multi layer grids connections can be rigid, semi rigid or hinged.

Barrel Vaults: Barrel vault is a structure shaped by curving a grid to one single direction. Barrel vaults can be used in the form of single layer or double layer in different structures. In some structures barrel vaults are constructed without columns. They are placed in their corner edges connected to the supports.

Domes: Dome is a structure shaped by curving a grid to

two different directions. A dome's surface is usually a part of a sphere, cone or connection of multiple surfaces. Domes can be constructed in form of single layer, double layer or multi layer. The connections of domes are rigid, semi rigid or hinged.

Other Space Structures including: air field structures, folding structures, bridges, etc. [1, 2]

2. Capacity and Reflection Spectrum

2.1. Capacity

Total capacity of a structure depends on deformation capacity and strength of every single element of structure. To determine the post yield capacity of structure, nonlinear analysis such as pushover analysis is used. In this method, a series of consecutive elastic analysis is used to draw load-total displacement curve. In each step, the mathematical model is revised in a manner that the strength loss due to yielded elements is considered in the calculations. Vertical load distribution is increased repeatedly until one of the predefined limitations is reached.

2.2. Demand

Ground motion during an earthquake generates a complicated pattern which varies by time. It is not possible to consider these motions in each step to determine the demands of structural of design. The term "demand" usually mentioned as displacement demand, is the maximum expected response of a structure during an earthquake.

2.3. Performance

Having determined the capacity curve and displacement demand of structure, performance control is done. Performance control is carried out to assure that no damage is occurred in structural and nonstructural elements until achieving acceptable limitations of performance goal. [2, 3, 4]

3. Structural Capacity Determination and Capacity Spectrum Method

The nonlinear static analysis is used for determination of the structures capacity which is shown as a curve. In the capacity curve of a structure the vertical axis is structure's base shear and the horizontal axis is roof displacement. In this article this diagram is constructed of the structure's vertical earthquake force (vertical axis) versus the structure's vertical displacement (horizontal axis). It should be noted that capacity curve (which is usually generated to indicate the structure response in the first mode) is based on this assumption that the structure's main vibration mode is the dominant mode of response. This assumption is valid for the structures with one second main vibration mode or less, but with longer periods in very soft structures, the effects of next vibration modes should also be considered in analysis. [5, 6]

In this method, performance point of structure is determined by crossing the capacity spectrum and suitable demand response spectrum which is decreased due to nonlinear effects. [3, 4]

4. Determination of Performance Point Using Capacity Spectrum Method

Performance point indicates a condition in which structure capacity and seismic demand are equal. On this basis, the best method to determine performance point is crossing the capacity curve and the demand curve in which nonlinear effects are considered. The performance point is the crossing point of these two curves. It should be notice that the capacity curve is the base shear versus roof displacement and the demand curve is the spectral acceleration versus vibration period of the structure. In this paper the capacity curve is constructed by the vertical force of the earthquake versus the vertical maximum displacement and the demand curve is the spectral acceleration versus vibration period of the structure. Thus these two curves cannot be crossed in the same coordination system. Thus the demand and capacity curves are converted to acceleration response spectrum-displacement (ADRS) format. Then the curves are graphed in (S_a-S_d) coordinates to calculate the performance point. The calculation of performance point consists of the following steps:

4.1. Converting Capacity Curve to Capacity Spectrum

The capacity curve is constructed of the base shear axis versus roof displacement axis. The capacity spectrum expresses the capacity curve of the structure as the acceleration displacement-response spectrum (ADRS). This is shown as a diagram of S_a axis versus S_d axis. Each point on the capacity curve with (Δ_{roof},V) coordination is corresponded to (S_a,S_d) coordination on the capacity spectrum.

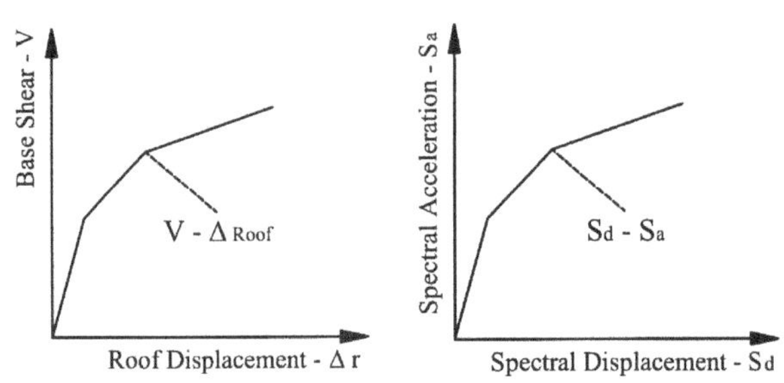

Figure 1. *Conversion of Structure's Capacity Curve to Demand Curve*

$$S_a = \frac{V/W}{\alpha_1} \tag{1}$$

$$S_d = \frac{\Delta_{roof}}{PF_1 * \phi_{roof,1}} \tag{2}$$

W: the building dead load plus a portion of the live load.
V/W: the base shear coefficient of the earthquake.

Δ_{roof}: the displacement in roof level.

$\Phi_{roof,1}$: the value of first mode shape in roof level.

α_1: the mode mass coefficient for the first natural mode of the structure.

PF_1: the mode collaboration coefficient for the first natural mode of the structure.

$$PF_1 = \left[\frac{\sum_{i=1}^{N} (W_i \phi_{i1})/g}{\sum_{i=1}^{N} (W_i \phi^2_{i1})/g} \right] \quad (3)$$

$$\alpha_1 = \frac{\left[\sum_{i=1}^{N} (W_i \phi_{i1})/g \right]^2}{\left[\sum_{i-1}^{N} W_i / g \right] \left[\sum_{i-1}^{N} (W_i \phi^2_{i1})/g \right]} \quad (4)$$

W_i/g: mass of the first mode in i^{th} level.

Φ_{i1}: the value of first mode shape in the ith level.

N: the total storey of the building.

i: the storey number building.

In this article the capacity curve is drawn by the vertical force of the earthquake versus the vertical displacement of the structure's center. Thus, the base shear force is transformed to the vertical displacement of the earthquake. The roof displacement is the vertical displacement of the structure's center, that is:

$$V \rightarrow \rightarrow P_{vertical} \quad (5)$$

$$\Delta_{roof} \rightarrow \rightarrow \Delta_{vertical} \quad (6)$$

Each point on the capacity curve with (Δ_{roof}, V) coordination is converted to $(\Delta_{vertical}, P_{vertical})$ and the corresponding point of (S_a, S_d) can be calculated using the formulas of next section.

$$(\Delta_{vertical}\text{-}P_{vertical}) \rightarrow \rightarrow (S_d\text{-}S_a) \quad (7)$$

Concerning the related references to this subject and the accepted calculations in space structures we have:

$$\alpha_1 = 1 \qquad PF_1 = 1 \qquad \Phi_{roof,1} = 1$$

Placing the coefficients and converting the parameters, the formulas in previous section are converted as below:

$$S_a = \frac{P_{vertical(i)}}{W_t} \quad (8)$$

$$S_d = \Delta_{vertical} \quad (9)$$

$P_{vertical(i)}$: the total applied force of vertical earthquake in each loading step.

W_t: the summation of the structure's total weight and the Dead and Live (Snow) forces applied to the structure.

$\Delta_{vertical}$: structure's middle vertical displacement due to total earthquake subjected to the structure.

4.2. Converting Demand Spectrum to ADRS Format

Standard response spectrum is drawn by spectrum acceleration (S_a) versus the structure's period and needs to be converted to the ADRS format. For each point of (T, S_a) on the standard spectrum, the spectrum displacement of the corresponding point on the demand spectrum in ADRS format is equal to:

$$S_d = \left[\frac{T}{2\pi} \right]^2 S_a \quad (10)$$

$$(T\text{-}Sa) \rightarrow \rightarrow (Sd\text{-}Sa) \quad (11)$$

Since there are not international valid codes for space structure based on available references, the 2/3 or 0.667 coefficient in earthquake codes components for frame structures can be used to define the demand spectrum. For this purpose, no changes are made to the horizontal axis components but the vertical components obtain by the code's terms is multiplied by 0.667 and then the result number is shown on the diagram.

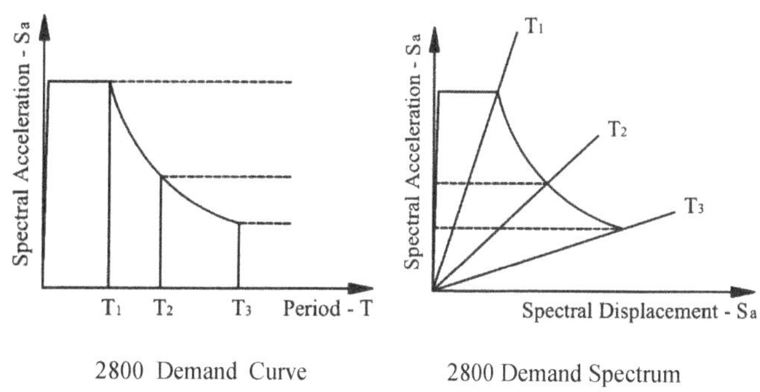

Figure 2. *Conversion of Structure's Demand Curve to Capacity Curve in 2800 Code*

To convert the earthquake acceleration response spectrum of 2800 standard design into ADRS format, the following formulas can be used:

$$0 \le T \prec T_0 \Rightarrow \begin{cases} \frac{S_a}{g} = AI(1 + S(T/T_0)) \\ S_d = AIg(1 + S(T/T_0)(T/2\pi)^2) \end{cases} \quad (12)$$

$$T_0 \le T \prec T_S \Rightarrow \begin{cases} \frac{S_a}{g} = AI(1 + S) \\ S_d = AIg(1 + S)(T/2\pi)^2 \end{cases} \quad (13)$$

$$T_S \le T \Rightarrow \begin{cases} \frac{S_a}{g} = AI(1 + S)(T_S/T)^{2/3} \\ S_d = AIg(1 + S)(T_S/T)^{2/3}(T/2\pi)^2 \end{cases} \quad (14)$$

4.3. Making a Bilinear Form for Capacity Spectrum

A bilinear form for capacity spectrum should be graphed to estimate the effective damping and consequently determining the suitable decreased demand spectrum. A point coordinated at (d_{pi}, a_{pi}) should be determined to construct a bilinear model.

To generate a bilinear form of capacity spectrum, first a straight line with a slope equal to the structure's elastic stiffness is drawn so that passes the origin. The second line is drawn from (d_{pi},a_{pi}) point so that crossing the first line at the point (d_{pi},a_{pi}), two approximately equal areas are generated between these lines and the capacity spectrum.

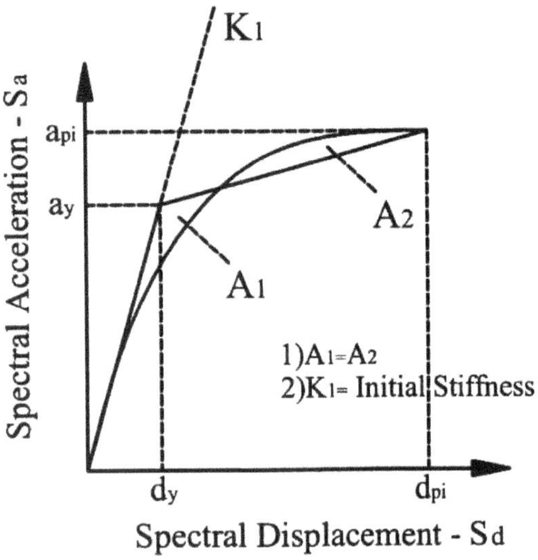

Figure 3. *Bilinear Form for Capacity Spectrum*

4.4. Approximating the Equivalent Damping and Equivalent Viscous

When a structure enters the nonlinear region due to earthquake force, its damping can be defined by a combination of natural viscous damping and hysteresis damping. Hysteresis damping is proportional to the area under curve of structure force-displacement curve. This damping can be stated in the equivalent viscosity damping (β_{eq}).

$$\beta_{eq} = \beta_0 \pm 0.05 \qquad (15)$$

β_0: the stated hysteresis damping as the equivalent viscosity damping is 5 percent of the structure's natural viscosity damping and is constant to the end of analysis.

$$\beta_0 = \left(\frac{1}{4\pi}\right)\left(\frac{E_D}{E_{S0}}\right) \qquad (16)$$

E_D: the amount of energy absorbed.
E_{S0}: the maximum strain energy.

$$E_D = 4*\left(a_y d_m - d_y a_{pi}\right) \qquad (17)$$

$$E_{S0} = a_{pi} d_m / 2 \qquad (18)$$

$$\beta_0 = \frac{63.7(a_y d_{pi} - d_y a_{pi})}{a_{pi} d_{pi}} \qquad (19)$$

$$\beta_{eq} = \beta_0 + 5 = \frac{63.7(a_y d_{pi} - d_y a_{pi})}{a_{pi} d_{pi}} + 5 \qquad (20)$$

The ideal hysteresis curve (parallelogram) is an acceptable estimation of the actual hysteresis curve of a flexible building that has been affected by the relatively short term earthquakes and has an equivalent viscosity damping less than 30 percent. For other conditions the ideal hysteresis curve results in over estimations for equivalent viscosity damping. This is because of that the actual hysteresis loops are practically incomplete due to phenomena such as buckling, fatigue, resistance drop or stiffness. The confined areas by these loops decrease in consequent cycles. By applying the concept of effective viscosity damping, we use damping modification factor (K) to enable damping coefficients (β) for simulating the incomplete hysteresis loops.

$$\beta_{eff} = K\beta_0 + 5 = k\frac{63.7(a_y d_{pi} - d_y a_{pi})}{a_{pi} d_{pi}} + 5 \qquad (21)$$

K: the modification factor that is used for converting equivalent viscosity damping in the ideal hysteresis curve to the corresponding equivalent viscosity damping in actual hysteresis curve.

K factor depends on the structure's actual behavior which in turn depends on quality of building lateral resistant system and the duration of earthquake. Three different type of structure behavior is concerned in ATC40 code:

Structure behavior type A: Indicates stable hysteresis loops. In this behavior the hysteresis loops are stable in all cycle and are very similar to ideal shape (parallelogram).

Structure behavior type B: Indicates relatively incomplete hysteresis loops. In this behavior the average area of hysteresis loops decrease.

Structure behavior type C: Indicates incomplete hysteresis loops. In this behavior involves significant decrease in the loops area in the consequent cycles.

Base on the above descriptions, the space structures behavior belongs to type B.

Table 1. *Modification Coefficients of Structural Behavior*

K	β_0	Structure Behavior Type
1.00	$\leq 16.25\%$	A
$1.130 - 0.008\beta_0$	$\succ 16.25\%$	
0.67	$\leq 25\%$	B
$0.845 - 0.007\beta_0$	$\succ 25\%$	
0.33	Other	C

4.5. Reduction of Demand Spectrum of 5% Damping

We consider β_{eff} as the effective viscous damping equivalent to the energy dissipation due to 5% structure's natural viscous damping and permanent plastic deformations (hysteretic deformations) up to the performance point. Therefore, the design elastic spectrum of 5% damping is decreased to the demand spectrum corresponding to β_{eff} damping. According to ATC40 bylaw, these portions are specified by SR_A and SR_V factors respectively which are called spectrum decrement factors:

$$SR_A = \frac{3.21 - 0.68Ln(\beta_{eff})}{2.12} \geq (SR_A)_{min} \quad (22)$$

$$SR_V = \frac{2.31 - 0.41Ln(\beta_{eff})}{1.65} \geq (SR_V)_{min} \quad (23)$$

$(SR_A)_{min}$ and $(SR_V)_{min}$, the minimum permissible values of spectrum decrement factor, are:

***Table 2.** Permissible Values of Spectrum Decrement Factors*

$(SR_V)_{min}$	$(SR_A)_{min}$	Structure Behavior Type
0.50	0.33	A
0.56	0.44	B
0.67	0.56	C

In most of space structures the spectrum reduction factors in both cases is more than 1 or so close to 1. Therefore, it can be concluded that spectrum reduction for most of space structures is not necessary and the crossing point of the demand spectrum and the capacity spectrum is the structure's performance point.

4.6. Standard Design Spectrum of Decreased New 2800 Standard in ADRS Format

If the space reduction factors are less than 1, it is necessary to apply damping effects on the structure. Based on the spectrum reduction factors SR_A and SR_V, the standard reduced design spectrum of Iran 2800 code is equal to:

$$S_a = S_a \times \begin{cases} SR_A \Rightarrow T \prec T_S \\ SR_V \Rightarrow T \geq T_S \end{cases} \quad (24)$$

Based on Euro code 8, in absence of precise studies in a region, the horizontal response spectrum and the horizontal earth acceleration are multiplied by 2/3 or 0.667 to calculate the corresponding vertical values.

4.7. Determination of Structure Performance Point

The most basic method to determine performance point position is to assume a performance point on the capacity spectrum. The structure's hypothetical β_{eff} is calculated based on the effective viscosity damping of the structure's natural viscosity. Then SR_V and SR_A are obtained. Finally the 5% damping elastic demand spectrum is reduced. This hypothetical point is the performance point if in an acceptable tolerance is passed by the reduced demand spectrum. If the spectrum acceleration on reduced demand spectrum is less than the spectrum acceleration on capacity spectrum, for spectral displacement of the hypothetical performance point, then the demand is more than the capacity of the structure. Therefore, the spectral displacement has not reached to its maximum level. For another try a performance point with a spectral displacement less than the previous assumed spectral displacement is chosen. This try and error is repeated until the reduced demand spectrum affected by the effecting viscosity damping of the hypothetical performance point passes the hypothetical performance point by an acceptable tolerance. In this case hypothetical performance point is the actual performance point of the structure. If the crossing point of the demand spectrum and the capacity spectrum is placed in the linear region of the spectrum, the displacement indicates the actual displacement of the structure. The acceptable tolerance of the performance point based on ATC40 code is 5%. It means that if the differences between the primary assumed point in the first steps with the crossing point is less than 5%, then obtained point is the structure performance point. Elsewhere try and error is continuously repeated. [3, 4, 6, 7, 8, 12, 13, 14, 15]

5. Different Levels of Structure Performance

To define the structure's performance, it is necessary to determine the allowable amount of damage caused by the earthquake level. The building behavior level is a combination of the behavior level of structural elements (beam, column, etc.) and non-structural elements (partition wall, interior finishing, windows, glasses and ...). According to this, a number of major performance levels are defined in FEMA and ATC40 codes:

1. Performance level of Immediate Occupancy (continuous use) (IO)
2. Life Safety Performance Level (LS)
3. Structural Stability Performance Level or Collapse Prevention (CP)

In seismic retrofitting of existing buildings code in Iran, Performance Levels are classified as below:

1. Continuous Service Performance Level (A-1)
2. Continuous Usability Performance Level (B-1)
3. Life Safety Performance Level (C-3)
4. Collapse Threshold Performance Level (E-5)

6. Acceptance Criteria in the Life Safety Performance Level

Two quantities are controlled to evaluate life safety performance level of the structures:

A) Plastic Deformation (resistance and flexibility control to study structural damages)

B) Story's Displacement (stiffness control to evaluate non-structural damages)

Allowed values for mentioned quantities are defined for steel and concrete structures and generally for all frame structures as boundaries of different performance levels in FEMA273 and ATC40 codes. In space structure, there is not a valid international code for desired problems. Thus, we cannot refer to any source for boundaries of different performance levels to be comparing the performance points with. In this paper, we have used a series of suggested definitions, some of which are extracted from international codes and some are used by changes in their application.

1. Displacement between 0 to 1/300 of span's length as non-stop use boundary
2. Between 1/300 of span's length to formation place of the

first plastic hinge as life safety boundary

3. From the first plastic hinge to 0.02 span's length as collapse threshold boundary

4. From 0.02 span's length afterwards as collapse boundary

Displacement for non-stop use boundary is: $0 \leq \Delta_{ver} \prec L/300$

Displacement for life safety boundary is: $L/300 \leq \Delta_{ver} \prec \Delta_{Pl}$

Displacement for collapse threshold boundary is: $\Delta_{Pl} \leq \Delta_{ver} \prec 0.02L$

Displacement for collapse boundary is: $0.02L \leq \Delta_{ver}$

L: structure's total length.

Δ_{ver}: maximum vertical displacement of structure's middle subjected to vertical earthquake load.

Δ_{Pl}: first vertical displacement of plastic hinge subjected to vertical earthquake load. [13, 14, 15]

7. Examination of Studied Samples

The analyzed models are single layer grid structure that plane elements shape of which are square (Shape A & B) or square-diagonal (Shape C & D). The length of every square members are assumed 1 meter. All connections are supposed rigid to maintain the stability structures.

In order to transmit the roof forces into foundation of structure, 2 kinds of supports are used which are actually the connection of column and roof.

Considering two 1 meter consul at each side, the total length of structure is supposed between 10 to 15 meters. The reason of selecting these spans is that single layer plane structures are usually used in structures with 10 to 15 meters span length and it is also economically reasonable. Since it is important in space structures to maintain their stability, all the connections in these structures are assumed to be rigid. Also, pipe shaped sections are used in these structures.

However the applied loads subjected to the space structures studied in this paper are:

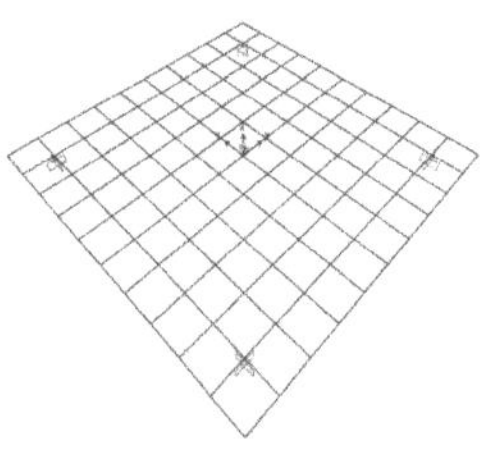

Figure 4. *Shape (A)*

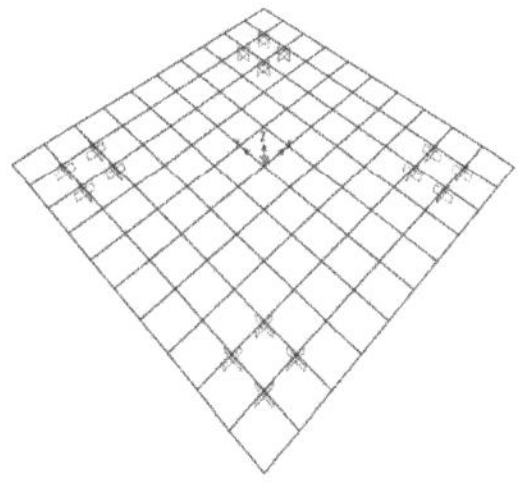

Figure 5. *Shape (B)*

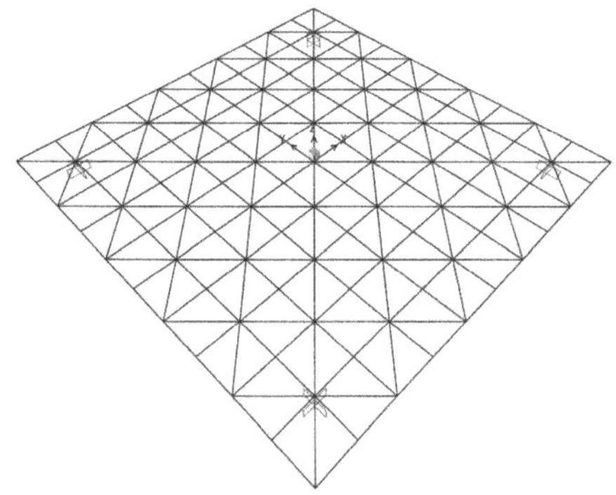

Figure 6. *Shape (C)*

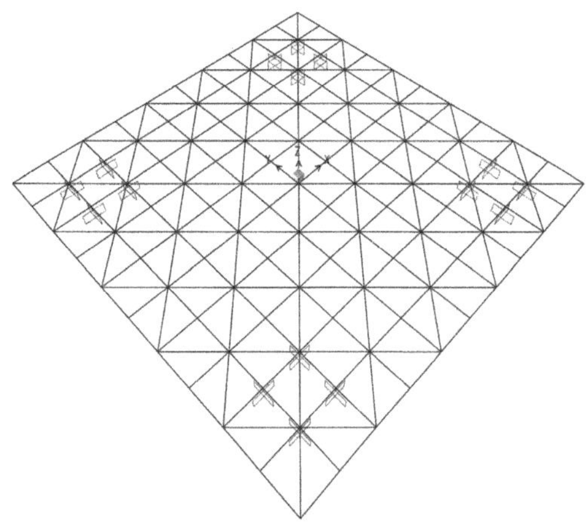

Figure 7. *Shape (D)*

7.1. Dead Loads

Surface unit weight of dead load for these structures is 50 kilograms per square meter and for applying it to the structure it should be converted to point load.

7.2. Live Loads

Since this type of structures is usually without roof or has a light non-structural coverage, the only live load subjected to these structures is snow load. Its load amount is extracted from the code according to the type of structure and place of project. In this paper live load (snow load) is assumed uniform and equal to 150 kilograms per square meter which should be applied to the structure as pint load in nodes.

7.3. Vertical Earthquake Load

This load is obtained by product of a coefficient by the structure's weight. It is applied to the structure locally in the structure's nodes. In this paper the structure's weight is assumed as the summation of dead loads and total live load.

After analysis, designing and optimizing the structures using SAP2000, we can have the following results to use in ANSYS software:

1. Structure's total weight
2. Structure's different sections specifications

Now, using SAP2000 software output and modeling the

structure in ANSYS software, we take none linear static analysis. In ANSYS software, three node BEAM189 elements are used for modeling the structures.

These elements have the rigidity condition of structure's connection elements. This analysis may be ended in one of these two ways:

1. Repeat the force increment until first point of destruction or instability of the structure.
2. Repeat the force increment until a point with specific condition.

In this paper we use the first way (point of destruction or instability of the structure) because collapse point or destruction point of the structure must be determined and we should find the maximum vertical displacement of the structure.

In this section, the diagrams and calculation steps for determining the performance point of a single-layer structure of size 10*10 meters is presented for instance. Since the parameters such as A (design base acceleration), I (structure importance factor), S (depends on the soil and condition of the region) are very important in determination of S_a, these parameters are assumed to be constant in all of the models. In this manner, the response of structure analysis is only depends on the structure itself. In below diagrams, the process of determining the performance point is presented. It is assumed that A=0.3, I=1.2 and region type=2)

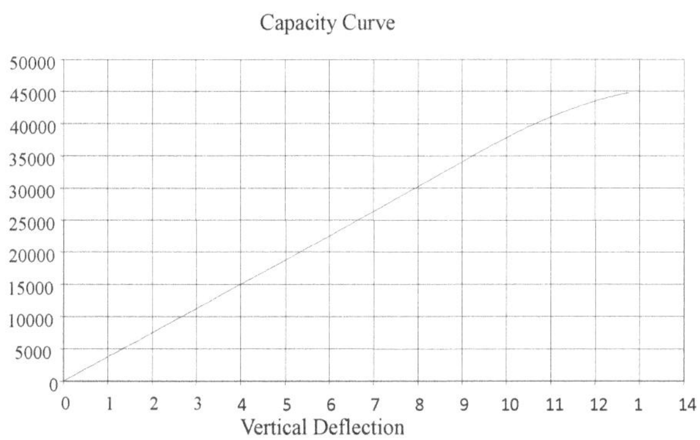

Figure 8. *Capacity Curve–Vertical Load*

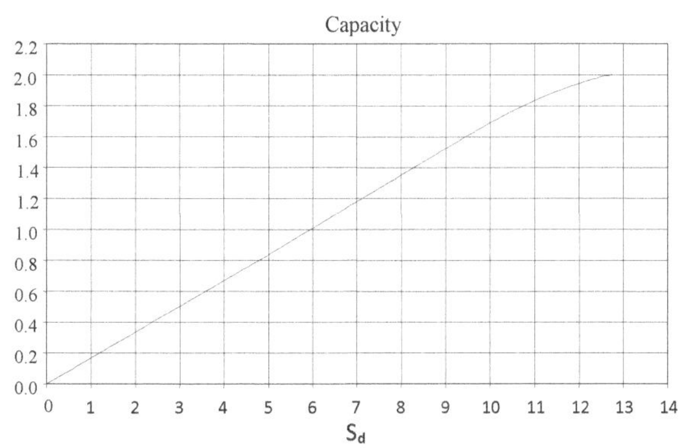

Figure 9. *S_d–S_a/g for Capacity Spectrum*

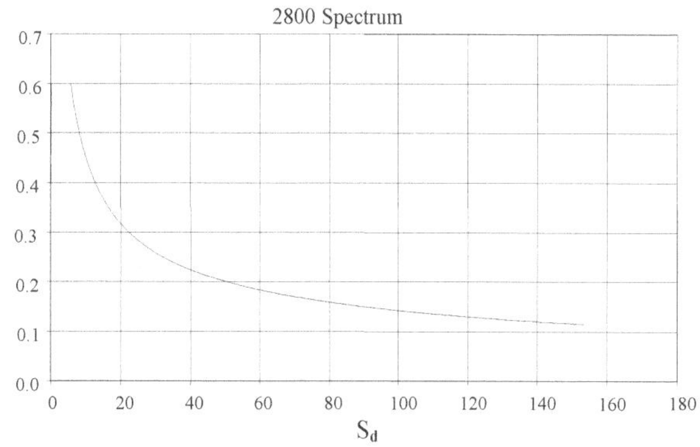

Figure 10. *S_d–S_a/g for 2800 Spectrum*

We hold the process for all considered structures with length of 10 to 15 meters and finally we find their performance points after drawing the needed diagrams.

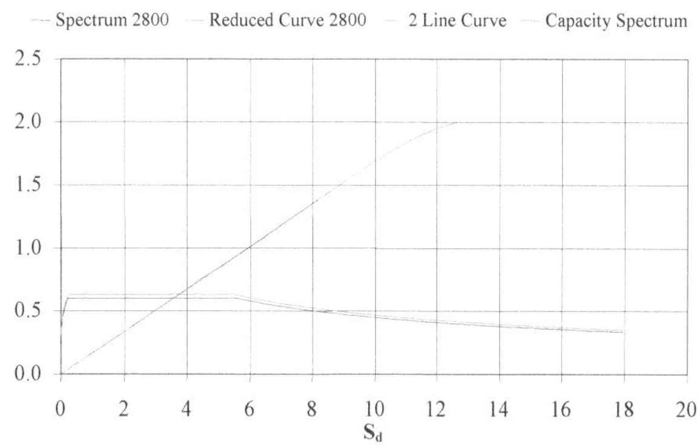

Figure 11. *S_d–S_a/g for ADRS format*

8. Discussion and Conclusion

Using the curves and the results of single layer plane space structure analyses, it can be concluded that the performance point in spans of about 15 meters are placed in life safety level and in small spans are placed in immediate occupancy level. In these structures, S_d is increased slightly by increasing the span length.

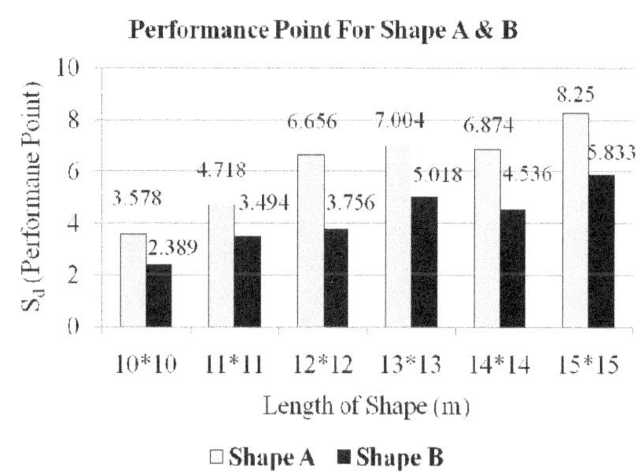

Figure 12. *Performance point for shape (A, B)*

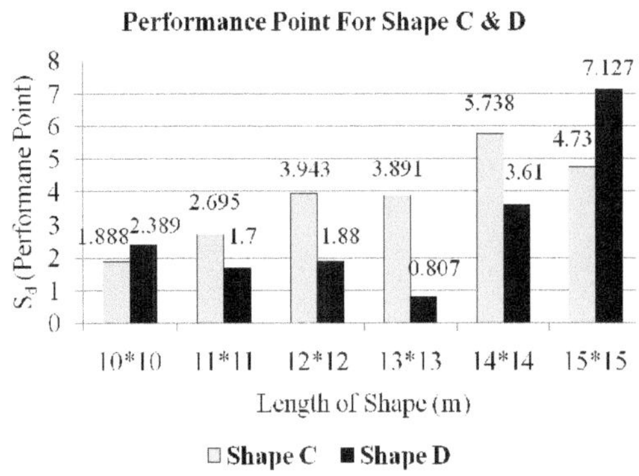

Figure 13. *Performance point for shape (C, D)*

Single supports in structure with square elements can perform better and stronger than in quadrate supports shape.

In structure with square-diagonal elements, excluding 10 and 15 meters spans, structure with single supports can perform better and stronger than structure with quadrate supports.

It seems that the performance can be better with fewer and smaller supports in single layer plane structure applying this type of connection and short length of spans.

References

[1] H. Nooshin, "Space Structure and Configuration Processing," Structural Engineering and Material, Vol. 1, No. 3, pp. 329-336, 1998.

[2] H. Nooshin, Z.S. Makowski, "Some Aspects of the Behavior of Square Diagonal Grids," International Journal of Building Science, pp. 267-275, Des 1974.

[3] M. Layegh Rafat, D. Layegh Rafat, M. Ghalehnovi, "The Examination of Performance Level of Barrel Vaults Space Structure by Capacity Spectrum Method," 1st International Conference on Seismic Retrofitting, Iran, Tabriz, 2008.

[4] P. Faifar, "Capacity spectrum method based on inelastic demand spectra," Earthquake engineering and structural dynamics, Vol. 28, pp. 979-993, 1999.

[5] A.K. Chopra, R.K. Goel, "A model pushover procedure to estimate seismic demands for building," Report No. PEER-2001/03, 1999.

[6] R.B. Malla, L. Serret, "Double Layer Grids: Review of Static and Thermal Analysis Methods," Journal of Structure Engineering, Vol. 122, NO. 8, 1996.

[7] R.B. Malla, L. Serret, "Double Layer Grids: Review of Dynamic Analysis Methods," Journal of Structure Engineering, Vol. 122, NO. 8, 1996.

[8] H.A. Moghaddam, "Seismic Behavior of Space Structure," International Journal of Space Structure, Vol. 15, No. 2, pp. 119-134, 2000.

[9] Sh. Malek, S.M. Ardakani, "Study Seismic Behavior of Double Layer Barrel Vaults Space Structure," 1st International Conference on Seismic Retrofitting, Iran, Tabriz, 2008.

[10] R. Kamali Salehabad, E. Salajegheh, "The Provision of Vertical Reflection Spectrum and Usage of them for Space Structure," 2nd National Conference on Spatial Structure, Iran, Tehran, 2007.

[11] G. Victor, "Instability Problems in Space Structure," Elsevier Applied Science Publisher, Vol. 7, No. 4, pp. 321-331, 1985.

[12] "Euro Code8: Design Provision for Earthquake Resistance of Structure – Part 1-1: General Rules-Seismic Action and General Requirements for Structure," European Committee for Standardization (CEN), 2002.

[13] FEMA273 (Federal Emergency Management Agency), Washington D.C, USA, Oct 1997.

[14] ATC40 (Applied Technology Council), Redwood City, California, USA, Nov 1996.

[15] "Instruction for Seismic Rehabilitation of Existing Buildings," Management and Planning Organization of Islamic Republic of Iran, No 360, 2007.

A New Approach to Teach Mechanics of Material Through Touching Experience

Yao George C.[1], Kuo Keng-Chang[2], Hsueh Cheng-Luen[3], Tang Yao[4]

[1]Department of Architecture, National Cheng Kung University, Tainan, Taiwan

[2]Department of Construction Engineering, National Kaohsiung First University of Science and Technology, Kaohsiung, Taiwan

[3]Department of Architecture, National Cheng Kung University, Tainan, Taiwan

[4]Institute of Education, National Cheng Kung University, Tainan, Taiwan

Email address:

gcyao@mail.ncku.edu.tw (Yao George C.), kkuo@nkfust.edu.tw (Kuo Keng-Chang), clhsueh@mail.ncku.edu.tw (Hsueh Cheng-Luen), alextang@mail.ncku.edu.tw (Tang Yao)

[*]Corresponding author

Abstract: Recent advancement in building structural analysis by computer has changed the structural analysis work in the building professional. Nowadays, professionals can read output from the computers and will need a good structural concept to judge the validity of the analysis and to verify the calculated result. This change of structural analysis work sequence prompted us to reconsider the pedagogy in the Strength of Material class taught in the Architecture schools at the university level. This paper presents a new pedagogy in teaching Strength of Material class in the Department of Architecture at National Cheng Kung University in Taiwan. This new pedagogy is experiment-oriented in teaching different units of the Strength of Material in contrast to the math-oriented approach popularly used in most of the Engineering schools. New experiments emphasize on visually observable mechanical behavior so that students can visually inspect the deformation of members when subjected to different types of loadings. Three modules, they are lecture, experiment, and public presentation, are designed into the weekly class to enhance learning efficiency.

Keywords: Experiment-Oriented, Structural Concept, Mechanics of Materials, Architecture

1. Introduction

The exterior form and structure of a building act like the two sides of a coin. In designing innovative spaces, the structure serves both as a pivotal agent of imagination and ingredient for the creation of beauty [1]. As famously stated by an esteemed Japanese structural engineer Tsuboi Yoshikatsu [2] "The beauty of architecture comes hand in hand with a rational structural system." For this reason, imparting an understanding of structural concepts has been an integral part of the architectural curriculum at universities worldwide.

The planning of teaching material for structural courses at architectural schools in Taiwan has always proven a challenge. Instructors of structural courses often struggle to design appropriate course content for mechanics that can familiarize students with structural concepts. Conventionally, structural courses have been delivered as class presentations. For each topic, the instructor applies mathematical models to work out the forces and deformation or the relationship between stresses and strain, from which equations are derived to calculate structural reactions such as deformation or cross-sectional strengths. This approach is pedagogically ideal for rigorous and comprehensive knowledge systems. The drawback is its unilateral nature makes it hard to engage students, while the lack of stimulus is also not conducive to drawing their attention.

There are other reasons the conventional approach for teaching structural mechanics at engineering schools is at odds with the current educational environment at Taiwanese architectural schools:

1) A heavily loaded design curriculum occupies most of students' energy, leaving them little room to study engineering courses and therefore limiting the depth of the content that can reasonably be taught in structural courses.

2) Structural courses at conventional engineering schools often use mathematical models to explain mechanical behavior, so students with an insufficient foundation in mathematics have a hard time using commonly available textbooks.

3) There is a gap between the structural competency required by students in architectural design and the equations they learn at structural mechanics courses during their junior years, therefore diminishing their interest in studying structural courses.

In an era where computer-aided structural analysis is readily used throughout all stages of design [3], we find the mathematically equations mentioned above practically irrelevant in real-life structural calculations for designers, because all the computing can be accomplished by software. What design practitioners need, given all the computed data from structural analysis, is a strong understanding of structural concepts to determine the rationality of the results and verify the basic assumptions of the model.

As such, the latest pedagogical philosophy in structural engineering is to equip students with sufficient structural concepts, rather than teaching them how to derive equations. [4] The same idea can be applied to structural courses in architectural education. Recently, different approaches are developed for this goal in different architectural school based on the needs of their students. [5] This study turns away from the teaching model that builds structural concepts through equation-driven teaching and substantial experience in structural calculations. Instead, we apply the idea of "seeing and touching structural concepts", proposed by University of Manchester professor Bell [6], in developing a series of visually accessible syllabuses for teaching structural models. Rather than learning structural concepts through conventional mathematical tools, students shape their understanding by engaging in visible structural maneuvers complemented by in-class explanations of real-life examples in applying equations. This pedagogy utilizes a sensory approach, i.e. vision, to intrigue students and develop their competence. A similar approach has been adopted in structural courses at some civil engineering faculties [7], demonstrating its growing acceptance at various faculties in universities.

2. Syllabus

The compulsory structural mechanics courses in the four-year architecture program at National Cheng Kung University (NCKU) are as follows: structural behavior (Year 1 Term 2), strength of materials (Year 2 Term 1), structures (Year 2 Term 2), structural systems (Year 3 Term 1), reinforced concrete and steel Design (Year 3 Term 2). Strength of materials appears to present the most challenge to students of all the courses. This course introduces the deformation of members and relationships between internal stresses and strains when external forces are applied.

Because the content is critical for familiarizing students with concepts in structural mechanics, we developed a new approach to teach this class by the theory introduced above. This study initially targets a few key units in strength of materials to design a set of simple and interesting experiments that students can operate on their own (Table 1). They can then familiarize themselves with the relationship between forces and deformation by actually seeing the deformation. Since the deformation is visible to the eye, students take away a strong mental impression. Then, examples in engineering applications of the equations involved in the chapters are discussed in class with the help of textbooks, captivating students with practical applications. Finally, students are asked to analyze and review the experiment data in order to reinforce their understanding [8, 9].

Table 1. *Lab units in the new strength of materials class.*

Unit	Topic	Unit	Topic
1	Internal forces, external forces and stresses	8	Horizontal shear stresses
2	Modulus of elasticity E	9	Bending moment experiment
3	Shear and shear strain	10	Cantilever beam: span and deflection at the center
4	Creep	11	Deflection of different connections
5	Plane stresses	12	Buckling
6	Torsional stiffness and cross-sectional shapes	13	BRB (Buckling Restrained Brace)
7	Torsional failure characteristics	14	LTB (Lateral Torsional Buckling)

The study materials comprise three modules as detailed below:

1) Lecture: The contents of the lectures are taught by the instructor. Focus is placed on the engineering application of equations already derived in conventional mechanics, such as external forces and deformation. Take creep for example: real-life engineering failures caused by excessive creep in chemical anchors and the serious resulting damage are used to supplement explanations on the common characteristics of creep. (Fig. 1)

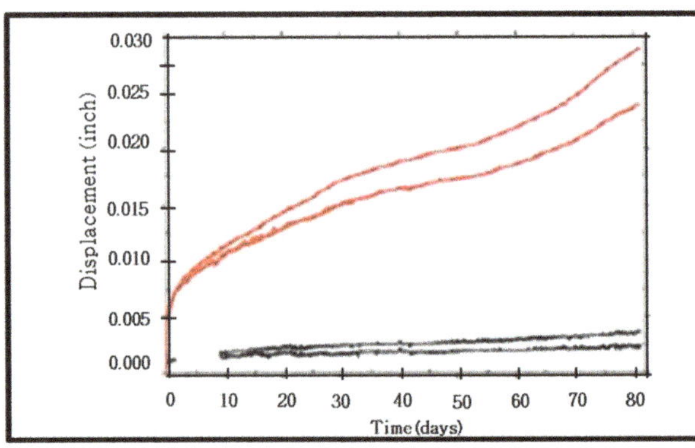

(a) Creep in different adhesives

(b) Damage to hanger due to adhesive failure

Figure 1. *Engineering failure caused by creep in adhesives.*

2) Laboratory: In the lab module, the teaching assistant leads students in conducting experiments, with a focus on deformation that is visible or can be photographed with a camera or measured with simple equipment. The teaching assistant must design the experiment according to the objective of the unit and select appropriate materials in order to demonstrate the characteristics covered in the unit. Take for example an experiment on buckling: we have designed an experiment using a plastic ruler and an electronic scale, where students can read the load values causing the buckling as they apply pressure to the top of the ruler (Fig. 2). An intermediate support is then added to stimulate higher-mode buckling, allowing students to actually measure and see the higher load values required in the second-mode buckling [10].

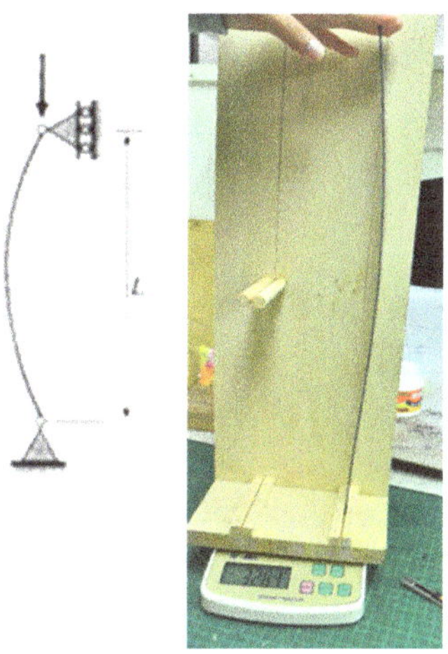

(a) Basic buckling

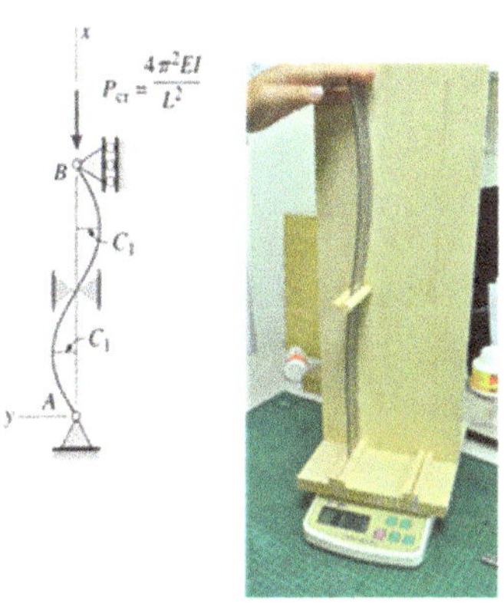

(b) Second mode buckling

Figure 2. *Experiment on buckling load.*

3) Public Presentation: Students conduct their own experiments, record data, analyze and compile reports in this module. Focus is placed on the processing of experiment data, discussing equations covered in the textbooks and the accuracy of the experiments, and examining possible causes of errors. Aside from submitting weekly reports, two teams of students must make an oral presentation followed by the instructor's comments and questions from their peers.

Hence, the course seeks to familiarize students with structural concepts by combining theoretical understanding and experiential learning. The textbooks and lab component for two units will be illustrated in the following chapter.

3. Examples and Analysis

1) Bending moments and deflection in beams

Since the relationship between bending moments and stress are is very important, key equations such as those for bending moments (Fig. 3a) and common deformation equation (Fig. 3b) will be explained to students first. As the main objective is to help them understand the close relationship between area moment of inertia and bending moment, it may be further explained in the follow-up examples that a deeper section should be used where the structure carries greater bending moment. This ensures efficient use of the section and reduced deformation (Fig. 4). [11]

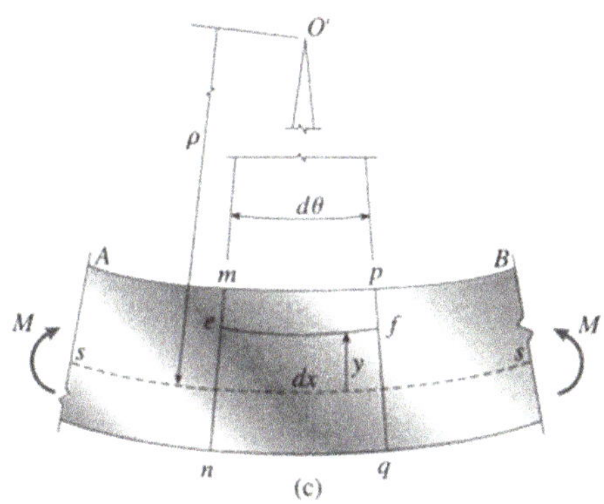

$$\sigma_x = \frac{M_z \cdot y}{I_z}$$

$$\sigma_x = E \cdot \varepsilon_x = (-1) \cdot E \cdot k \cdot y$$

(a) Relationship between stress and external forces

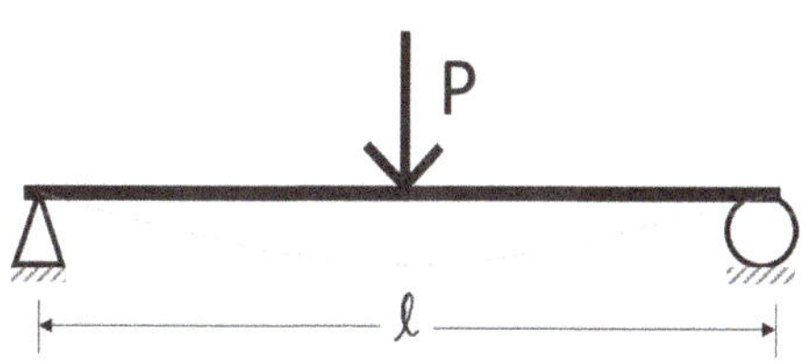

$$\delta_c = \frac{P\ell^3}{48EI}$$

(b) Equation for simple beam deformation

Figure 3. *Examples of equations and relationships for bending moment.*

Figure 4. *Examples of designing for bending moment --a deeper cross section is used on the cantilevered structure where the bending moment is greater.*

The lab component for this unit is divided into two parts. Part 1 compares the I values and deflections at center for different cross sections of beams and then compare them with theoretical stress values. Part 2 calculates the radius of curvature for bending deformation. The experiment in Part 1 compares two members with different cross sections (Fig. 5a). By actually measuring the deformation of simply supported members built of balsa wood and under loading (Fig. 5b), students gain an insight into the close relationship between members with different moments of inertia and their deformation. To facilitate measurement, we used a non-contact laser displacement meter to achieve higher resolution.

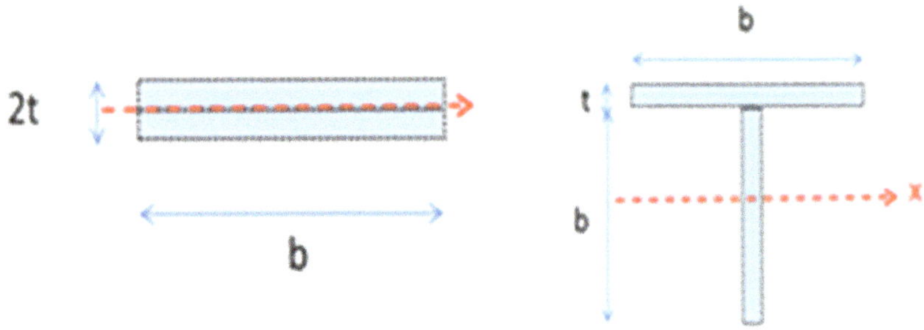

(a) Members with different cross sections

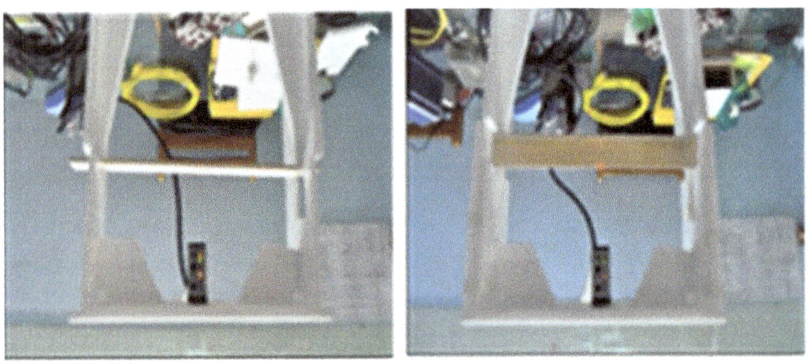

(b) Measuring deformation under equal loads

Figure 5. *Relationship between the moment of inertia and deformation of members.*

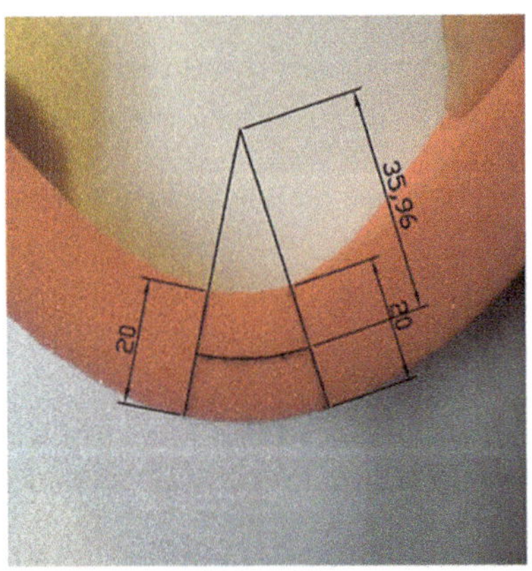

$$\varepsilon_x = \frac{y}{\rho} = \frac{-(20 \div 2)}{35.96} = -0.2781$$

(a) Calculating strain by measuring the deformation (curve radius)

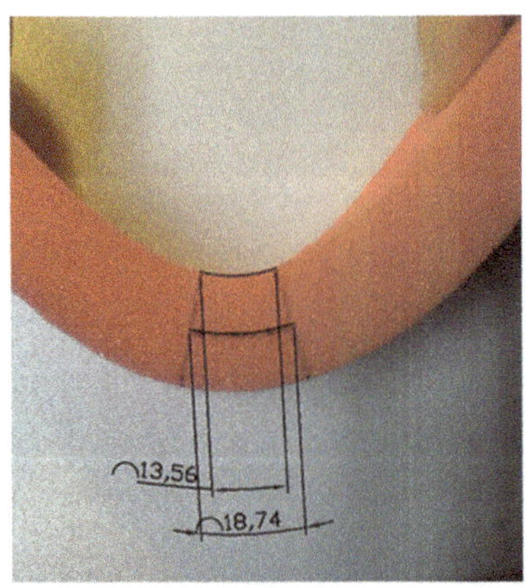

$$\varepsilon_x{}' = \frac{13.56 - 18.74}{18.74} = -0.2764$$

(b) Measuring the strain directly

Figure 6. *Experiment on the bending deformation of a sample sponge.*

In the experiment in Part 2, the central axis is drawn on a piece of bendable sponge; the test sample is bent with hands, and the photographs are imported into AutoCAD for processing and analysis. Students find the radius of curvature, ρ, and then calculate the strain at the top surface of the member after compression, using the equation (Fig. 6a). They then sketch the curve of the top surface, measure its actual length and then find the strain. Finally, they calculate the error between the number they get from the equation and actual measurement. (Fig. 6b)

2) Horizontal shear in beams

Shear acting on a beam includes vertical and horizontal shear stresses. The main objective of this unit is to help students understand the relationship between these two types of shear stresses. Therefore, the cause of horizontal shear stresses will be explained first (Fig. 7) and the distribution of shear stresses on a rectangular beam and the corresponding equation will be detailed.

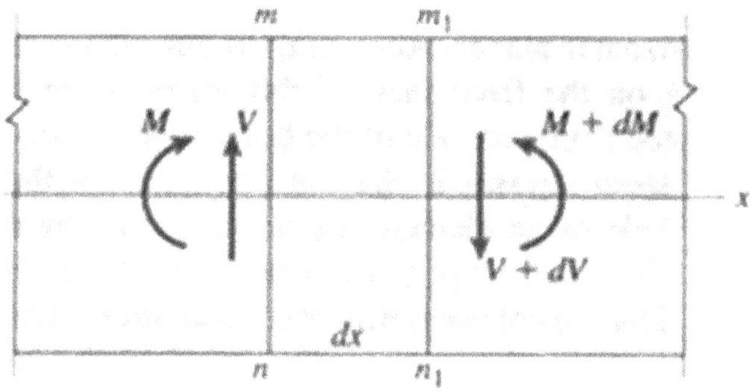

$$\tau = \frac{V \cdot Q}{I \cdot b}$$

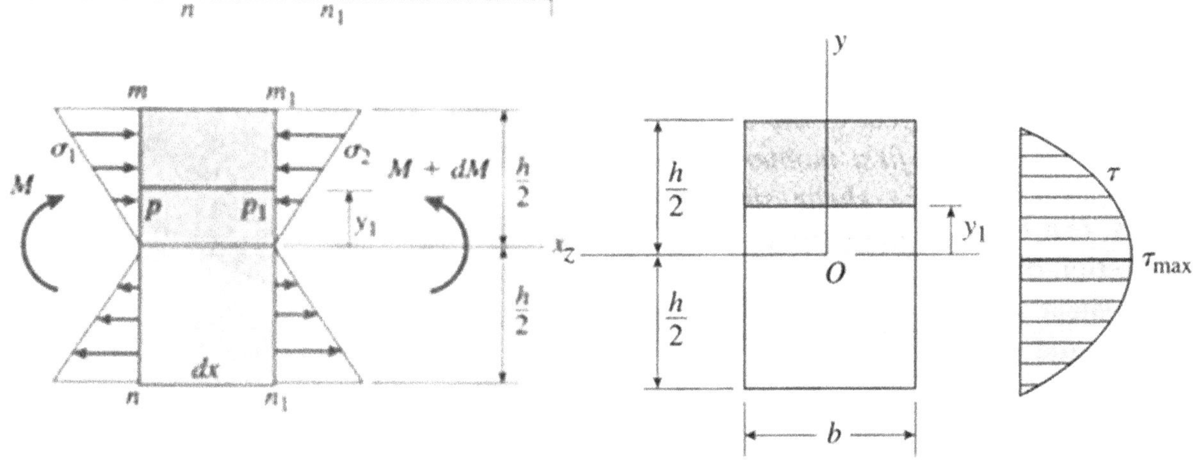

Figure 7. Introduction of the theory on horizontal shear stresses.

The engineering example used in this unit is shear design for steel I-beams. The reason why the web area of the cross section is used in the design of most steel beams (Fig. 8a) and the role played by shear stud is explained (Fig. 8b)

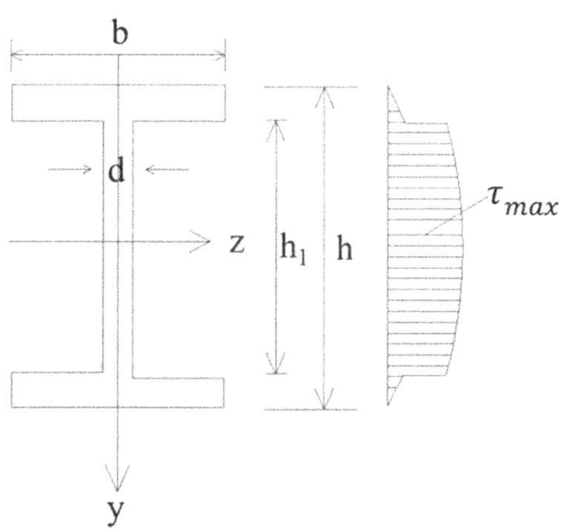

(a) Distribution of horizontal and vertical shear stresses

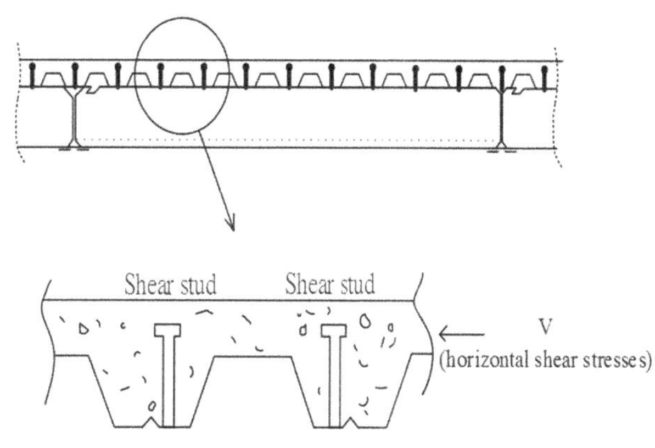

(b) Shear studs resist horizontal shear stresses

Figure 8. *Engineering example of horizontal shear stresses.*

This unit includes two experiments. The first aims to help students realize horizontal shear stresses cause relative movement and result in relative deformation if there is not a sufficient resisting mechanism. A simple beam structure made from two erasers is loaded at center to generate deformation. Erasers are used because their low stiffness results in visible deformation. (Fig. 9)

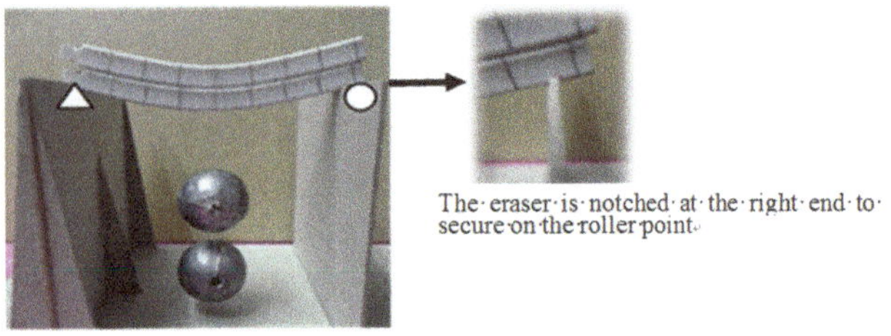

Figure 9. Horizontal shear causing deformation.

In the second experiment, foam board is used to make two sets of double-layered beams. One set is fastened with screws that act as shear studs, and the other set has no screws (Fig. 10). Equal loads are applied to the center of both sets, students observe the deflection of the center points of both beams, and deformation on both types of beams is explained.

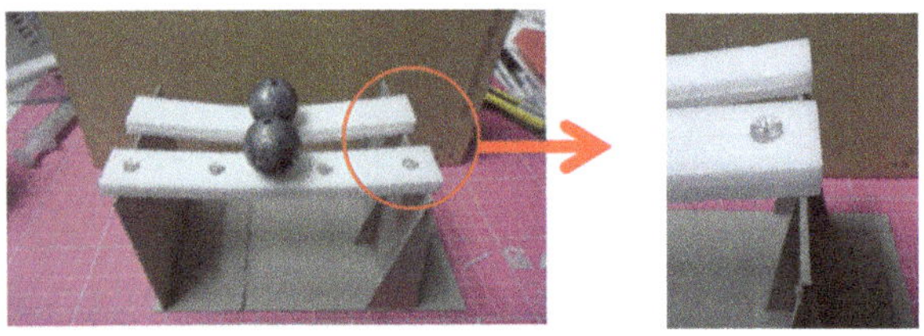

Figure 10. Comparing deformation between plates with and without connections.

4. Conclusions

Strength of materials is a very important structural course in the architectural curriculum. It is critical in familiarizing students with structural concepts because different types of external forces (e.g. axial force, bending moment and torque), member deformation and the relationships between internal forces are introduced. However, the traditional pedagogy of explaining mechanical behavior using mathematics often impedes architectural students' understanding of the underlying concepts in mechanics. Thus, this study proposes using experiments with visible models to help achieve this goal.

The lab content for 14 units in strength of materials is proposed herein. It has been tested over two years and the design for Year-3 courses is now underway. The students are first engaged in class through introduction to equations and relevant examples on the textbooks. Then, the lab component with visible features enables them to examine the applicability of the equations and secure a strong impression. Finally, students conduct oral presentations and peer discussions, gaining full insight into strength of materials.

We intend to share the lecture notes for this course online, with the hope that teaching professionals interested in developing this curriculum will collaborate and promote this new pedagogy. Beginning in 2016, we plan to compile new textbooks for various structural courses, and we cordially invite your contribution in this endeavor.

Acknowledgements

Special thanks for generous support from National Science Council's Research Program No. MOST 103-2511-S-006-009.

References

[1] Charleson, A. W. (2005), Structure As Architecture: A Source Book For Architects And Structural Engineers. Architectural Press, Oxford.

[2] Architectural Institute of Japan (2001), Spatial Structures Seismic Design and Realization in Japan, Maruzen Publishing.

[3] Computers & Structures, Inc. (2015), CSI Analysis Reference Manual, Computers & Structures, Inc.

[4] An Innovative Tool for Teaching Structural Analysis and Design, IBK, vdf Hochschulverlag AG, Swissland.

[5] Dytoc, Bronne, Graphic Learning Strategies to Motivate Structures Education in Architecture, 5th Annual International Conference on Architecture, 6-9 July 2015, Athens, Greece.

[6] Ji, T. and Bell, A. (2008), Seeing and Touching Structural Concepts, Higher Education Press, Beijing.

[7] Tao, Y., Liu, Z., Chen, J., Li, G. and Wang, Q. (2012), Teaching Concrete Structural Design Theories by Visualizing and Feeling of Structural Concepts, Journal of Southeast University (Philosophy and Social Science), 14 (S2), 191-194.

[8] Ján Ilkovi. (2014), To think in architecture, to feel in structure: Teaching Structural Design in the Faculty of Architecture. *Global Journal of Engineering Education.* 16: 59-65.

[9] Kolb, D. A. (1984), Experiential learning: Experience as the source of learning and development, Englewood Gliff, NJ.

[10] Hibbeler, R. C. (2011), Mechanics of Materials, Prentice Hall.

[11] Gere, J. M. and Goodno B. J. (2013), Mechanics of Materials, Cengage Learning.

Effect of Calcium Chloride on the Compressive Strength of Concrete Produced from Three Brands of Nigerian Cement

S. O. Odeyemi, M. A. Anifowose, M. O. Oyeleke, A. O. Adeyemi, S. B. Bakare

Department of Civil Engineering, Federal Polytechnic Offa, Offa, Nigeria

Email address:
odeyemiso@yahoo.ca (S. O. Odeyemi)

Abstract: Concrete is a common material used in the Nigerian building industry. It is obtained by mixing cementitious materials, water, fine and coarse aggregates and sometimes admixtures in required proportions. Admixtures are added to concrete to modify its properties so as to make it more suitable for any situation. In recent times, building collapse in Nigeria has been a source of concern particularly to those associated with the building industry. This has necessitated the need to check the effect admixtures have on some properties of concrete. This study investigated the effect of calcium chloride ($CaCl_2$) as an admixture on the compressive strength of concrete produced from Dangote, Elephant and Burham brands of cement available in Nigeria. The impact of calcium chloride admixture on the compressive strength of concrete made from these brands of cement was compared with the compressive strength of concrete without calcium chloride. Sieve analysis, natural moisture content and specific gravity tests were carried out on the aggregates used in the production of the concrete and the results were recorded. Slump tests were carried out on the fresh concrete containing $CaCl_2$ and without $CaCl_2$ and the results were also recorded. Cubes produced from the concrete were cured for 7, 14, 21 and 28 days respectively. The mean densities of the concrete cubes made from the three brands of cement with and without admixture were computed. The corresponding mean compressive strength for the concrete cubes was also computed. The results showed that concrete with calcium chloride have higher compressive strength compared with those without calcium chloride.

Keywords: Concrete, Admixture, Calcium Chloride, Cement, Compressive Strength

1. Introduction

Concrete has been the most common building material for many years [1]. It is obtained by mixing cementitious materials, water, aggregate (usually sand and gravel or crushed stone) and sometimes admixtures in required proportions. Fresh concrete or plastic concrete is freshly mixed material which can be moulded into any shape for it to harden into a rock-like mass known as hardened concrete. The hardening is as a result of chemical reaction between water and cement. Portland cement is the most common type of cement in general usage. It is a basic ingredient of concrete, mortar and plaster [2]. Anything other than cement, water and aggregate (fine and coarse aggregate), if added in concrete either before or during mixing to alter the properties to our desired requirement are termed as admixtures [3].

In recent times, building collapse in Nigeria has been a source of concern to so many people particularly those associated with the building industry. This is so because there are so many cases of building collapse all over the world and particularly Nigeria. Most of these cases had resulted into colossal economic losses in terms of lives and property [4]. This has necessitated the need to check the effect admixtures have on some properties of concrete.

This research was aimed at investigating the effect of calcium chloride ($CaCl_2$) as an admixture on the compressive strength of concrete produced from Dangote, Elephant and Burham brands of cement in Nigeria.

2. Collapse of Building Structures in Nigeria

In recent times in Nigeria, there have been several cases of structural collapse of buildings, especially in Abuja, Lagos and Port Harcourt where several construction works are presently going on. Several of these incidents occurred in buildings that had outlived their life spans, some were still undergoing construction, some had been declared unsafe and undergoing demolition while others were already completed and in use when the sudden collapse occurred. Figures 1 and 2 show some of the collapsed buildings in Nigeria.

Figure 1. *Collapsed Synagogue building in Nigeria [5].*

Figure 2. *Collapsed Naval building in Gwarinpa, Abuja, Nigeria [4].*

3. Compressive Strength of Concrete

The quality of concrete is judged largely on the strength of that concrete. Strength is usually the basis for acceptance or rejection of the concrete in a structure. A deficiency in the strength of concrete can lead to expensive and difficult repairs or at worse a failure of the structure. Concrete is an excellent material for resisting compressive loading. Compressive strength of concrete is usually determined by means of test cylinders made of fresh concrete on the job and tested in compression at various ages of the concrete. The requirement is a certain strength at an age of 28 days or such earlier age as concrete is to receive its full service load or maximum stress. Additional tests are frequently conducted at earlier ages to obtain advance information on the adequacy of strength development where age-strength relationship have been established for the materials and proportions used [6].

4. Cement Manufacturers in Nigeria

The following companies manufacture cement in Nigeria:

1. Cement Company of Northern Nigeria Plc: They are the producer and marketer of Sokoto BUA brand of cement in Nigeria. They are located at Km 10, Kalambaina Road, Sokoto, Sokoto State, Nigeria.
2. Dangote Group: The Head office of Dangote Group is located at 1 Alfred Rewane Road, Falomo Ikoyi, Lagos, Lagos State, Nigeria. They are the manufacturer of Dangote brand of cement in Nigeria.
3. Eastern Bulkcem Company Limited: Their Head office is located at 11, Awolowo Road, Flat 5 Block 2, Ikoyi, Lagos, Lagos State. They are the manufacturer of Eagle Cement in Nigeria.
4. Lafarge Cement WAPCO Nigeria Plc: They are the manufacturer of Elephant brand of cement in Nigeria. Their Head office is located at Km 64, Lagos/Abeokuta Expressway, Ewekoro, Abeokuta, Ogun State.
5. Flour Mills of Nigeria Plc: They are the producers of Burham cement. Flour mills began the production of Burham cement in Nigeria in 1975. They are located at No 2, Old dock road, Apapa, Lagos, Nigeria.

5. Admixtures

Admixtures are materials other than cement water and aggregate that is used as an ingredient of concrete and added to the batch immediately before, after or during mixing. They are used to modify the properties of concrete so as to make it more suitable for any situation. Admixtures, although not always cheap, do not necessarily represent additional expenditure because their use can result in concomitant savings, for example, in the cost of labour required to effect compaction, in the cement content which would otherwise be necessary, or in improving durability without the use of additional measures [7].

5.1. Chemical Admixtures

Chemical admixtures are added to concrete in very small amount mainly for air entrainment reduction of water or cement content plasticization of fresh concrete mixtures, or control of setting time [8]. Chemical admixtures are materials in form of powder or fluid that are added to concrete to give it certain characteristics not obtainable with plain concrete mixes. In normal use, admixture dosages are less than 5% by mass of cement and are added to concrete at the time of batching/mixing.

Calcium chloride ($CaCl_2$) is a chemical admixture and a by-product of the Solvay process in the manufacturing of sodium carbonate. Calcium chloride is available in two forms: (i) Regular flake calcium chloride and (ii) concentrated flake, pellet, or granular calcium chloride [9].

Calcium chloride has been used in concrete since 1885 and finds application mainly in cold weather, when it allows the strength gain to approach that of concrete cured under normal curing temperatures [8]. In normal conditions, calcium chloride is used to speed up the setting and hardening process for earlier finishing or mould turnaround.

Aside from affecting setting time, calcium chloride has a minor effect on fresh concrete properties. It has been observed that addition of $CaCl_2$ slightly increases the workability and reduces the water required to produce a given slump [10] and reduces bleeding. Initial and final setting times of concrete are significantly reduced by using calcium chloride. The total effect of adding calcium chloride depends on dosage, type of cement used, and temperature of the mix.

6. Methodology

The material constituents, mix proportions, mixing, curing process are paramount important factors that determine the strength of concrete.

6.1. Materials Used

The following materials were used in producing the concrete used in this research work:

6.1.1. Coarse and Fine Aggregates

Granite and Sharp sand which are extremely needful materials for production of concrete are the product of natural or artificial disintegration of rocks and minerals. Clean aggregates, free from waste stone and impurities were used in the production of the concrete used in this research work.

6.1.2. Cement

Ordinary Portland cement (OPC) – Dangote and Elephant and Burham cement brands which conformed to NIS 444 – 1: 2003 were used.

6.1.3. Water

Portable water which is free from suspended particles, salts and oil contamination were used throughout this study as specified by [11].

6.1.4. Calcium Chloride

Calcium chloride was added to the concrete mix as an admixture. 4% of calcium chloride was used.

A mix ratio of 1:2:4 was adopted for the concrete production. Batching was done by weight.

6.2. Tests on Materials

The following tests were carried out:

6.2.1. Sieve Analysis

The test samples were dried to a constant weight at a temperature of $110 \pm 5^{0}C$ and weighed. The samples were sieved by using a set of IS Sieves and electric sieve shaker. On completion of sieving, the materials on each sieve were weighed. Cumulative weight passing through each sieve was calculated as a percentage of the total sample weighed. Fineness modulus was obtained by adding cumulative percentage of aggregates retained on each sieve and dividing the sum by 100.

$$\text{Fineness modulus} = \text{Total cumulative \% retained}/100 \qquad (1)$$

6.2.2. Natural Moisture Content

Sets of container and wet soil were used. The container was clean and dried before use. The weight of the empty container is taken as W_1. The required quantity of the wet soil specimen was placed in the container and weigh as W_2 (weight of empty container + wet soil). The containers with sample were oven dried in a controlled temperature for 24 hours to a constant weight. After attaining the constant weight, the specimen was removed, allowed to cool, and the reading was taken as W_3 (weight of empty container + dry soil). Equation (2) was used to determine the moisture content of the aggregate.

$$Moisture\ content(Mc) = \frac{weight\ of\ water}{weight\ of\ dry\ sample} = \frac{W_2 - W_3}{W_3 - W_1} \times 100\% \qquad (2)$$

6.2.3. Specific Gravity Test

The samples were first screened on BS sieve thoroughly to remove grass particles and other deleterious materials. The weight of the empty density bottle was recorded as W_1. The sample used was filled into the density bottle and weighed; the weight was recorded as W_2 (weight of bottle + dry sample). The density bottle was gradually filled with distilled water to gauge mark, soon after the end of soaking, air entrapped and bubbles on the surface of the aggregate sample were removed by heating the density bottle on a water bath and the weight was recorded as W_3 (weight of bottle + dry sample + distilled water), after which the bottle was emptied and allowed to dry. The density bottle was then filled with distilled water to the gauge mark and weighed as W_4 (weight of bottle + distilled water). The equation used to determine the specific gravity of aggregate is given in Equation 3:

$$Specific\ Gravity(Gs) = \frac{W_2 - W_1}{(W_4 - W_1) - (W_3 - W_2)} \qquad (3)$$

6.2.4. Slump Test

The internal surface of the cone mould used was thoroughly cleaned and applied with a light coat of oil. The mould was placed on a smooth, horizontal, rigid and non- absorbent surface. The mould was then filled in three layers with freshly mixed concrete, and each layer was tamped 25 times with tamping rod. After the top layer was rodded, the concrete was struck off the level with a trowel. The mould was removed from the concrete immediately by raising it slowly in vertical direction. The difference in level between the height of the mould and that of the highest point of the subsided concrete was measured using a meter rule. This difference in height in mm was recorded as the slump of the concrete.

6.2.5. Density Test

At the end of each curing age (7, 14, 21 and 28 days respectively), the concrete cubes were removed from the curing tank and allowed to drain off. The weight of each cube was obtained by a weighing balance to determine the density of the concrete cube. Equation (4) was used to obtain the density of each cube.

$$Density, \rho = \frac{Weight\ of\ Cube}{Volume\ of\ Cube}\left(kg/m^3\right) \qquad (4)$$

6.2.6. Compressive Strength Test

The compressive strength test was carried out immediately after obtaining the density of the concrete cube. The weighed cube was carefully placed in the compression test machine with the smooth face in contact with the plates of the machine. Then, the compression testing machine was turned on from the power source and the gear of the machine turned clockwise. At failure, the dial gauge of the compression testing machine stops and the gear was reversed. The load at which failure occurs was recorded. 72 cubes (36 cubes for concrete with calcium chloride and 36 cubes for concrete without calcium chloride) were crushed.

7. Results and Discussion

7.1. Physical Properties of Aggregates

Sieve analysis, natural moisture content and the specific gravity of the aggregate were determined and recorded.

7.1.1. Sieve Analysis

From the sieve analysis, the following fineness moduli were obtained:

Fineness modulus for Fine Aggregate = 3.08

Fineness modulus for Coarse Aggregate = 6.31

These values fall within the acceptable limits for fine and coarse aggregates [12]

7.1.2. Natural Moisture Content

Average moisture content of Fine Aggregate (Sand) = 0.94%

Average moisture content of Coarse Aggregate (Crushed Granite) = 0.16%

These values fall within the acceptable limits for fine and coarse aggregates [12].

7.1.3. Specific Gravity

Average Specific Gravity (Gs) of Fine Aggregate (Sand) = 2.6

Average Specific Gravity (Gs) of Coarse Aggregate (Crushed Granite) = 2.65

These values are within the acceptable limits for fine and coarse aggregates [13].

7.2. Workability

Workability is the ability of fresh concrete mix to mould or form properly with the desired work without reducing the concrete quality.

7.2.1. Slump Test Results

Table 1 shows the slump test results for the concrete with and without calcium chloride. True slump was exhibited by the concrete in the fresh concrete mix. The slump was satisfactory as segregation was minimized. The highest slump height was obtained using Dangote cement without $CaCl_2$ and the least height was obtained using Burham cement with $CaCl_2$.

7.3. Tests on Hardened Concrete

Hardened concrete assumes important properties that are retained for the life of the concrete and these properties include density, strength and deformation under load and durability among others.

7.3.1. Density Test Results

The mean densities of concrete cubes made from dangote, elephant and burham cement with and without calcium chloride for age 7, 14, 21 and 28 days hydration period are given in Tables 2 and 3. All concrete cubes produced falls within the range 2300Kg/m^3 – 2600Kg/m^3. The densities of all the samples tested fell within the normal range of concrete [14]. It was noticed that the densities of the concrete increased with days irrespective of whether it contains $CaCl_2$ or not.

7.4. Compressive Strength Test

Tables 4 and 5 shows the mean compressive strength test results of the concrete cubes cast from the three brands of cement with calcium chloride admixture and without calcium chloride admixture respectively. The Tables reveal that the compressive strength of concrete cubes made from Dangote, Elephant and Burham cement with calcium chloride were greater than the compressive strength of concrete cubes made without calcium chloride at the different curing ages. The highest compressive strength was obtained from the concrete made from Dangote brand of cement containing calcium chloride while the lowest compressive strength was obtained from the concrete made from Elephant brand of cement without calcium chloride. The results are represented graphically in Figure 3 and Figure 4.

Table 1. *Result of Slump Test for Concrete Made with different brands of Cement*

Serial Number	Brands of cement with and without $CaCl_2$	Slump Height (mm)
1	Dangote with $CaCl_2$	70
2	Dangote without $CaCl_2$	73
3	Elephant with $CaCl_2$	60
4	Elephant without $CaCl_2$	63
5	Burham with $CaCl_2$	60
6	Burham without $CaCl_2$	65

Table 2. *Results of Mean Densities of Concrete Cubes with Calcium Chloride*

Age (Days)	Mean Densities of Concrete Cubes (Kg/m^3)		
	Dangote Cement	Elephant Cement	Burham Cement
7	2397.93	2377.38	2397.93
14	2414.02	2419.16	2450.47
21	2441.48	2488.69	2556.35
28	2419.16	2520.29	2556.74

Table 3. *Results of Mean Densities of Concrete Cubes without Calcium Chloride*

Age (Days)	Mean Densities of Concrete Cubes (Kg/m^3)		
	Dangote Cement	Elephant Cement	Burham Cement
7	2419.16	2350.72	2397.93
14	2430.91	2430.91	2450.47
21	2437.14	2474.77	2409.77
28	2441.48	2491.65	2556.35

Table 4. *Mean Compressive Strength of Concrete Cubes Made from the three brands of Cement with Calcium Chloride*

Age of cubes (Days)	Mean Compressive Strength of concrete cubes (N/mm²)		
	Dangote Cement	Elephant Cement	Burham Cement
7	17.93	17.48	18.89
14	21.56	19.48	21.78
21	23.13	22.22	22.81
28	25.23	24.67	25.29

Table 5. *Mean Compressive Strength of Concrete Cubes Made from the three brands of Cement without Calcium Chloride*

Age of cubes (Days)	Mean Compressive Strength of concrete cubes (N/mm²)		
	Dangote Cement	Elephant Cement	Burham Cement
7	13.17	11.91	12.90
14	16.99	16.38	16.60
21	18.80	17.78	18.51
28	23.54	22.25	22.67

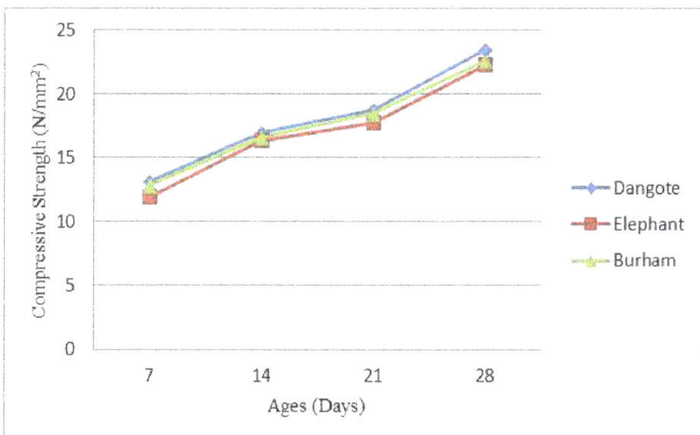

Figure 3. *Graph of Compressive Strength against Ages of concrete without Calcium Chloride.*

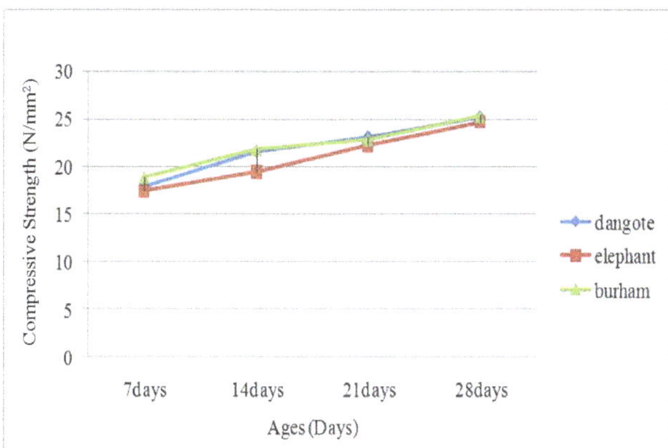

Figure 4. *Graph of Compressive Strength against Ages of concrete with Calcium Chloride.*

5. Conclusion

The results of the mean compressive strength test at 7, 14, 21 and 28 days shown in Tables 4 and 5 shows that the compressive strength of concrete cubes made from Elephant cement with calcium chloride is lesser than that of Dangote and Burham cement with calcium chloride. It was noticed that the compressive strength of concrete cubes made from Dangote, Elephant and Burham cement with calcium chloride increases as the curing age increases at almost the same rate, but concrete made with Burham cement with calcium chloride has the highest compressive strength. Conclusively, the compressive strength of concrete cubes made from Dangote, Elephant and Burham cement with calcium chloride is greater than compressive strength of concrete cubes made from Dangote, Elephant and Burham cement without calcium chloride at different curing ages.

References

[1] W.F. Chen and J.Y. Richard Liew, The Civil Engineering Handbook, 2nd ed., CRC Press, 2003.

[2] M.S. Shetty, Concrete technology, theory and practice, S. Chand & company ltd., New Delhi, 2011, pp. 1-31.

[3] K. Nitish, Admixtures used in concrete and their effects, retrieved from www.nersp.nerdc.ufl.edu on 18th November, 2014.

[4] A.A. Fakere, G. Fadairo and R.A. Fakere, Assessment of building collapse in Nigeria: A case of Naval building, Abuja, Nigeria, International Journal of Engineering and Technology, Vol. 2, No. 4, 2012, pp.585-588.

[5] www.nigeriancurrent.com, retrieved on the 7th January 2015.

[6] www.iccsafe.org, retrieved on the 7th January 2015.

[7] A.M. Neville, Properties of concrete, 5th ed., Pearson Education Ltd., 2011.

[8] M.R. Rixom and N. P. Mailvaganam, Chemical admixtures for concrete, Cambridge, England: The University Press, 1986.

[9] ACI Committee 212, Admixtures for concrete, ACI Journal proceedings 60(11), 1963, pp. 1481-1524.

[10] V.S. Ramachandran, Calcium chloride in concrete, Science and technology, 2nd ed., Applied Science Publishers Ltd., Essex, England, 1996.

[11] NIS 2000. NIS 87:2000. Nigerian Industrial Standard: Standard for Sandcrete blocks. Standard Organisation of Nigeria, Lagos, Nigeria.

[12] IS 383, Indian Standard Specification for coarse and fine aggregates from natural sources for concrete, Bureau of Indian Standards, New Delhi 110002, 2nd revision, 9th reprint, 1993, pp. 4-18.

[13] IS 2386-3, Indian Standard Methods of test for aggregates for concrete, Part III: Specific gravity, density, voids, absorption and bulking , Bureau of Indian Standards, New Delhi 110002, 8th reprint, 1997, pp. 3-17.

[14] K. Jones, Density of concrete, The Physics Factbook, 1999 (retrieved from www.hypertextbook.com/facts/1999/katrinajones.shtml on 7th January, 2015).

Remotely sensed ore mineralization potentials in Ramand altered region, north of Iran

Abolfazl Ezzati[1,*], Reza Mehrnia[2], Kimiya Ajayebi[1]

[1]Department of Geology, Islamic Azad University, Science and Research Branch, Tehran, Iran
[2]Department of Geology, Payam Noor University, Tehran, Iran

Email address:
Ezzatiabollfazl@gmail.com (A. Ezzati), R_Mehrniya@pnu.ac.ir (R. Mehrnia), Kimiya.ajayebi@kiau.ac.ir (K. Ajayebi)

Abstract: In some regions the main geological occurrences contain igneous pyroclastics formations such as rhyodacite and rhyolithic tuffs. An extended region with dominant alterations has been detected by remote sensing techniques for sampling purposes according to field observation activities. Enhancement of alterations has been done in Ramand region after applying a Crosta-based selective principle component analysis technique on ETM photomaps. Advanced Silicification haloes are most important evidence indicate to Base and Precious metals potentials in postmagmatic environments. For identifying the Silica content and related variations in epithermal systems, Crosta technique is adequate. This method introduces to recognition some clay and iron-oxides aggregations as the main part of gold bearing hydrothermal alterations in prospected regions. Concluded photomaps of which signals have been improved by mathematical analysis contained number of unique digital numbers related to mineralization potentials in Ramand region, Qazvin province, Iran.

Keywords: Remote Sensing, Crosta Technique, Silicic Alteration, Ramand

1. Introduction

Remote Sensing is a modern technology which without physical contacts with the phenomenon, we are able to measure and analyze spectral characteristics. Not only determining the alteration zones (and variations), but also separating mineralization anomalies is an important application of this technology as well. remote sensing techniques allows the identification of a large area with high accuracy, high speed and low cost.

Ore deposits are often produced by fluid flow processes that alter the mineralogy and chemistry of the country rock. A diversity of previous researches has proved the reliability of multispectral data analysis in the field of alteration detection [1]. Many ore deposits are first detected in the field by the recognition of hydrothermally altered host rocks, and are typically zonally distributed. Economic mineralization is often produced by fluid processes that substantially alter the mineralogy and chemistry of the host rocks. This alteration can produce distinctive assemblages of minerals that vary according to the location, degree, and longevity of those flow processes. When exposed to the surface of the Earth this alteration can sometimes be mapped at a zonal pattern [2]. By using remote sensing techniques these zones can be detected in a regional scale. In this study, Processing was performed on the ETM satellite imagery data of the Ramand area to map spectral signatures associated with the hydrothermal alterations. Ramand region in North-West of Iran is an importance region for, presence of good mineralization of, copper, lead, zinc and valuable metals such as Gold and Silver.

2. Spectral Optimization by IHS Filtering

In other to have primary display of the area, the boundary of layers and lineaments in ETM image are detected by using Intensity Hue Saturation (IHS) filter. This filter is a high-pass quality filter that despite other ENVI filters, relates to the intensity and quality. However, it is not relevant to DN values. This filter is used to optimize the image quality and separate the geological layers from alterations zones by ENVI 4.7. The advantage of this filter is to identify geological phenomenon specially faults structures [3]. In figure 2 the RGB (7, 4, 1) is filtered by IHS. It enhances and separates the different lithological units better not revise any of the current designations.

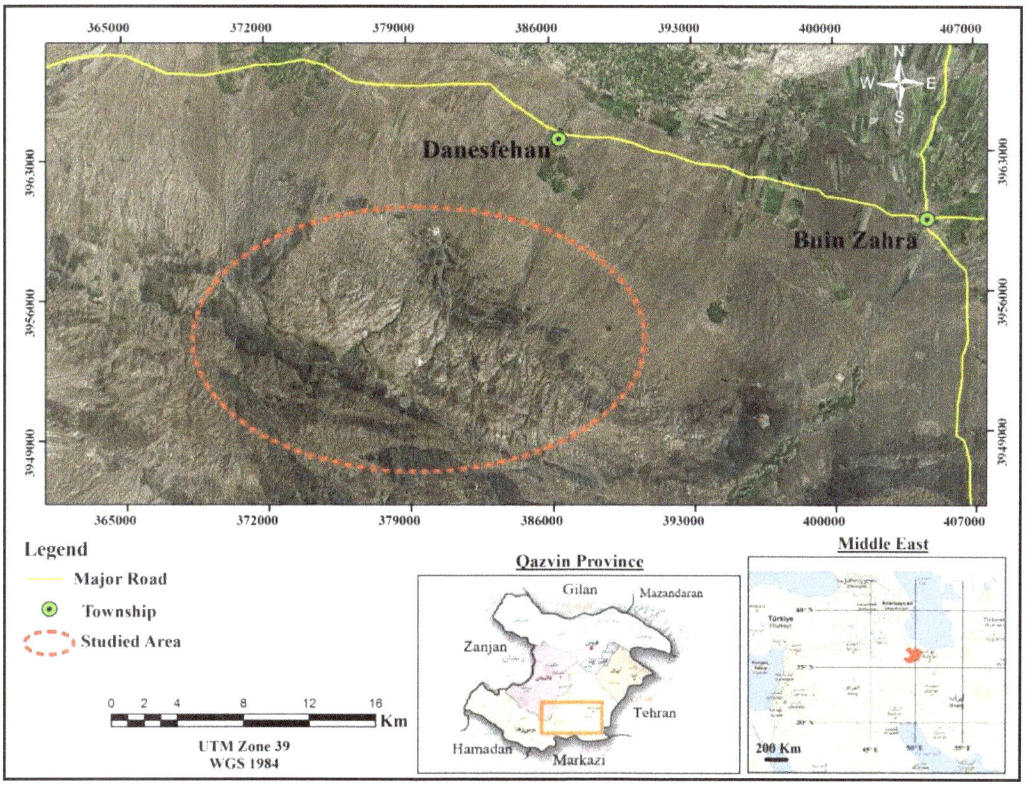

Figure 1. Study area

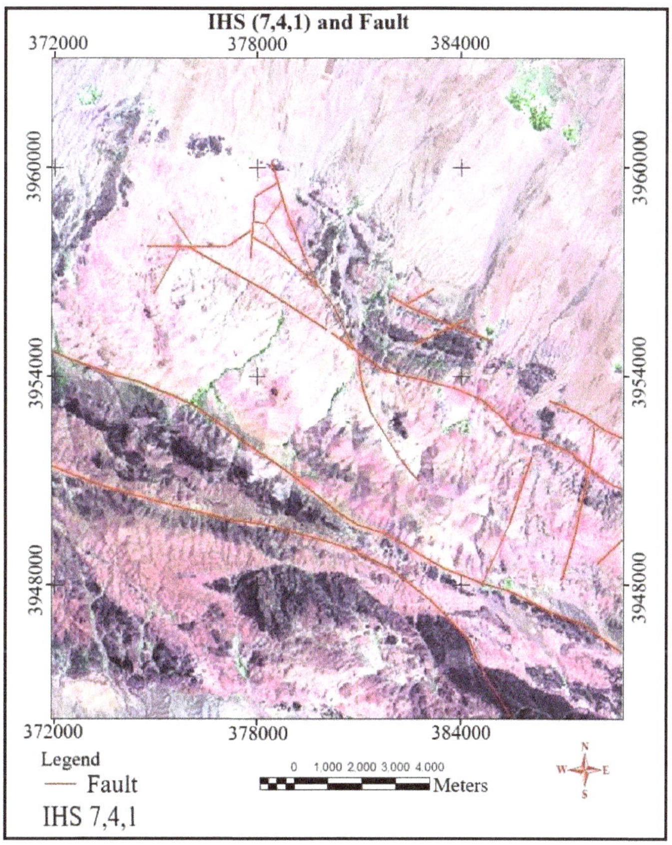

Figure 2. Enhancement of lineaments and boundary layers by IHS Filtering.

3. Implementation of Crosta Technique (Crosta PCA) for Enhancement of Alterations in Ramand Region

The principal component transformation is a multivariate statistical technique that selects uncorrelated linear combinations (eigenvector loadings) of variables in such a way that each successively extracted linear combination or principal component (PC) has a smaller variance. The main aim of PC analysis is to remove redundancy in multispectral data. Principal component analysis is widely used for mapping of alteration in metallogenic provinces. Crosta technique is also known as feature oriented principal component selection. Through the analysis of the eigenvector values it allows identification of the principal components that contain spectral information about specific minerals, as well as the contribution of each of the original bands to the components in relation to the spectral response of the materials of interest [4].

According to present experiences and previous studies and also geological structure of region, appropriate bands for PCA analysis can be recognized.

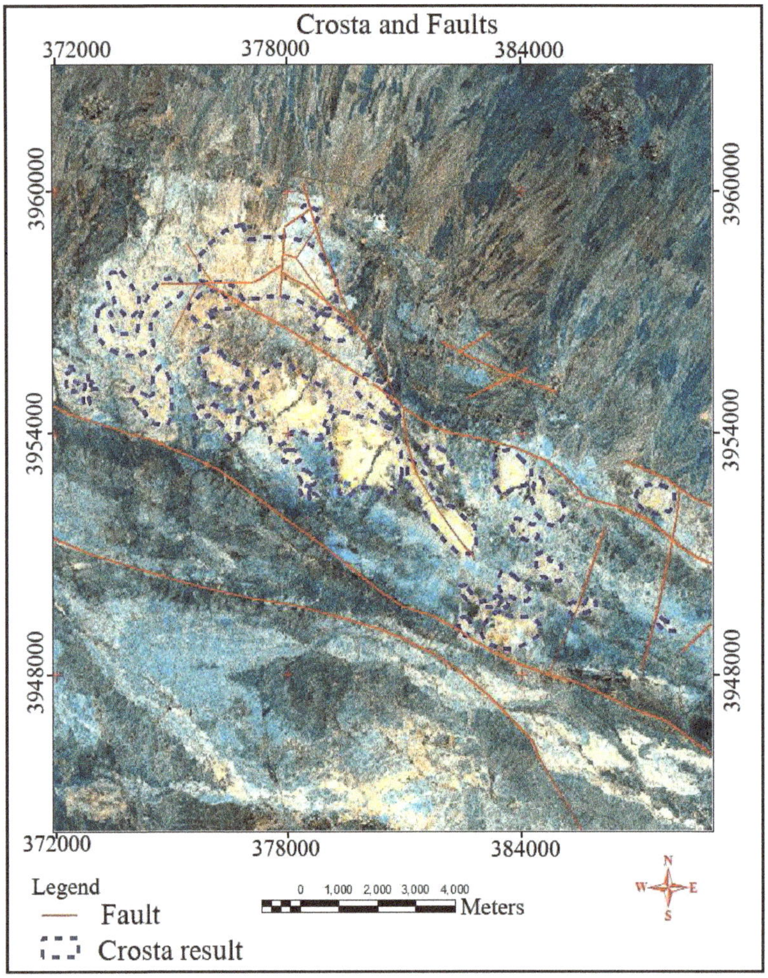

Figure 3. *Hydrothermal Alterations according to Crosta PCA*

In this method, reducing of the bands usually increases the probable mapping of definite processes according to discriminations between Eigen values (principal components-PC) [5]. Hydroxyl-bearing minerals are important because of their abundance in the alteration zones. Crosta technique defines the nature of alteration halos surrounding most of magmatic intrusions. Selective 7,5,4,1 ETM bands for mapping hydroxyls and selective 5,4,3,1 bands for mapping iron oxides have been proposed. The component analysis results of the first selections known as F-series and the second known as H-series. Studying Eigenvector of F and H-Series, the PC4 values made the most differential spectrums for revealing iron oxides and clayey minerals respectively (Table1 & 2).

Applying Band Math technique caused adding H to F components for obtaining the new H+F spectrums. RGB visible panels were constructed by these components as: R= H, G= H+F and B= F for mapping hydrothermal alterations within yellowish-buff to orange colors of photomap (fig. 3).

A Selective Principle Component Analysis (SPCA) technique for identifying clay + iron oxide aggregation has been used and correlated with geochemical haloes for introducing young (Neogene) alteration facies related to epithermal mineralization processes in Ramand region.

Table 1. The result of PCA for Enhancing Iron oxides

Eigenvector	Band 1	Band 3	Band 4	Band 5
PC 1	0.44759	0.597438	0.33271	0.57623
PC 2	-0.0279	0.635341	0.7262	0.261136
PC 3	-0.7294	0.045967	-0.2202	0.646037
PC 4	0.51656	0.487131	-0.5599	0.427087

Table 2. The result of PCA for Enhancing Hydroxyls.

Eigenvector	Band 1	Band 4	Band 5	Band 7
PC 1	0.47284	0.354945	0.61472	0.522072
PC 2	-0.1644	0.784691	0.07611	0.592812
PC 3	0.8517	0.270296	0.4443	-0.06449
PC 4	0.15489	0.430365	0.64726	-0.6098

5. Detection of Siliceous Facies Using Crosta Technique

Crosta PCA is a recently developed technique for identifying alteration halos related to magmatic hydrothermal activities. This technique was proposed for the first time by Crosta and Moore (1989) [6], to indicate a specific purpose as a lighter pixel than the other phenomenon, in one of the principal component images. This is an improved technique for revealing hydrothermal alterations containing iron oxides and clayey matrixes with hydroxyl ions combinations mineralogicaly [7]. In practice, numbers of aggregated halos have close relationships with hydrothermal systems. Therefore it can be determined by Crosta technique. In addition, due to large amount of paragenesis of silica mineralization, the electromagnetic responses of Crosta photomap are coincided with clay, iron oxides and silicification aggregations in this region. Our field observations "Fig. 4", plus instrumental (X-ray) and micrographic studies "Fig. 5", confirmed the reality of remotely sensed mineralization content in close associations with quartz, illite and jarosite paragenesis "Table 3", [8]. It means that, for the first time in Silica bearing facies, a Crosta based technique has been carried out for separating Si-alterations successfully.

In a prone area of hydrothermal deposits with large scale amounts of siliceous and iron oxide mineralization, Crosta technique can be easily introduced these areas among the hydrothermal alteration halos.

Table 3. X-ray result (Zarazma Mineral Studies Company).

Sample	Major Phase	Minor Phase	Trace Phase
M37	Quartz,Muscovite-Illite	Albite, Natrojarosite	_
M38	Quartz, Illite	Orthoclase, Jarosite	Kaolinite

Figure 4. *Jasperoid facies in the study area (see to the South West).*

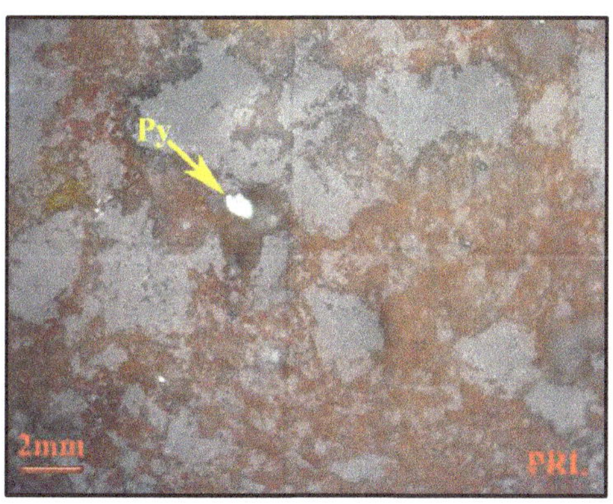

Figure 5. *The higher values of iron oxides cause to red colors of matrix. (probably presence of jasperoid).*

5. Laboratory Studies

Ore Microscopy and Petrographic studies have been well done after polished and thin sections preparation for focusing on mineralogical peculiarities of the samples. Figure 6 shows two micrographs indicate to paragenesis of available ore minerals (right) with nonmetallic aggregations as gangue traces (left). Pyrite and Hematite are important known mineralization at the surface of remotely sensed alteration (figure 6 - right). Also a massive silicified texture (fine grained) has been intruded by quartz and Fe-hydroxides according to petrographic evidences (figure 6- left). From mineralogical points of view, a massive Silica facies in paragenetic association with Fe-Oxide and Fe-Sulfide minerals have enough potential for increasing in gold traces anomalies. Therefore an atomic absorption analysis (AA) has been well done and summarized for Au analysis in table 4.

As depicted, Au-grade varies between 131-145 ppb in adjacent of alteration occurrences which are mostly contained by argillic-silicific sequences. Individual shiny pyrites with bright appearance have additional reflections (comparing with pale types) and therefore selected for detailed quantitative analysis by Scanning Electron Microscopy (SEM).

Both mineralogical analytical evidences indicate to this fact that Ramand altered region not only has potentials for basics but considered as gold-bearing prospect according to SEM performances.

Increasing in Silica content has a direct relationship with pyrite and invisible gold traces associate with quartz-agate veinlets in post magmatic differential stages of hydrothermal environments [9].

Table 4. *Atomic Absorption Analysis (AA) result for Ramand Au-bearing sample. (Zarazma Mineral Studies Company).*

Sample	M 28	M 37	M 31	M 22
Au(ppb)	131	133	125	145

Figure 6. *Ore microscopy (right) and ore petrography (left) results for remotely sensed altered samples- Ramand region. Right: two main ores are Hematite and Pyrite. Left: Silica masses plus quartz and feldspars are common gangue minerals in gold-bearing samples.*

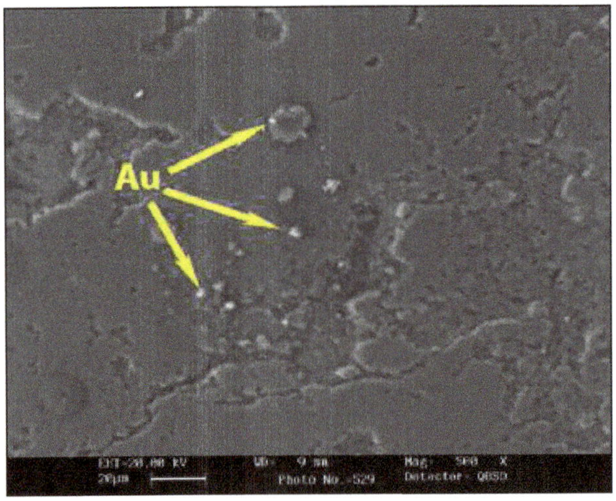

Figure 7. *Shiny appearances in gold particles (Zarazma Mineral Studies Company).*

6. Discussion and Conclusions

This research study focused on extended altered regions in Ramand mineralized prospects by processing ETM images according to Crosta technique. An aggregated alteration containing clay and iron oxides has been identified due to investigations. Sampling and instrumental analysis results showed that a clayey matrix within silica based mineralization is dominant in this region. Also our thin section studies have been well done and confirmed the hydrothermal origination of ore bearing solutions in a post magmatic environment.

From remote sensing points of view, Ramand prospected regions have enough potentials for epithermal mineralization because of extending number of solution related alterations on surfaces and existing evidence which indicate continuous alterations in depth of region. Therefore, a deep investigation on altered features and corresponding mineralization processes are highly recommended for further exploration activities in future with emphasizing on precious metals accumulations in the silica veinlets.

References

[1] AN. Rensz, "Remote Sensing For the Earth Sciences," John Wiley & Sons, USA, 1999.

[2] G. Ferrier, K. White, G. Griffiths, R. Bryant, M. Stefouli, "The mapping of hydrothermal alteration zones on the island of Lesvos,Greece using an integrated remote sensing dataset," National Journal of Remote Sensing, ISSN 0143-1161, 2001.

[3] R. K. Vincent, "Fundamentals of Geological and Enviromental Remote Sensing," 1st edition Prentice Hall, 131p, (1997).

[4] H. Kaufman, "Mineral exploration along the Agaba-Levant structure by use of TM-data concepts, processing and results," International Journal of Remote Sensing, 1630–1658, 1998.

[5] W. P. Loughlin, "Principal Component Analyses for Alteration Mapping," Photogrammetric Engineering and Remote Sensing, Vol. 57, pp. 1163_1169, 1992.

[6] A. P. Crosta, J. Moore, "Enhancement of Landsat Thematic Mapper imagery for residual soil mapping in SW Minais Gerais State," Proceedings of the Seventh ERIM Thematic Conference: Remote Sensing for Exploration Geology, pp. 1173–1187, Brazil, 1989.

[7] S. R. Mehrnia, "Using fractal filtering technique for processing ETM data as criteria for evaluating of Au-indices in north-west of Iran," 13th Australian Remote Sensing Conference, Canberra, Australia, 2006.

[8] A. Ezzati, R. Mehrnia, K. Ajayebi, "Detection of Heydrothermal potential zones using remote sensing satellite data in Ramand region Qazvine Province Iran," Journal of Tethys, vol. 2, pp. 93–100, April 2014.

[9] G. Morison, "AMIRA Project, Revised version : Evaluating of Gold Mineralization Potentials in Queensland Epithermal Systems," Queensland J.C Univ. press, Queensland, Australia, 249 p, 2003.

Effect of Drying-Wetting Cycles on Saturated Shear Strength of Undisturbed Residual Soils

Sayem Hossain Md[1, 2, *], Kong Ling-wei[1], Yin Song[1]

[1]State Key Laboratory of Geomechanics and Geotechnical Engineering, Institute of Rock and Soil Mechanics, Chinese Academy of Sciences, Wuhan, Hubei, China

[2]Department of Geological Sciences, Jahangirnagar University, Dhaka, Bangladesh

Email address:

sayem_geo@yahoo.com (Sayem H. M.), lwkong@whrsm.ac.cn (Kong L. W.), flysong1314@126.com (Yin S.)

[*]Corresponding author

Abstract: Due to world-wide distributions and extensively used as construction materials, geotechnical engineers are interested in understanding the mechanical behavior of residual soils which are sometimes referred in the literature as problematic soils. The climatic zones where residual soils occur are often experienced by many drying-wetting cycles due to seasonal variations. This seasonal variation in the water content termed as drying-wetting cycle is regarded as the most destructive environmental factor that may cause many foundation diseases. Considering these, the purpose of this study is to understand the effects of drying-wetting cycles on saturated shear strength characteristics of undisturbed residual soil. A series of consolidated drained (CD) triaxial tests are conducted on multiple drying-wetting soil specimens to analyze the saturated shear strength. The test results indicate that the stress-strain relationships appear to be strain-hardening. The deviatory stress and initial stiffness of saturated soils increase and the volume of soils becomes contractive as the net normal stress increases but decrease with increasing drying-wetting cycle numbers. The cohesion (c_d) and internal friction angle (φ_d) decrease with increasing cycle number (N) but the attenuation rate of φ_d is less than c_d. The variations of c_d and φ_d with respect to drying-wetting cycles can be expressed by exponential function. The saturated shear strength and it's attenuation rate due to drying-wetting cycles are analyzed. The effect is more significant for the first cycle and decreases with subsequent cycles and finally reaches to a constant state after 4 cycles. Furthermore, a mathematical function is proposed in this paper which can describe the saturated shear strength attenuation rate of drying-wetting cycle samples. Such studies are useful to understand the possible changes in shear strength behavior of residual soils below the engineering structures that are subject to periodic drying and wetting from climatic variations.

Keywords: Drying-Wetting Cycle, Residual Soil, Shear Strength, Attenuation Rate

1. Introduction

Residual soils are the weathering product of their parent materials. The engineering properties and behavior of residual soils vary widely from place to place even within depth depending upon the rock of origin and the local climate during their formation. These soils are found in many parts of the world (for example in Bangladesh, China, India, Malaysia, Singapore, Thailand etc) and are used extensively in construction, either to build upon, or as construction material of both geotechnical and geoenvironmental structures such as embankments, pavements, earth fills and soil barriers. Therefore, geotechnical engineers are interested in understanding the mechanical behavior of residual soils which are sometimes referred in the literature as problematic soils.

Due to world-wide distributions and extensively used as construction materials, many research worked have already been done on residual soils. For example, the engineering characteristics, microfabric and mineralogical composition, degree of weathering, compressibility and shear strength properties and their controlling factors, bonding behavior and its loading impact, stiffness or yielding behavior of both natural and compacted residual soils are examined by numerous scholars [1-11]. Several geo-engineering problems associated with residual soils such as landslides, subsidences, damage of road and railway tract, building collapse in many parts of the world have also been reported.

Under extreme climatic conditions such as heavy rainfall, persistent drought and multiple drying-wetting cycles will cause many engineering problems including the weakening of the mechanical properties in consequence foundation diseases. Based on a large amount of investigations, climatic condition, namely drying-wetting, is regarded as the most destructive environmental factor that can induce damage to infrastructures such as highways and pavements [12-13].

A significant number of investigations have been conducted to understand the effects of drying-wetting cycles on soil physical and mechanical properties. It is found that drying-wetting cycle is one of the important factors to impact soil fabric, particle cementation, water content, and void ratio of soils [14-20]. This effect leads to the formation of cracks as well as the development of fissures in soils which significantly increase soil compressibility [21-23]. The hydraulic conductivity or permeability coefficient increases with increasing drying-wetting cycles of compacted clay cover but decreases if the soil is improved with the addition of lime [24-25]. The strength change of expansive soil under drying-wetting cycles plays an important role in the slope stability analysis. The shear strength of expansive soils decreases with the increasing number of cycles, the cohesion of expansive soil decreases obviously and the friction angle keeps a stable value. Furthermore, the slope stability and safety factor decrease with increasing number of cycles [26-27]. Several authors showed that the influence of the first drying-wetting cycle on soil structure is greatest and decreases with subsequent cycles [28-30]. The study on lime or fly ash stabilized expansive soils or solidified sludge found that drying-wetting cycles may lead to increase in swelling percent, liquid limit and plasticity index, and a reduction in plastic limit and soil strength [31-32].

Recently, the effects of drying-wetting cycles on soil strength profile of a silty clay are investigated using a micro-penetrometer and found that the strength tends to decrease with increasing cycles and the reduction between the 1st and 2nd drying paths is much higher than that between the 2nd and 3rd drying paths. The pattern of the penetration curve changes from typical mono-peak to multi-peak pattern after the 3rd drying-wetting cycles which reflects that the drying-wetting cycles create more defects in soil microstructure and intensify the heterogeneity of strength in profile [33-34].

Most of the previous investigations on the effects of multiple drying-wetting cycles on engineering properties are based on expansive soil or artificial soils (soil mixed with fly ash, cement, lime or organic polymers) or reconstructed soils for the purpose of slope stability analysis or soil solidification or stabilization. But the effects of multiple drying-wetting cycles on shear strength of undisturbed residual soils are not well understood yet. Therefore, the main objective of this study is to evaluate the effects of drying and wetting cycles on saturated shear strength of undisturbed residual soils.

2. Materials and Methods

Undisturbed residual soil samples of 5.0-7.0m depths collected at areas around Kaiping, Gaungdong, China are used for this study. According to X-ray diffraction analysis, the studied soils are mainly composed of kaolinite with small amount of illite. The non-clay minerals include quartz, pyrite and gibbsite. The basic material properties are measured in the laboratory and given in table 1 and the grain size distribution is shown in Fig. 1.

Table 1. *Basic material properties of the studied residual soils.*

Depth (m)	P_d (g/cm^3)	NMC %	Atterberg limits			Free swell (%)	Grain size distribution (%)			
			L_L %	P_L %	I_P %		Gravel	Sand	Silt	Clay
5.7~6.0	1.30	40.2	57.1	30.7	26.4	9.75	4.6	33.4	45.9	16.1

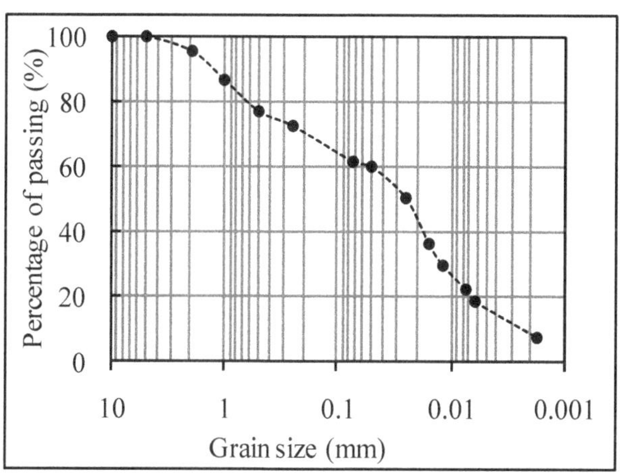

Figure 1. *Grain size distribution of the studied residual soil.*

The preparation procedures include cutting appropriate size of soil using cutting ring from the core samples. After cutting and weighted, the samples are dried up to moisture content 20% at a constant temperature of 40 ± 2°C (the highest temperature under simulated natural conditions). Remove the specimens from oven and allow 1-2 h for cool at room temperature. The samples are covered by filter paper and water-permeable stone both on top and bottom to prevent disturbance during the saturation process. Then the samples are saturated with distilled water after vacuum seeding about 2 hours and then submerged about 24h. Later the samples are dried up to 20% moisture content oven at a constant temperature of 40 ± 2°C. This is one drying-wetting cycle. The process is repeated until the desired numbers of drying-wetting cycles (0, 1, 2, 4 and 8) are completed.

A conventional triaxial test apparatus is used to measure the shear strength parameters of saturated drying-wetting soil specimens under consolidated drained (CD) conditions following the ASTM D7181-11 [35] standard test method. Prior to the tests, the soil specimens are saturated from the top until a value of B (the pore-water parameter) is 0.95 (± 0.02) and it takes 2-4 days. For this purpose, the cell pressure and saturation water pressure (back pressure) are applied and then increased gradually. A difference of 10 kPa between cell pressure and back pressure is maintained to prevent swelling or consolidation during the saturation process. After completing the saturation process, the soil specimens are consolidated under a confining pressure of σ_3, and pore water is allowed to drain out. The intervals of elapsed time (0.25, 0.5, 1, 2, 3, 4, 5, 10, 15, 30 min and at 1, 2, 4, and 8 h, and so forth) and the corresponding volume change are recorded. The consolidation process is allowed to continue until a steady value of volume change is obtained. At the end of the consolidation, the specimens are sheared at a strain rate of 0.011%/min. For all the samples, the net confining pressures are of 50, 100, 200 and 300 kPa and shear strain upto 20%.

3. Results and Discussions

The drained deviator stress versus axial strain curves for the drying-wetting cycle samples are shown in Fig. 2-6 and it can be seen that the stress-strain relationships appear to be strain-hardening. The curves show a maximum stress level in each case and after reaching the maximum stress level, there is a slight reduction or tends to reach a steady or equilibrium state in deviator stress with increasing strain. The results show that the peak deviatory stresses reduce with increasing drying-wetting cycles. The reduction is more pronounced in the 1st cycle and decreases with subsequent cycles and finally reaches to a constant state after 4 cycles. The axial strains to attain maximum deviator stress for these drying-wetting samples show higher values range from 14.57-18.94% which is a reflection of highly plastic nature of the soil.

The volumetric strain versus axial strain curves are shown in Fig. 7-11. It can be observed that all the samples show contraction behavior because of reduction in volume during drained shearing. It can also be seen that the amount of contraction varied from sample to sample.

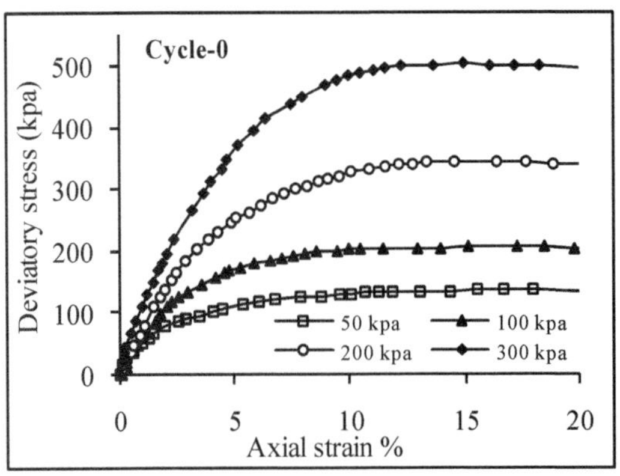

Figure 2. Deviatory stress versus axial strain curve of initial soil samples.

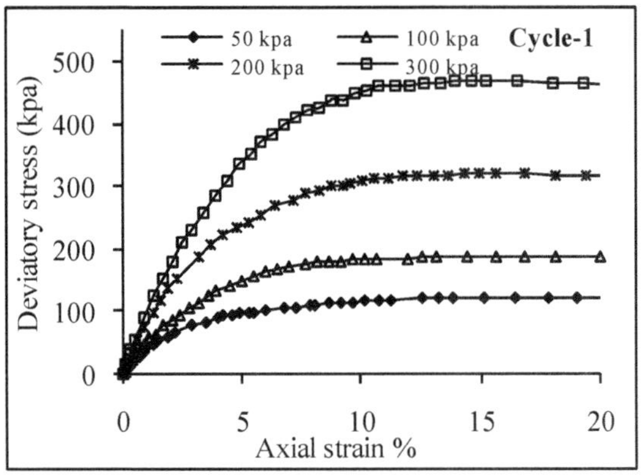

Figure 3. Deviatory stress versus axial strain of 1st drying-wetting cycle.

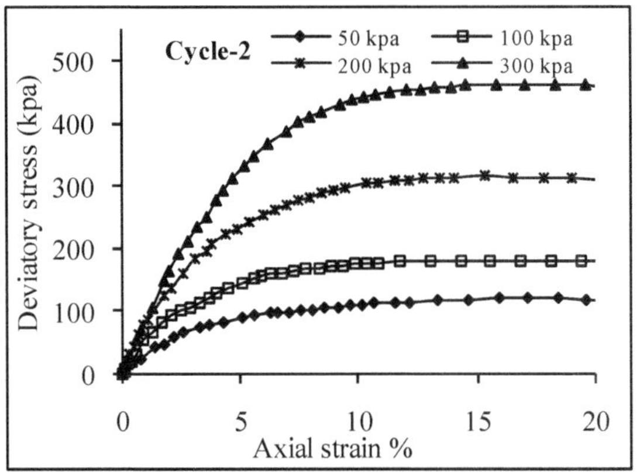

Figure 4. Deviatory stress versus axial strain of 2nd drying-wetting cycle.

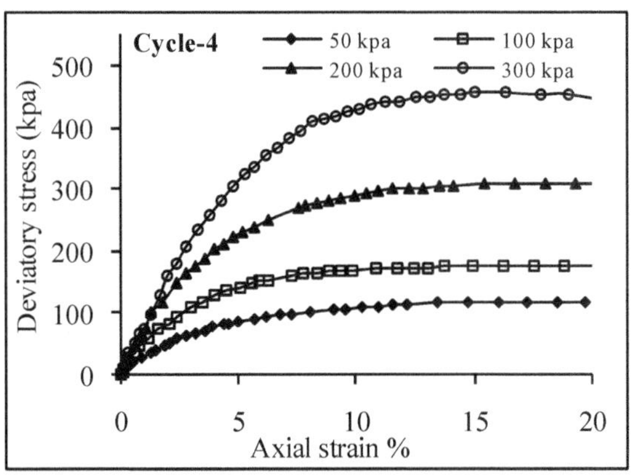

Figure 5. Deviatory stress versus axial strain of 4th drying-wetting cycle.

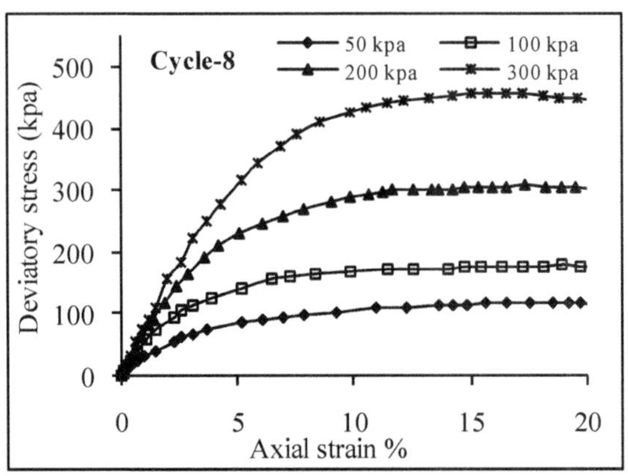

Figure 6. Deviatory stress versus axial strain of 8th drying-wetting cycle.

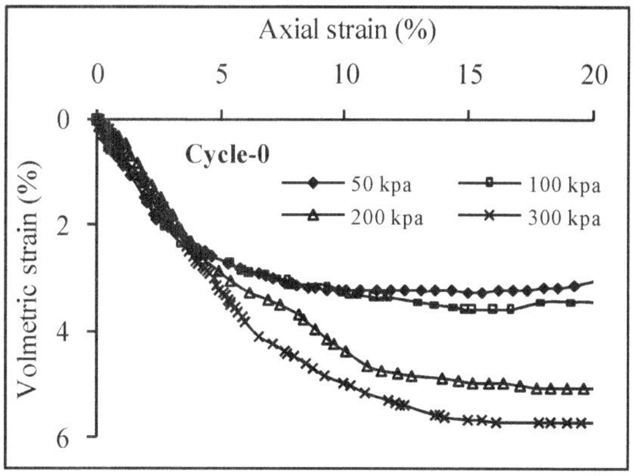

Figure 7. *Volumetric strain versus axial strain curve of initial soil samples.*

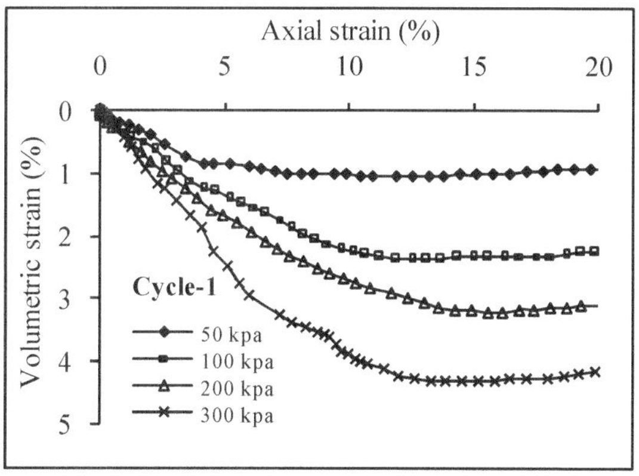

Figure 8. *Volumetric strain versus axial strain of* 1st *drying-wetting cycle.*

After attaining a maximum contraction, the volumetric strain of each sample tends towards a steady value of the volumetric strain at very large strain and there is no distinct dilation behavior which suggesting that they might approach the critical state condition. The effective stress paths of those drying-wetting cycle samples are shown in Fig. 12-16. It is observed that the failure point and end point almost close to each other indicating the hardening behavior of the soil. All the samples failed with a number of indistinct shear planes and by bulging. At high net normal stress, the sample failed with distinct shear planes and prominent bulging. The possible critical state lines (CSL) and the slope of CSL (M) in the plane of deviator stress against mean effective stress of the saturated specimen are also given in Fig. 12-16.

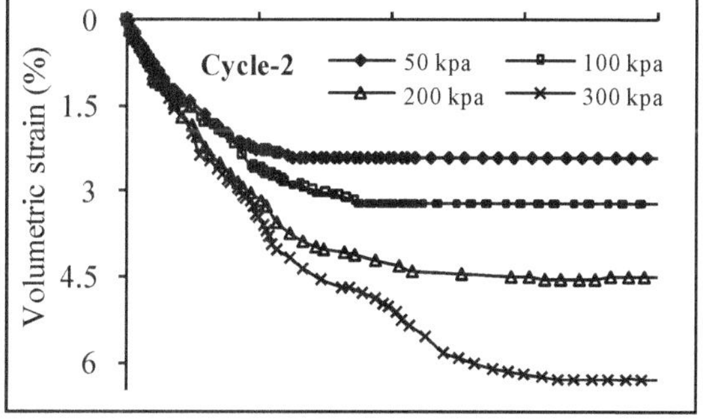

Figure 9. *Volumetric strain versus axial strain of* 2nd *drying-wetting cycle.*

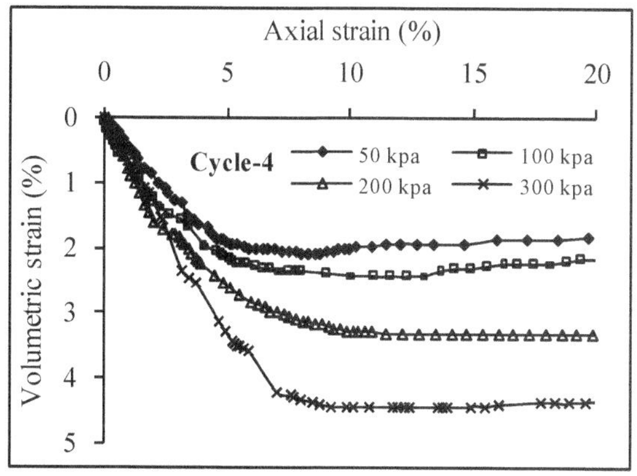

Figure 10. *Volumetric strain versus axial strain of* 4th *drying-wetting cycle.*

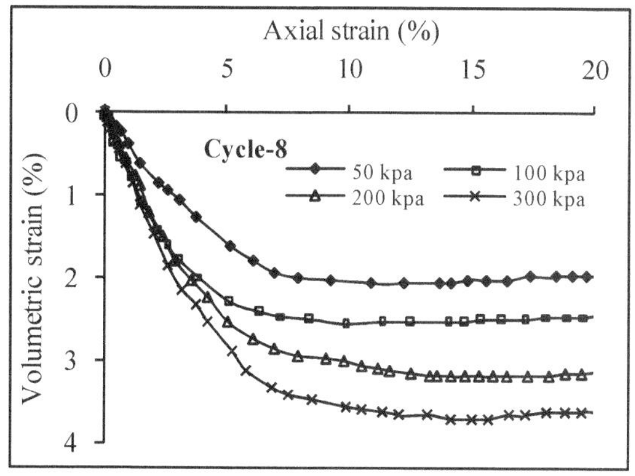

Figure 11. *Volumetric strain versus axial strain of* 8th *drying-wetting cycle.*

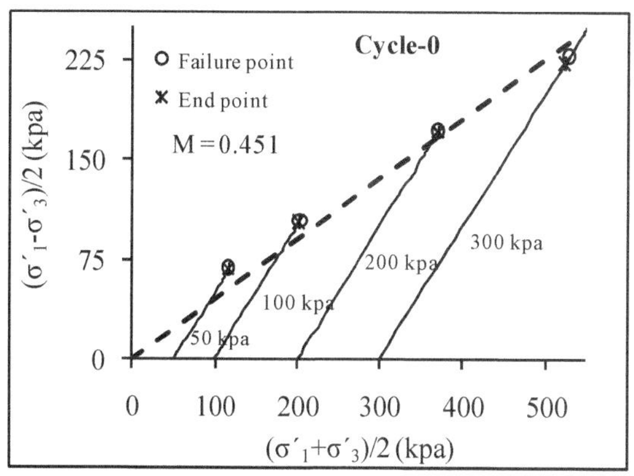

Figure 12. *Effective stress paths of initial soil samples.*

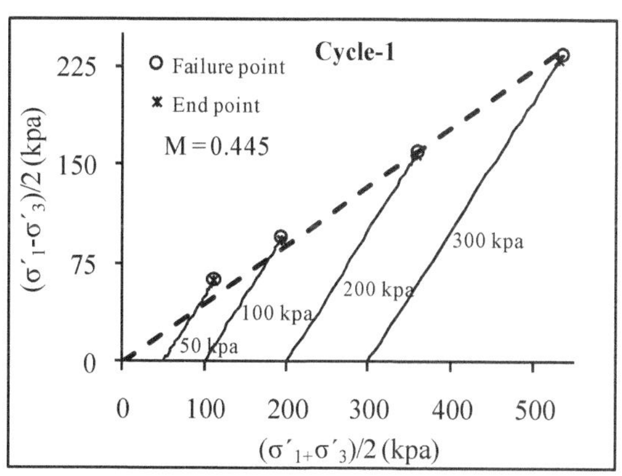

Figure 13. *Effective stress paths of* 1st *drying-wetting soil samples.*

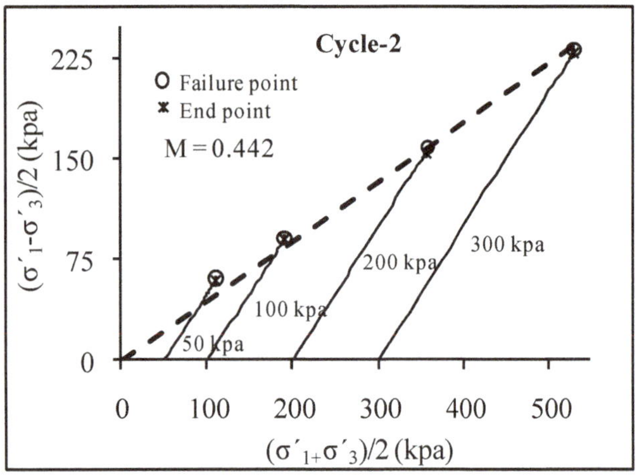

Figure 14. Effective stress paths of 2nd drying-wetting soil samples.

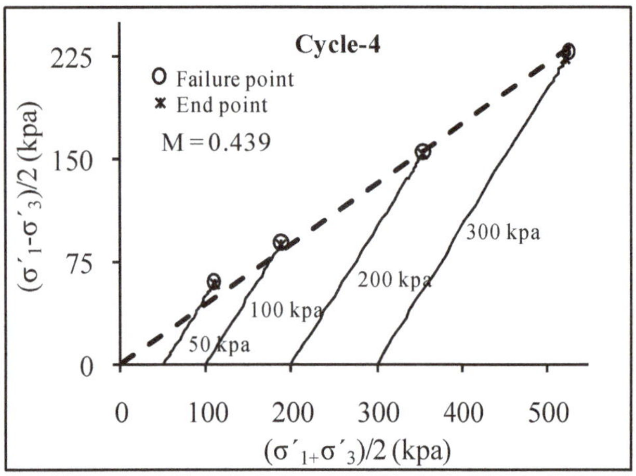

Figure 15. Effective stress paths of 4th drying-wetting soil samples.

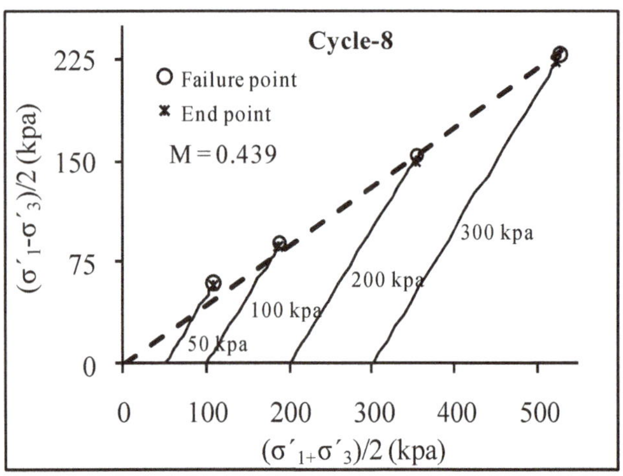

Figure 16. Effective stress paths of 8th drying-wetting soil samples.

The deviatory stress and initial stiffness of saturated soils increase and the volume of soils becomes contractive as the net normal stress increases. The increase in the net normal stress may force soil particles to close pack arrangement (volume contraction) during shearing.

This volume contraction makes the sample denser, eventually the shear resistance and the initial stiffness increase. But the deviatory stress and initial stiffness of saturated soils decrease with increasing drying-wetting cycle numbers. The slope of the critical state line also decreases with increasing cycle numbers.

The saturated strength parameters (c_d and φ_d) of the drying-wetting cycles specimens are determined by the classical Mohr-Coulomb approach which are given in table 2. The failure envelope of the drying-wetting cycles is shown in Fig. 17. The cohesion (c_d) and internal friction angle (φ_d) decrease with increasing cycle number (N) but the attenuation rate of φ_d is less than c_d (Fig. 17). The variations of c_d and φ_d with respect to drying-wetting cycles can be calculated using exponential function and the obtained results are fitted well with the measured c_d and φ_d data (Fig. 18).

Table 2. Saturated shear strength parameters of drying-wetting cycles.

Shear strength parameters	Drying-wetting cycle numbers (N)				
	0	1	2	4	8
Cohesion c_d	19.35	16.6	15.5	15	15
Friction angle φ_d	24.95	24.25	23.97	23.8	23.78

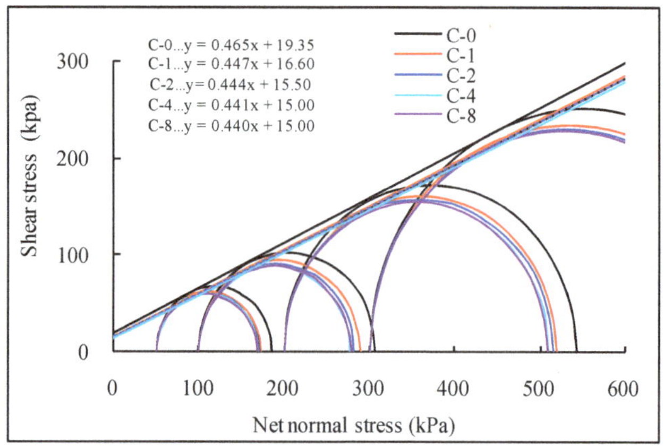

Figure 17. Mohr-Coulomb failure envelop of different drying-wetting cycles.

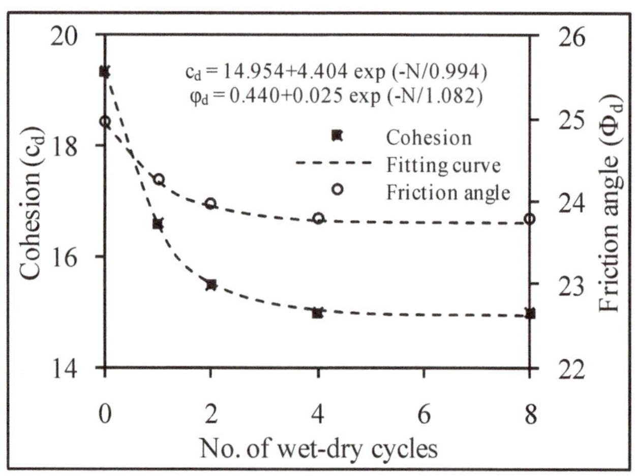

Figure 18. Variation of c_d and φ_d with respect to drying-wetting cycles.

The variations of saturated shear strength under different net normal stress and drying-wetting cycles are given in table 3 and found that the saturated shear strength increases with increasing net normal stress. It can also be seen that at same net normal stress, the shear strength decreases with increasing drying-wetting cycles.

Table 3. τ_{sat} with different wetting-drying cycles and net normal stress.

$(\sigma-\mu_a)$/kpa	τ_{sat} (kpa) with dry-wet cycles				
	0	1	2	4	8
50	42.6	39.1	37.7	37.05	37
100	65.85	61.6	59.9	59.1	59
200	112.4	106.6	104.3	103.2	103
300	158.9	151.6	148.7	147.3	147

The attenuation rate (%) of saturated shear strength with respect to drying-wetting cycles is calculated using following equation-

$$\Delta_\tau = \frac{|\tau_i - \tau_0|}{\tau_0} \times 100\% \quad (1)$$

Where, Δ_τ is the attenuation rate (%), τ_0 & τ_i are the saturated shear strength of initial and after different drying-wetting cycles respectively.

The variations of saturated shear strength attenuation rate (%) under different net normal stress and drying-wetting cycles are given in table 4. For example, at 50 kpa confining pressure, the attenuation rate is 8.22%, 11.50%, 13.03%, 13.15% for 1, 2, 4 and 8 cycles respectively. The attenuation rate from cycle 0→1, 1→2, 2→4 and 4→8 are about 8.22%, 2.78%, 1.53% and 0.12% respectively. It is stated that the attenuation rate is more pronounced in the first cycle and decreases with subsequent cycles and finally reaches to a constant state after 4 cycles. The attenuation rate also decreases with increasing net normal stress. This reduction of shear strength with drying-wetting cycles might be due to the changes of particle arrangement and the grain size distribution, decrease of pore volume and void ratio or due to the formation of micro-cracks and fissures.

Table 4. τ_{sat} attenuation rate (%) with different wetting-drying cycles and net normal stress.

$(\sigma-\mu_a)$/kpa	τ_{sat} attenuation rate (%)			
	Δ_{c-1}	Δ_{c-2}	Δ_{c-4}	Δ_{c-8}
50	8.22	11.5	13.03	13.15
100	6.45	9.04	10.25	10.4
200	5.12	7.17	8.14	8.32
300	4.56	6.39	7.27	7.46

Based on regression analysis, a mathematical function is proposed to obtain the best fitting saturated shear strength attenuation rate with drying-wetting cycle samples. The mathematical equation is-

$$\Delta_\tau = J_0 - \frac{J_0}{\left(1+\frac{N}{J_1}\right)^{J_2}} \quad (2)$$

Where, Δ_τ is saturated shear strength attenuation rate, N is the drying-wetting cycle number and J_0, J_1 & J_2 are fitting parameters. The variations of fitting parameters (J_0, J_1 & J_2) with respect to different net normal stress $(\sigma-\mu_a)$ can be calculated by the following equations-

$$J_0=7.2128 + 10.7652 \exp[-0.012(\sigma-\mu_a)] \quad (3)$$

$$J_1=-2.5494 - 5.36 \exp[0.0066(\sigma-\mu_a)] \quad (4)$$

$$J_2=-2.1357 - 5.1727 \exp[0.0065(\sigma-\mu_a)] \quad (5)$$

Using the above mathematical function, the predicted attenuation rate of saturated shear strength with respect to different drying-wetting cycles and net normal stresses are fitted well with the calculated data (Fig. 19). The correlation coefficients are more than 99%, indicating that the proposed mathematical function can describe the attenuation rate of drying-wetting cycle samples saturated shear strength.

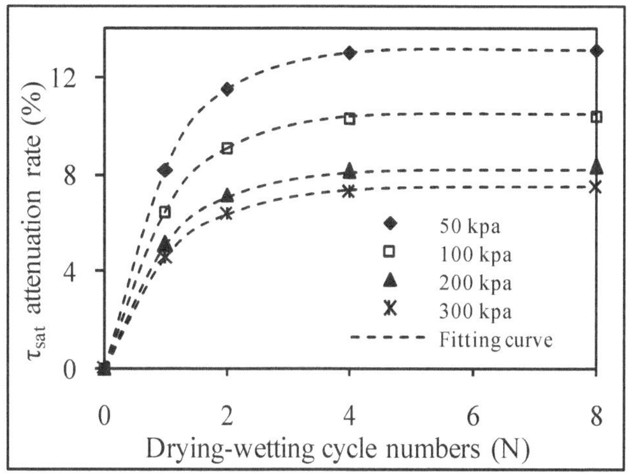

Figure 19. Fitting curves of τ_{sat} attenuation rate (%) under drying-wetting cycles.

4. Conclusion

In this study, the effects of multiple drying-wetting cycles on saturated shear strength of undisturbed residual soil are investigated using a series of consolidated drained triaxial tests. The stress-strain curves for the drying-wetting cycle samples appear to be strain-hardening. The deviatory stress and initial stiffness of saturated soils increase with increasing net normal stress but decrease with increasing drying-wetting cycle numbers. The axial strains to attain maximum deviator stress for these drying-wetting samples show higher values range from 14.57-18.94% which is a reflection of highly plastic nature of the soil. The slope of the critical state line decreases with increasing cycle numbers. The cohesion (c_d) and internal friction angle (φ_d) decrease with increasing cycle number (N) but the attenuation rate of φ_d is less than c_d. The variations of c_d and φ_d with respect to drying-wetting cycles can be expressed by exponential function. The saturated shear strength of residual soil decreases with increasing number of cycles. The effect is more significant for the first cycle and decreases with subsequent cycles and finally reaches to a constant state after 4 cycles. The attenuation rate of shear strength due to drying-wetting cycles is also analyzed. Furthermore, a mathematical function is proposed in this paper which can describe the saturated shear strength attenuation rate of drying-wetting cycle samples. Such studies are useful to understand the possible changes in shear strength behavior of residual soil that are subject to periodic drying and wetting from climatic variations and may provide the reference to the future geotechnical engineers and disaster reduction professionals to build up engineering structures safely and economically.

Acknowledgements

The authors would like to acknowledge financial support from the National Natural Science Foundation of China (Grant No. 41372314) and the Science and Technology Service Network Initiative of the Chinese Academy of Sciences (Grant No. KFJ-EW-STS-122). The first author gratefully acknowledges the PhD scholarship from CAS-TWAS President's Fellowship Programme.

References

[1] P. R. Vaughan, M. Maccarini and S. M. Mokhtar, "Indexing the engineering properties of residual soil," Quarter. Jour. Eng. Geol., vol. 21, 1988, pp. 69-84.

[2] L. D. Wesley, Geotechnical Engineering in Residual soils, Jonh Wiley & Sons, Inc., USA, 1990.

[3] P. G. Fookes, Tropical residual soils (Ed.), The Geological Society Publishing House, U. K., 1997.

[4] A. T. M. S. Hossain, "The engineering behaviour of the tropical clay soils of Dhaka, Bangladesh," Durham theses, Durham University, 2001 (*Available at Durham E-Theses Online: http://etheses.dur.ac.uk/3792*).

[5] H. Rahardjo, K. K. Aung, E. C. Leong and R. B. Rezaur, "Characteristics of residual soils in Singapore as formed by weathering," Eng. Geol., vol. 73 (1–2), 2004, pp. 157–169.

[6] H. X. Lana, R. L. Hu, Z. Q. Yue, C. F. Leek and S. J. Wang, "Engineering and geological characteristics of granite weathering profiles in South China," Jour. Asian Earth Sci., vol. 21, 2003, pp. 353-364.

[7] I. Jworchan, "Mineralogy and chemical propertries of residual soils," IAEG, No. 21, 2006.

[8] Y. E. -A. Mohamedzein and M. H. Aboud, "Compressibility and shear strength of a residual soil," Geotech. Geol. Eng., vol. 24, 2006, pp. 1385-1401.

[9] H. Rahardjo, F. Melinda, E. C. Leong and R. B. Rezaur, "Stiffness of a compacted residual soil," Eng. Geol., vol. 120, 2011, pp. 60-67.

[10] M. Kumruzzaman and J. Yin, "Stress-Strain Behaviour of Completely Decomposed Granite in both Triaxial and Plane Strain Conditions," Jordan Jour. Civil Eng., vol. 6 (1), 2012, pp. 83-108.

[11] F. A. M. Marinho, O. M. Oliveira, H. Adem and S. Vanapalli, "Shear strength behavior of compacted unsaturated residual soil," Inter. Jour. Geotech. Eng., vol. 7 (1), 2013.

[12] M. M. Allam and A. Sridharan, "Effect of wetting and drying on shear strength," Jour. Geotech. Eng., vol. 107 (GT4), 1981, pp. 421–438.

[13] K. Sobhan and B. M. Das, "Durability of soil–cements against fatigue fracture," Jour. Mater. Civ. Eng., vol. 19 (1), 2007, pp. 26–32.

[14] O. Cuisinier and F. Masrouri, "Hydromechanical behavior of a compacted swelling soil over a wide suction range," Eng. Geol., vol. 81, 2005, pp. 204-212.

[15] S. M. Rao and K. Revanasiddappa, "Influence of cyclic wetting drying on collapse behavior of compacted residual soil," Jour. Geotech. Geol. Eng., vol. 24, 2006, pp. 725-734.

[16] S. Tripathy, S. Kanakapura and S. Rao, "Cyclic swell-shrink behavior of a compacted expansive soil," Geotech. Geol. Eng., vol. 27, 2009, pp. 89-103.

[17] J. C. B. Benatti, M. G. Miguel, R. A. Rodrigues and O. M. Vilar, "Collapsibility study for tropical soil profile using oedometric tests with controlled suction," In: Alonso, E. E. & Gens, A. (Eds.), Proceedings of the 5th international conference on unsaturated soils, Taylor & Francis Group, London, 2011, pp. 193-198.

[18] R. D. Tovar and J. E. Colmenares, "Effects of drying and wetting cycles on the shear strength of argillaceous rocks," In: Alonso, E. E. & Gens, A. (Eds.), Proceedings of the 5th international conference on unsaturated soils, Taylor & Francis Group, London, 2011, pp. 1471-1476.

[19] M. Kholghifard, K. Ahmad, N. Ali, A. Kassim and R. Kalatehjari, "Collapse/Swell Potential of Residual Laterite Soil Due to Wetting and Drying-wetting Cycles," Natl. Acad. Sci. Lett., vol. 37, 2014, pp. 147-153.

[20] D. A. Sun, and D. J. Huang, "Soil-water and deformation characteristics of Nanyang expansive soil after wetting-drying cycles," Rock Soil Mech., vol. 36, 2015, PP. 115-119 (*in Chinese*)

[21] P. H. Morris, J. Graham and D. J. Williams, "Cracking in drying soils," Can. Geotech. Jour., vol. 29, 1992, pp. 263–277.

[22] B. A. Albrecht and C. H. Benson, "Effect of desiccation on compacted natural clays," Jour. Geotech. Geoenviron., vol. 127, 2001, pp. 67-75.

[23] K. Harishkumar and K. Muthukkumaran, "Study on swelling soil behaviour and its improvements," Int. Jour. Earth Sci. Eng., vol. 4, 2011, pp. 19-25.

[24] Y. Wan, Q. Xue, L. –Y, Zhao, Y. –J. Du and L. Liu, "Effects of wetting-drying cycles on permeability of compacted clay cover at landfill site" Rock Soil Mech., vol. 36 (3), 2015, pp. 679-693 (*in Chinese*).

[25] F. Akcanca and M. Aytekin, "Impact of wetting-drying cycles on the hydraulic conductivity of liners made of lime-stabilized sand-bentonite mixtures for sanitary landfills," Environ. Earth Sci., vol. 72, 2014, pp. 59-66.

[26] Z. T. Zeng, H. B. Lu and Y. L. Zhao, "Wetting-Drying Effect of Expansive Soils and its Influence on Slope Stability," Appl. Mecha. Material, vol. 170-173, 2012, pp. 889-893.

[27] J. Zhou, H. Z. Xu and W. J. Hu, "Impact of wetting-drying cycle effects on stability of expansive soil slopes," Chin. Jour. Geotech. Eng., vol. 35 (zk2), 2013, pp. 152-156 (*in Chinese*).

[28] A. A. Basma, A. S. Al-Homoud, A. I. Husein-Malkawi and M. A. Al-Bashabsheh, "Swelling-shrinkage behavior of natural expansive clays," Appl. Clay Sci., vol. 11, 1996, pp. 211-227.

[29] S. Tripathy, K. S. Subba Rao and D. G. Fredlund, "Water content-void ratio swell-shrink paths of compacted expansive soils," Can. Geotech. Jour., vol. 39, 2002, pp. 938-959.

[30] F. J. Leij, T. A. Ghezzehei and D. Or, "Modelling the dynamics of the soil pore-size distribution," Soil Tillage Res., vol. 64, 2002, pp. 61-78.

[31] L. Li, W. Zhu, C. Lin and T. Ohki, "Study of wet and dry properties of solidified sludge," Rock Soil Mech., vol. 30 (10), 2009, pp. 3001-3004 (*in Chinese*).

[32] C. Yang, F. Zha and K. Cui, "Effect of cyclic wetting and drying on the engineering properties of stabilized expansive soils," Indust. Constr., vol. 42 (1), 2012, pp. 98-103 (*in Chinese*).

[33] C. -S. Tang, D. -Y. Wang, B. Shi and J. Li, "Effect of wetting-drying cycles on profile mechanical behavior of soils with different initial conditions," Catena, vol. 139, 2016, pp. 105-116.

[34] D. –Y. Wang, C. –S. Tang, Y. –J. Cui, B. Shi and J. Li, "Effects of wetting–drying cycles on soil strength profile of a silty clay in micro-penetrometer tests," Eng. Geol., vol. 206, 2016, pp. 60-70.

[35] ASTM D7181-11, Method for Consolidated Drained Triaxial Compression Test for Soils, ASTM International, West Conshohocken, PA, 2011.

Quantitative and Qualitatively Evaluation of New Technologies in Perspective of Construction Optimization

Hassan Saghi*, Nima Arefizadeh

Department of Civil Engineering, Hakim Sabzevari University, Sabzevar, Iran

Email address:
h.saghi@hsu.ac.ir (H. Saghi), mohandes.n.arefi@gmail.com (N. Arefizadeh)

Abstract: Novel research studies to find approaches to reduce the cost and time in technology and materials of construction has resulted in modern technologies including insulating concrete formwork, hebelex blocks materials, roof technology of metal deck, Light Steel Framing and Rebar butt forge welding process. Investigation of the new technologies for construction in terms of the cost and time is the purpose of this paper. At first, these technologies are introduced. Then elements, restrictions and some unique special features of this system are presented. In addition, these technologies are studied in an important regarded for the optimization parameters including the cost and time. This paper provides a quantitative, qualitative and relative comparison between different techniques. Furthermore, new optimized technologies are determined and other new technologies for the investigating factors of optimization are suggested.

Keywords: New Technologies, Construction, Construction Optimization, Optimized Technology

1. Introduction

Creative searching to find approaches to reduce the cost and time of production of construction materials has led to the introduction of new technologies in construction. New technologies for construction include insulating concrete formwork system, Hebelex block materials, roof technology of metal deck, light steel framing system and Rebar butt forge welding process. The purpose of this paper is to investigate these technologies for construction in term of cost and time. Previous researches presented results in order to introduce and evaluate new technologies for manufactures. According to this research, lack of awareness of the creators from modern methods of construction was reason of not using these technologies in large scale [1].

Also, Hebelex is considered as a good alternative for traditional materials such as brick [2]. Considering the advantages and disadvantages of the light steel framing system, the need to develop manufacturing and machine tools and use of these systems has been known. Furthermore, training of the skilled manpower in the field of installation and working with materials, developing practical regulations and supporting research centers such as universities to cover and eliminate the disadvantages of these systems have been recommended [3]. Therefore, this study aimed to investigate the construction of new technologies from the perspective of the important optimizing factors, i.e., cost and time.

2. New Technologies for Construction

A number of new technologies including Insulating concrete formwork system, Hebelex blocks materials, Metal deck roof technology, and Rebar butt forge welding process are investigated, separately and cost and time of each technology are evaluated.

2.1. Insulating Concrete Formwork System

Insulating concrete formwork system is a building system including reinforced concrete and polystyrene panels in which the reinforced concrete acts as load-bearing and polystyrene panels act as concrete formwork and thermal and acoustic insulation [1]. As it could be understood from name of this system, after concreting, molds remain. The system includes load-bearing components for walls, ceiling panels and wall separators [1]. Main ingredients in this system are expanded polystyrene that is available with relatively acceptable price. The initial cost of this technology is significant, but for some reasons such as saving transportation cost, speed of implementation of the system, and wall thermal insulation at the same time, eventually cost of the product is reduced. Saving transportation cost and increasing implementation speed is achieved due to some reasons such as low weight, easy displacement of parts, and not needing to heavy machinery [5].

Limitation: Due to the use of polystyrene in the system, it is necessary to consider a series of safety cases [6], including that foam polystyrene blocks must be valid in accordance with standards; For example, the block must be fire retarded [6]. This system is used for small residential buildings and villas [6].

2.2. Hebelex Blocks Materials

Hebelex is trade name of Aerated Concrete that is invented and introduced for the first time in 1924 by Swedish architect engineer. It is widely used in the United States and Europe [2]. This product (Figure. 1) is now presented with the name of Autoclaved Aerated Concrete (AAC) or Gas concrete [5, 6]. In this production process, bubble and porosity are caused by chemical reaction due to materials such as Aluminum powder [6]. In comparison with other kinds of concrete, these are more suitable in term of thermal insulation [7]. Considering different ways of production it is provided with different brand names, such as SYPOREX, YTON and ARECON [4]. Its ingredients include silicium oxide, cement, calcium oxide and some aluminum powder that are and cooked in 200 °C and they are produced in sizes 10*25*15, 25*25*60, 60*25*60, and 30*25*60 cubic centimeters or manufactured with arbitrary and custom made sizes.

Fig. 1. Production of Hebelex.

Block parts can be made with Portland cement- sand mortar with additional materials, calcium oxide-Portland cement-sand mortar (Batard) and special adhesive. It is recommended that traditional cement and sand mortar (1:6) to be used as leveling mortar under [4]. Since this product is simply cut, wastage of these materials is low. The suitable cutting ability and Navigability accelerate implementation [4]. Cement mortar in the Hebelex wall is 25 percent of the needed materials for producing brick and hence, this product consumes less cement [2]. The relative lightness of this product (Table 1) reduces volume of steel or concrete sections. Not needing to run the liners of Chalk soil on the wall (without intermediary, layer of chalk can be implemented on the Hebelex wall) in case of the proper execution of the work and relative high speed of installation (Table 2) will results in reducing the cost and time of implementation of the wall picking.

Table 1. *Comparison of the weight of walls with different materials (kg/m^2) [4].*

Wall Thickness(cm^2)	Type of wall			
	Brick	earthenware block	3D Panel	Hebelex
10	185	85	110	65
20	360	165	130	130

Table 2. *Comparison of the implementation speed of the various walls by an executive team at a working day (m^2) [4].*

Wall Thickness(cm^2)	Type of wall		
	Brick	earthenware block	Hebelex
10	15	25	45
15	10		40
20	5	20	30

Hidden costs of building related to the use of this product in regard to energy (Table 3) and fire resistance, resistance against frost due to the porosity, as well as high durability, and the possibility of recovering the product in the production cycle should be considered. The shortage of the silicium oxide in the future may lead to an increase in the price of this product. Usage limitation of this concrete is determined by its resistance. In this regard, these kinds of concrete are divided into three categories: 2.5, 5, and 7.5 MPa. [8]

Table 3. *Compare the Coefficient of Thermal Conductivity (w/(m *k)) [4].*

Wall Thickness(cm^2)	Type of wall			
	Brick	Earthenware block	3D Panel	Hebelex
20	1.81	1.52	0.62	0.13

2.3. Metal Deck Roof Technology

In this technique, galvanized sheets with ribbed trapezium sections which are cold-formed are used as the permanent concrete mold and replacement of the major rebar. These galvanized sheets are used up to ceiling span of 4 meters without temporary pillar and to ceiling span of 7 meters with temporary pillar in the midspan (Fig. 2). The materials of the ceiling include metal deck sheets, pegs or stud, rebar, concrete and spacers (rebar to the metal deck) [9]. Restriction of using the metal deck plates is in the condition of severe wind blowing (before connection of structural elements). With considering time and cost criteria, we can mention the following points:

Fig. 2. Galvanized sheets used in temporary pillar in the Midspan.

- Improving the performance and reducing the total cost.
- Reducing the cost of concrete curing with water conservation.
- Relatively reduction of the consumable materials cost (Table 4).

Table 4. Comparison of the materials consumption between metal deck roof and composite per square meter [14].

Cases	Composite roof technology	Metal deck roof technology
Steel	An average of 40 Kg	An average of 28 Kg
Rebar	8 ϕ 25@ (cm)	8 ϕ 30 @(cm)
Concrete with 350 (kg/m^3) cement content	0.1 m^3	0/085 m^3
UNP steel profile	2 Kg	
Plastic	1 m^2	
Welding electrode	On average, metal deck consumes 0.5 Kg lesser than composite.	
Equal angles (L Steel profile)	On average, metal deck consumes 0.5 Kg lesser than composite.	
Foundation rebar	On average, metal deck is 2 Kg lighter than composite.	
Shear wall rebar	On average, metal deck is 1.2 Kg lighter than composite.	

In order to improve the performance and reduce the total cost, the following methods can be used:

1. The ability to run several roof formatting simultaneously (albeit, in the case of the suitable temporary side bracing) or the ability to install a quick platform in all levels [10]
2. The ability to install elevator on the roof (assuming proper bracing).
3. The ability of faster installation of the stairways [10] (Fig. 3).

Fig. 3. The ability of faster installation of the stairways [13].

2.4. Lightweight Steel Frame

Lightweight steel (LFS) frame (Fig. 4) is made from cold rolled steel sections [11]. Using of this system is one of the suitable ways to build additional floor in the structure. The main components include cold-formed galvanized steel profile that are shaped in Z, C and U sections. These components are used horizontally or vertically and at specific intervals and also self-dilling screws or nuts and bolts are employed for connecting them. [11]

Fig. 4. Light weight steel.

Peripheral components include gypsum board or cement board or polystyrene foam [3]. This type of roof is usually sloped which is made of metal trusses. The trusses are made of cold-rolled profiles. Also, other parts of the building are implemented by using cold-rolled steel and with a variety of boards (gypsum, cement, wood) [12]. Restriction of the usage of the LSF in the windy areas is that the number of stages is limited to 5, and maximum to 8 in some special cases [15]. By studding the cost and time criteria, it can be concluded that using this technique the speed is increased and cost of the materials is reduced (Table 5).

Table 5. *Comparison of the features of traditional buildings and LSF system .The average specifications of Buildings 2, 3 and 5 floors [3].*

The row number	Quantity	Concrete	Steel	LSF
1	Steel weight (kg/m^2)	37.2	62.5	28.6
2	Concrete weight (kg/m^2)	1197	529	245
3	Skeleton weight (ton)	213	19.4	7.1
4	Total weight (ton)	489	340.9	160.7
5	Cutting the earthquake (ton)	57	39.2 and 43.9	28.8
6	Displacement (cm)	2.1	6.5 ,3.15	1.2
7	Frame stage (day)	80	70	25

2.5. Head to Head Forging Welding

This technology was invented in 1952, having the aim to use in the rail transportation industry. Over time, it is expanded into the construction industry due to its advantages (Fig. 5) [16].

Fig. 5. *Head to head forging welding.*

In this way, speeding up the construction could be achievable, by taking advantage of the followings [17]:

1. Decreasing consumed rebar ratio, with respect to the rebar diameter and its specified brokerage.
2. Waste rebar could be appropriately connected, to be re-used in construction process.
3. Reducing latent cost by increasing the resistance and persistence of the structure [17].
4. Increasing time and cost of operating mistakes, such as: when length of an armature root being short.

It is noteworthy to mention that reducing final cost are effected by time and cost, will be done with regards to above mentioned methods.

To complement the aforementioned method, the following points should be considered:

1. For instance, decreasing 25-35 % of vertical consumed rebar's taking place is sheer walls with respect to the following assumption:

Firstly, similar vertical reinforcement all over the height of the sheer wall from the foundation elevation up to the highest floor, is done. Moreover, vertical rainforest are typically connecting on each floor.

2. These advantages will be provided due to removing the armature patch, it is noteworthy, according workshop executive formula this length is 60 times longer than nominal diameter of the larger rebar, however, the maximum nominal diameter is residential constructions are limited to 28 millimeter.
3. According to [16] consultant Company reports using forging methods in case of rebar with nominal diameter less than 18 has no economic justification[16]. Regarding aforementioned points, an interval for decreasing vertical consumed rebar, will be predicted. For instance, suppose a building structure with height of 324 cm for sheer walls, the mentioned interval will be calculated by the following formula (Equation 1):

$$(18*60/10)/[(18*60/10)+324] \leq \text{interval} \leq (28*60/10)/[(28*60/10)+324] \quad (1)$$

$$0/25 \leq \text{interval} \leq 0/35$$

This formula is based on the proportion of armature patch length to the brokerage length.

In this method nominal diameter is limited to lower band 18 and upper bond 28. In addition, the coefficient of 60 is related calculating patch length of the rebar. Additionally, number of 324 is the height of the sheer walls for each story.

Clearly, this formula is expressed in SI. Not using this method, the waste rebar must be sold cheaper than its buying costs. This benefit will be achieved providing better condition for vibration due to the decrease of weight of extra rebar in place of joints rather than mechanical patch method. If one does not use the proposed method for solving the problem of length of the rebar root being short, another method have to be used. Such as implanting bolts which are relatively more expensive and wasting time.

3. Discussion, Conclusion and Recommendations

This article presented a quantitative and qualitatively evaluation of new technologies in perspective of construction optimization using relative parameters analysis method in terms of the cost and time. It seems that the presented new technologies provide the possibility to a quicker return of the investment with increasing the speed of construction, and thus reducing the time. Also, they will eventually reduce cost of building with reducing amount of consumed materials. As a result, these new technologies are innovative technologies to optimize manufacturing.

It is suggested to investigate optimization criteria of other new technologies such as dry wall technology (Dry Wall), polymeric or sprayed plaster, and *Concrete-Filled Steel Tubular Columns* (CFT).

References

[1] M. Mansouri and F. Hatami Bargh, " Investigating Insulating concrete formwork (ICF)," First National Congress on Construction Engineering and Projects Assessment, Gorgan, Iran, 2014.

[2] R. Zare'ian, A. R. Raeisi Vanani, M. R. Ghasemian, "performance review and comparison Heblex as an alternative to the brick", *2nd Conference National of new findings in Civil Engineering*, Islamic Azad University of Najaf Abad, 2012.

[3] A. Rahman, "The advantages and disadvantages of the light steel frame", *National Conference on Structure, Road, and Architecture*, Chalus Azad University, 2011.

[4] www.hebelexbirjand.com.

[5] M. Golabchi and H. Mazaherian, "New Architectural Technologies," Tehran University, Tehran, 1388.

[6] B. Mohammadkari, et al, "Insulating Concrete Formwork system (ICF), Collection of research reports, *Review and evaluation of the proposed system in residential buildings,* Mass housing projects, Building and housing research center of Iran, Tehran, 1387; (G-493): 52.

[7] M. Monzavi, M. Kermani, H. sobhani, "Evaluation and requirements for lightweight concrete in industrialization, *Construction industry of new technologies journal,* 1390; (2): 59.

[8] Iran's National Building Regulations Office, "Section 5, National Building Regulations- Building materials and products," 4th, Tehran, Iran's development, 1393; 62.

[9] www.bornaista.com.

[10] H. Hatefi, "Comparison of the roof (composed of steel – concrete) with common combined ceilings" *Construction industry of new technologies journal*, 1390; (1): 13.

[11] www.ista-lsf.ir.

[12] B. Mohammadkari, et al, " Lightweight Steel Frame (LSF)," Collection of research reports, Review and evaluation of the proposed system in residential buildings, mass housing projects, building and housing research center of Iran, Tehran, 1387; (G-488): 3,65.

[13] www.padabgostar.comR9.

[14] www.arshdamoun.com.

[15] A. R. Raeisi Vanani, R. Zare'ian, E. Rezaei, "Studding the light steel frame structures (LSF), causes and mechanisms of its development in developing countries", *2nd National Conference of new findings in Civil Engineering*, Islamic Azad University of Najaf Abad, 2012.

[16] http://forjing.sts-weld.com.

[17] www.parsasaze.com.

Comparison between Strain-Based vs. Stress-Based Criteria in Seismic Performance Evaluation of High Arch Dams

Masood Heshmati[1,*], Seyed Mohammad Seyed Kolbadi[2], Seyed Mahdi Seyed Kolbadi[1]

[1]Department of Civil Engineering, K. N. Toosi University of Technology, Tehran, Iran
[2]Departmet of Civil Engineering, Lame'i Gorgani University, Tehran, Iran

Email address:
masood_heshmati@sina.kntu.ac.ir (M. Heshmati), mohamad.kolbadi.pwut@gmail.com (S. M. S. Kolbadi), mahdi_kolbadi@sina.kntu.ac.ir (S. M. S. Kolbadi)

Abstract: In the present paper, seismic performance of a high concrete arch dam is evaluated based on both the stress and strain criteria. For this purpose, the finite element model of the selected arch dam-reservoir-foundation system was provided. Reservoir was modeled using Eulerian approach as a compressible domain and the foundation rock was assumed to be massless. Dynamic equilibrium equations for the coupled system were solved using Newmark's time integration algorithm. Seismic performance evaluation of dam-foundation-reservoir systems were performed considering parameters such as demand-capacity ratio, cumulative inelastic duration and extension of overstressed (or overstrained) areas obtained from linear elastic analyses and compared with the real crack profile from nonlinear analysis. It was found that although results obtained for the stress and strain rules have similarities, performance evaluation based on the strain gives different results which can be lead to different decision making in dam safety related projects.

Keywords: Arch Dam, Cumulative Inelastic Duration, Demand-Capacity Ratio, Seismic Performance Evaluation, Strain-Based Criteria

---

1. Introduction

Performance of arch dams maybe evaluated in linear or nonlinear phase considering various assumptions for material and loading. Generally, it's common to evaluate seismic behavior in linear phase considering various parameters and; in the case of extensive damage within the dam body, nonlinear analysis of the system should be conducted.

Several researchers such as Ghanaat [Ghanaat, Y. 2004, Ghanaat, Y. 2002]; Fok [Fok and Chopra, 1986]; Yamaguchi et al. [Yamaguchi, Y. et al. 2004], Bayraktar [Bayraktar, A., et al. 2009] and Hariri et al [Hariri Ardebili and Mirzabozorg and Ghaemian, 2011] have investigated seismic performance of concrete arch dams. Hall et al. [Hall and Matheu, and Liu, 1999] proposed some indices for systematic comparison of various ground motions effects. Ghanaat [Ghanaat, Y. 2002] proposed a methodology for damage estimation in concrete dams which can be found in USACE guideline [US Army Corps of Engineers (USACE), 2007]. Bayraktar [Bayraktar, A., et al. 2009] evaluated seismic performance of concrete gravity, arch, roller-compact-concrete (RCC) and concrete-face-rock fill-dam (CFRD) dams using indices proposed by USACE subjecting to near and far fault ground motions. Hariri et al. [Hariri Ardebili and Mirzabozorg and Ghaemian, 2011] investigated the effect of water level on dynamic performance of arch dams.

Wieland and Fan [Wieland and Fan, 2004], Wieland et al. [Wieland and Brennerand Sommer, 2003] investigated behavior of concrete dams under recent earthquake. Studer [Studer, 2004] was studied seismic performance of new and existing dams using methods proposed by international committee of large dams and expressed that the general procedure is clear and straightforward while practically detailed information are needed. Yamaguchi et al.

[Yamaguchi, Y. et al. 2004] discussed the role of nonlinear dynamic analyses in seismic evaluation problems of 2D concrete gravity dams. They examined the application of linear analyses in term of demand-capacity ratio (DCR) to provide qualitative estimates of potential level of damage under moderately severe excitations. Also, Hariri and Mirzabozorg [Hariri Ardebili and Mirzabozorg, 2011] studied seismic performance of concrete arch dams using real ground motions and also Endurance Time Acceleration Functions (ETAFs).

All previous researches utilize the stress to determine seismic behavior of concrete dams. However, it's important to note that behavior of concrete is based on the strain. In the present paper the common criteria for seismic assessment of concrete arch dams based on the stress are substituted by similar criteria, which are based on the strain rule. A high concrete arch dam is modeled for this purpose and evaluated using stress-based and strain-based rules. The results are compared with each other using the parameters such as the DCR, cumulative inelastic duration (CID) and percentage of overstressed (or overstrained) areas within the dam body in various performance levels.

2. Behavior of Mass Concrete

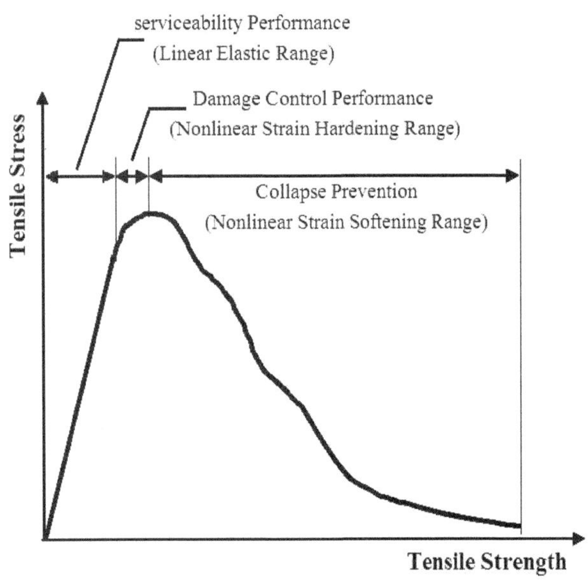

Figure 1. Conventional stress-strain relation for mass concrete structures (USACE, 2007)

Actual response of massive concrete structures to earthquake ground motions is very complicated. Loading histories and rapid seismic strain rates have important roles in structural performance (USACE, 2007). As it is known, mass concrete has limited ductile behavior. This behavior is characterized by a stress-strain relation composed of elastic and inelastic strain ranges followed by a complete loss of strength (USACE, 2007). Typical tensile stress-strain diagram of mass concrete is shown in Figure 1. The curve is divided into the three parts. In the first section, in which concrete behaves as a linear elastic (LE) material, the dam called to have serviceability performance. The second part is inelastic-strain hardening range which called as damage control phase and provides only limited inelastic behavior in the dam body. In this situation damage may be significant but all cracking and joint openings are limited and discrete [Ghanaat, Y. 2002]. Except for unlikely MCE events, it is desired to prevent damage in main elements such as foundation and other inaccessible structural elements. A LE analysis combined with a predefined performance evaluation criteria can be used to assess the dam response in the damage control phase. The dam response beyond the damage control range is governed by complete loss of strength, sliding, and nonlinear response behavior of discrete blocks bounded by opened joints and cracked sections and called as collapse prevention performance. This behavior must be evaluated using nonlinear time-history analysis [US Army Corps of Engineers (USACE), 2007].

3. Methodology of Performance Evaluation

Figure 2 describes the proposed flowchart for seismic evaluation of concrete arch dams using stress-based and strain-based methodology. Seismic performance of concrete arch dams is evaluated in accordance with displacements, stresses, strains, demand-capacity ratio, cumulative inelastic duration and spatial extension of overstressed (or overstrained) areas ($A^{overstress}$ or $A^{overstrain}$) on the upstream (US) and downstream (DS) faces of the dam body.

3.1. Demand-Capacity Ratio

For arch dams where high stresses and strains are usually oriented in the arch and cantilever directions, DCR refers to the ratio of calculated arch or cantilever stress (or strain) to the tensile strength of mass concrete or its equivalent strain, but it can also be developed for principal stresses (or strains) [Ghanaat, Y. 2002]. Tensile strength of mass concrete used in computation of DCR is obtained from uni-axial splitting tension tests or from Raphael proposed diagram (Raphael, 1984).It should be noticed that the DCR used in this method is the ratio of dynamic stresses (or strains) to static tensile strength (or its corresponding strain) and should be used in conjunction with other criteria for safety evaluation of concrete dams. In the method proposed by USACE, the mass concrete is assumed as homogeneous isotropic material and so its properties in three principal directions are the same. Also the static tensile strain of the concrete is calculated at the end of the linear part of stress-strain curve showed in Figure 1 and in fact just considers the serviceability performance range. The strains in dynamic range are approximately time-independent and are calculated using dynamic modulus of elasticity and Poisson's ratio of concrete.

3.2. Cumulative Inelastic Duration

Cumulative inelastic duration, which is a measure of energy, accounts for magnitudes as well as duration of stress (or strain) excursions. It refers to the total duration of stress (or strain) excursions above a stress (or strain) level associated with a certain DCR (Ghanaat, 2004). As shown in the flowchart introduced in Figure 2, the behavior of the dam is evaluated in three zones based on the estimated damage severity. The Performance Threshold Curve (PTC) for arch dams is shown in Figure 3 (Ghanaat, 2004 and USACE, 2007).

3.3. Spatial Extension of Overstressed (or Overstrained) Areas

In addition to foregoing performance criteria, introduced damage criteria require to be bounded in the limited areas, so that evaluation based on LE analysis is still valid. If spatial extent of damage or nonlinear response is limited to 20% of the total areas on the upstream or downstream faces, LE analysis is valid (USACE, 2007).

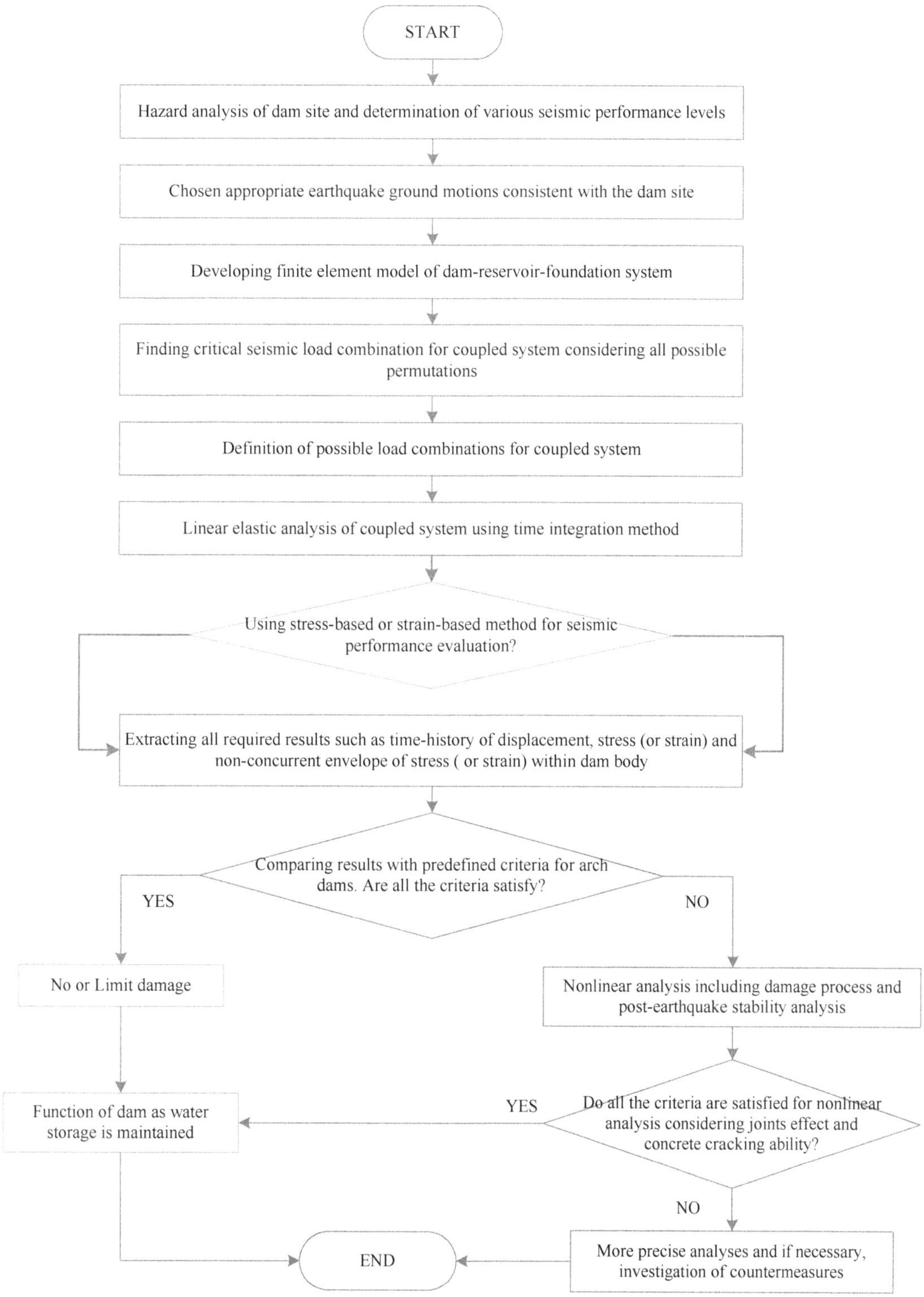

Figure 2. *Flowchart for seismic performance evaluation of arch dams using stress- (or strain-) based*

Table 1. *Quantifying the limit-states*

Limit-States	DCR		DCR-CID Diagram		$A^{overstress}/A^{overstrain}$
Minor or No Damage	DCR≤1.0	&	Zone I	&	0.0%
Acceptable Level of Damage	1.0<DCR<2.0	&	Zone II	&	≤20.0%
Severe Damage	DCR≥2.0	or	Zone III	or	>20.0%

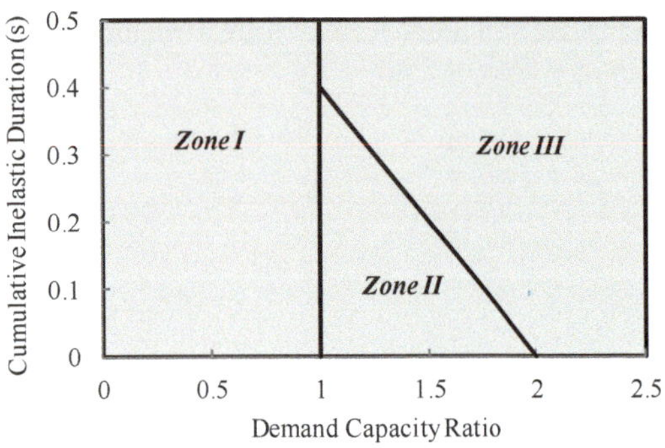

Figure 3. *Zoning the CID-DCR diagram and Performance Threshold Curve (PTC) for arch dams*

3.4. Quantification of Limit-States

Herein, the utilized methodology for quantifying the defined limit-states in the previous section is introduced. This method uses a combination of all previously defined criteria in conjunction with LE analysis for both stress-based and strain-based rules. Table 1 represents the tabular form of the introduced criteria for performance evaluation of arch dams.

4. Case Description

4.1. Finite Element Model of DEZ Dam

Dez high double curvature arch dam is selected as the numerical example. Total height of the dam is 203m but the height above its concrete plug (the simulated dam) is 194m. Also, the height of the main body sitting on the concrete saddle is 186m. Thickness of the dam at the crest is 4.5m and its maximum thickness at the base is 21m. Rectangular shaped mass-less foundation is used in this case while reservoir length modeled is about five times of the dam height. The provided finite element model is shown in Figure 4, which consists of 792 eight-node solid elements in the body and 3770 elements in the surrounding foundation rock. In this model, hexagonal elements are used in the dam body, while prism elements are utilized at some localized regions with irregular geometry. Also, 3660 eight-node Eulerain fluid elements are used to model the reservoir domain.

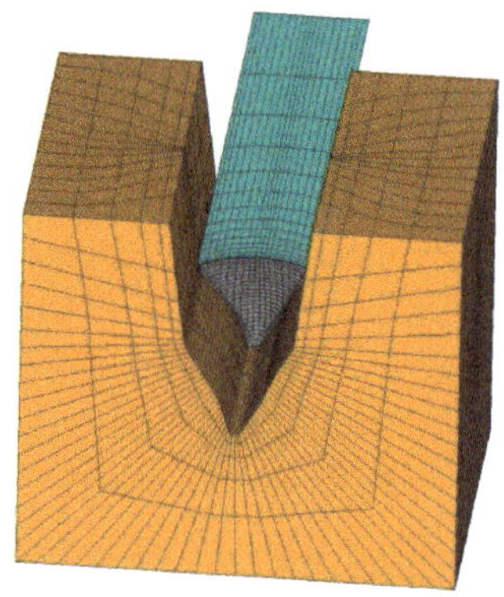

Figure 4. *Finite element model of dam-reservoir-foundation system;Dez Dam*

4.2. Material Properties

Material properties for mass concrete and foundation rock are given in Table 2 [Hariri Ardebili et al., 2011].The reservoir water density is taken as 1000kg/m^3, the sound velocity is 1440m/s in water and the wave reflection coefficient for the reservoir around boundaries is taken 0.8, conservatively.

Table 2. *Material properties for the dam body and its surrounding foundation*

	Properties	Static Values	Dynamic Values
Mass concrete	Modulus of Elasticity	40GPa	46GPa
	Poisson's Ratio	0.2	0.14
	Mass Density	2400 kg/m^3	
	Compressive Strength	35MPa	36.5MPa
	Tensile strength	3.4MPa	5.1MPa
	Thermal expansion coefficient	6×10^{-6}/°C	
Foundation rock	Deformation Modulus-Saturated	13GPa	
	Deformation Modulus-Unsaturated	15GPa	
	Poisson's Ratio	0.25	

5. Loading the Coupled System

The applied loads are the dam body self-weight, hydrostatic pressure in summer condition, thermal loads and finally seismic loads based on seismic hazard analysis of the dam site. The applied load combinations are presented in Table.3. In this table, each load combination is shown with an abbreviation in which the first letter represents the dam name, the second one represents the thermal load conditions (S indicates summer), the three other letters are the level of the seismic excitation and the final two numbers represents No. of earthquake ground motions. Nine ground motions are used for analysis of Dez dam in each excitation levels (Table 4).

Table 3. *Applied load combinations*

Dam name	Abbreviation	Definition
DEZ Dam	D-S-MDE09	$W+h_s(NWL)+T_s+MDE^1$

W: Dam Self-Weight
h_s: Hydrostatic Pressure
NWL: Normal Water Level
T_s: Summer Temperature
[1] Maximum Design Earthquake

Table 4. *Characteristics of ground motions*

Excitation Level	NO	Earthquake Name	Station
	1	DUZCE	At 1061 Lamond station
	2	LOMA-PRIETA	At 47006 Gilroy-GalivanColl station
	3	MANJIL	At Abbar station
	4	NORTH RIDGE 1	At 24088 Pacoima Kagel Canyon station
MDE	5	NORTH RIDGE 2	At 90059 Burbank Howard Rd station
	6	QAEN	At Qaen station
	7	SAN FERNANDO	At 128 Lake Hughes # 12 station
	8	SPITAK	At Gukasyan station
	9	TABAS	At Tabas station

All ground motions are scaled based on the horizontal and vertical components of the design response spectrums' considering 5% for damping as shown in Figure 5 (USACE, 2007; Zhang et al., 2009; Chen et al., 2012) .The β-Newmark time integration method is utilized to solve the coupled problem of dam-reservoir-foundation model and finally, the system is excited at the foundation boundaries using the scaled earthquake records. Moreover, structural damping is taken to be 5% of critical damping in the conducted analyses. The thermal distribution within the dam body is obtained from the thermal transient analysis of the dam considering water temperature, air temperature and solar radiation recorded at the site. The temperatures obtained on the upstream and downstream faces of the dam body along the crown cantilever for the winter and summer conditions are shown in Figure 6.

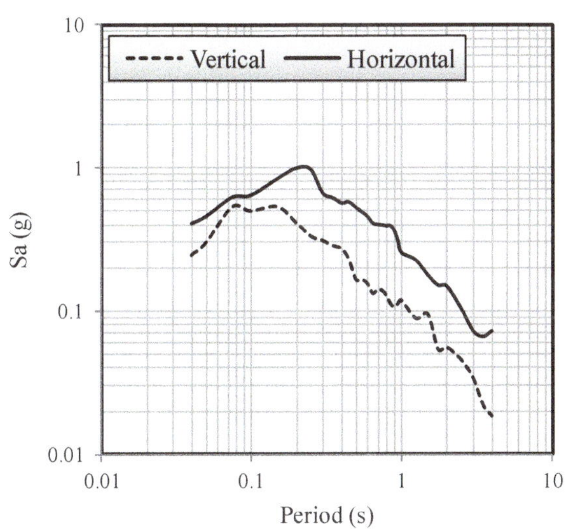

Figure 5. *Design response spectrum MDL*

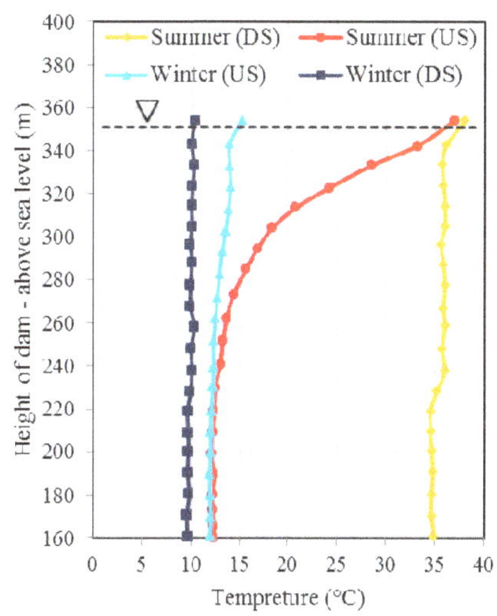

Figure 6. *Temperatures distribution on the upstream and downstream faces of the dam body along the crown cantilever*

6. Results and Discussion

6.1. Performance in MDE

All records are scaled to maximum design level. Figure 7 represents performance curves for the most critical node within the dam body. In both the stress-based and strain-based approaches, the dam expected to experience nonlinear behavior (some observed DCRs are greater than 2). Generally, the calculated inelastic durations using the stress-based method are higher than those in the strain-based method. The averages of cumulative inelastic durations at DCR=1.0 and DCR=2.0 for the stress- based method are 0.70s, 0.11s and these values for the strain-based method are 0.58s and 0.07s, respectively. So, using the strain-based approach decreases the cumulative inelastic duration about 17.1% for DCR=1.0

and 36.4% for DCR=2.0 in MDE.

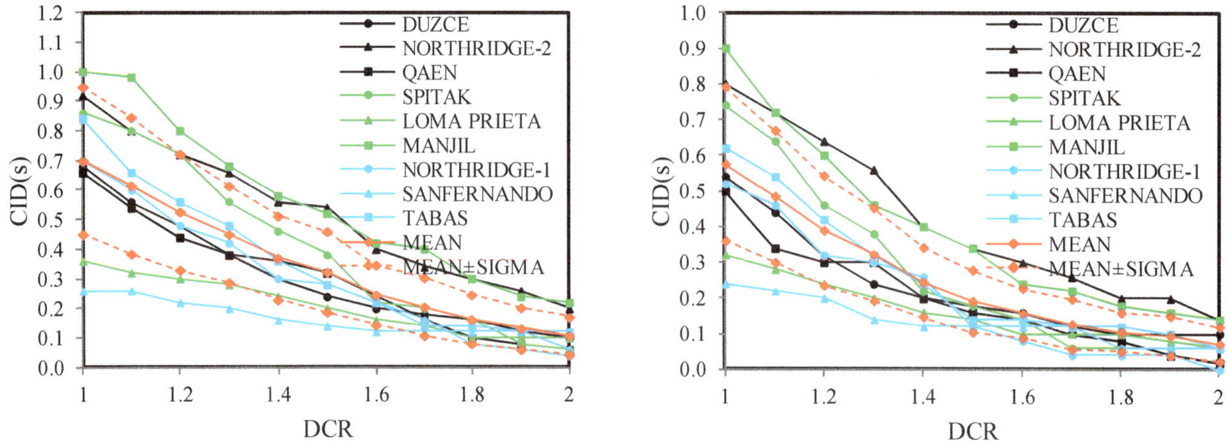

(a) (b)

Figure 7. *Performance curves in term of CID-DCR for critical nodes in the dam in MDE; (a) stress-based approach; (b) strain-based approach*

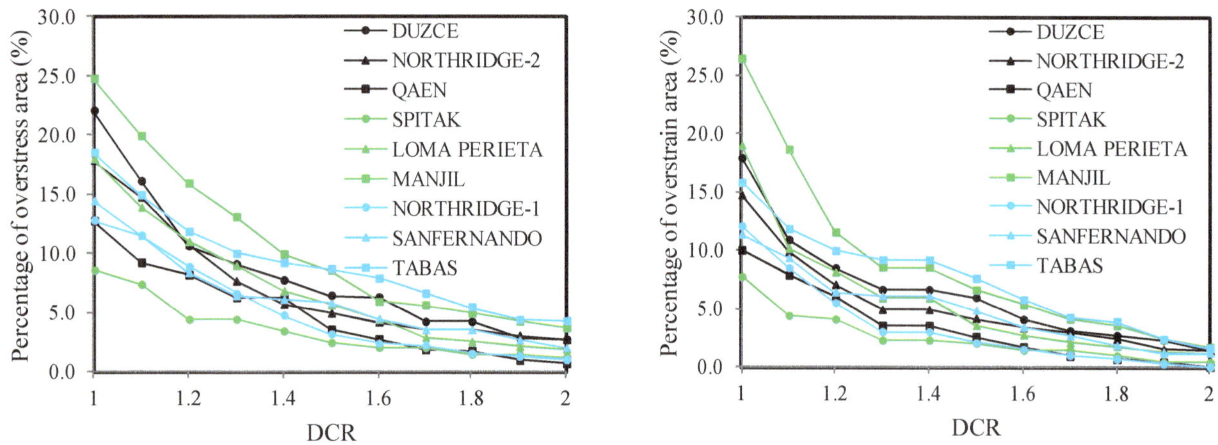

(a) (b)

Figure 8. *Percentage of (a) overstressed; (b) overstrained areas on the upstream face of the dam body in MDE*

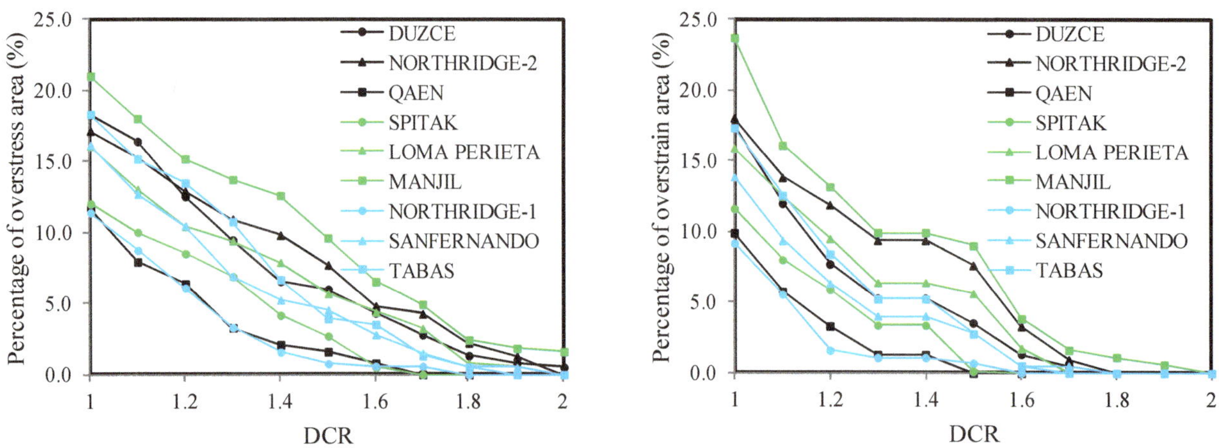

(a) (b)

Figure 9. *Percentage of (a) overstressed; (b) overstrained areas on the downstream face of the dam body in MDE*

Figures 8 and 9 represent the percentages of overstressed and overstrained areas on both the upstream and downstream faces of the dam body. analysis of the system in MDE shows that using the strain-based method increases the percentage of critical areas on both the upstream and downstream faces for some of ground motions in lower DCRs, while decreases percentage of critical areas in higher DCRs.

6.2. Discussion on the Results

Comparing stress- (or strain-) based approaches show that using the strain-based method leads to lower extension of critical areas in the higher DCRs. On the downstream face there is no overstrained areasin DCR=2.0 while at least three ground motions generate overstressed areas with DCR=2.0. Based on Table 1, all of ground motions lead to the results exceeding the predefined threshold curve for overstressed areas, while only six of them lead to the results above the predefined overstrained areas on the upstream face. So, the dam body experiences severe damage in the forms of joint opening and concrete cracking for all ground motions when the stress-based approach is used for results interpretation, while using strain-based approach reduces the probability of severe damage.

7. Conclusion

As pointed, all the guides and criteria for structural performance assessment of concrete arch dams in the literature are based on the stress. However, the behavior of mass concrete is governed by the strain. In the present paper, the seismic performance assessment of concrete arch dams was considered using the criteria based on both the stress and the strain. For this purpose Dez dam, which is a high double curvature arch dam, was selected and the numerical model of the dam-reservoir-foundation was provided using the finite element technique. Nine earthquake records were selected and all of them were scaled using the site response spectra in MDE. Linear analyses of the system in this level were conducted

Based on the conducted linear analyses, it was found that in spite of some similarities between the results obtained for the stress and strain based approaches. Interpreting the results based on the stress-based criteria can lead to different decision making in the dams' safety related projects.

Abbreviations

USACE	US Army Corps of Engineers
RCC	Roller Compact Concrete
CFRD	Concrete Face Rock-fill Dam
ETAF	Endurance Time Acceleration Functions
DCR	Demand Capacity Ratio
CID	Cumulative Inelastic Duration
LE	Linear Elastic
US	Upstream
DS	Downstream
PTC	Performance Threshold Curve
DBE	Design Base Earthquake
MDE	Maximum Design Earthquake
MCE	Maximum Credible Earthquake

References

[1] Bayraktar, A., et al. (2009), "Comparison of near and far fault ground motion effects on the seismic performance evaluation of dam-reservoir-foundation systems", Dam Engineering, XIX (4), 201-239.

[2] Chen, D.H., Du, C.B., Yuan, J.W. and Hong, Y.W., (2012), "An Investigation into the Influence of Damping on the Earthquake Response Analysis of a High Arch Dam", Journal of Earthquake Engineering, 16(3), pp.329-349.

[3] Fok, K.L. and Chopra, A.K. (1986), "Hydrodynamic and foundation flexibility effects in earthquake response of arch dams", Journal of Structural Engineering, 112 (8), 1810-1828.

[4] Ghanaat, Y., (2004), "Failure modes approach to safety evaluation of dams", 13th World Conference on earthquake engineering, Vancouver, B.C., Canada.

[5] Ghanaat, Y., (2002), "Seismic performance and damage criteria for concrete dams", Proceedings of the 3rd US-Japan Workshop on Advanced Research on Earthquake Engineering for Dams. San Diego, California.

[6] Hall, R. L., Matheu, E. E., and Liu, T. C., (1999), "Performance evaluation of the seismic response of concrete gravity dams", International Conference on Health Monitoring of Civil Infrastructure Systems, Chongqing, China.

[7] Hariri-Ardebili, M.A. and Mirzabozorg, H., (2011), "Investigation of Endurance Time Method Capability in Seismic Performance Evaluation of Concrete Arch Dams", Journal of Dam Engineering, XXII (1), 35-64.

[8] Hariri-Ardebili, M.A., Mirzabozorg, H., Ghaemian, M., Akhavan, M. and Amini, R., (2011),"Calibration of 3D FE model of DEZ high arch dam in thermal and static conditions using instruments and site observation", Proceeding of the 6th International Conference in Dam Engineering, Lisbon, Portugal.

[9] Hariri-Ardebili, M.A., Mirzabozorg, H. andGhaemian, M., (2011),"Seismic performance evaluation of high arch dams considering reservoir fluctuation", Proceeding of the 6th International Conference in Dam Engineering, Lisbon, Portugal.

[10] Hariri-Ardebili, M.A., Kolbadi, S.M., Heshmati, M. and Mirzabozorg, M., (2012), "Nonlinear Analysis of Concrete Structural Components using Co-axial Rotating Smeared Crack Model", Journal of Applied Science, 12(3), pp.221-232.

[11] Raphael, J. M. (1984), "The Tensile Strength of Concrete", ACI Journal Proceedings, 81, pp. 158-165.

[12] Studer, J.A., (2004), "Evaluation of earthquake safety of new and existing dams: Trends and experience", 13th World Conference on Earthquake Engineering, Vancouver, B.C., Canada, August, Paper No. 233.

[13] US Army Corps of Engineers (USACE), (2007)."EM 1110-2-6053: Earthquake design and evaluation of concrete hydraulic structures", Washington, D.C.

[14] Yamaguchi, Y., Hall, R., Sasaki, T., Matheu, E., Kanenawa, K., Chudgar, A., and Yule, D., (2004), "Seismic performance evaluation of concrete gravity dams", Proceedings of the 13th World Conference on Earthquake Engineering, Vancouver, B.C., Canada, August, Paper No. 1068.

[15] Wieland, M., and Fan, B.H., (2004), "The activities of the international commission on large dams (ICOLD) in the earthquake safety of large dams", Proceedings of the 13th World Conference on Earthquake Engineering, Vancouver, B.C., Canada, August, Paper No. 5051.

[16] Wieland, M., Brenner R.P. and Sommer, P., (2003), "Earthquake resiliency of large concrete dams: Damage, repair, and strengthening concepts", Proceedings of the 21st International Congress on Large Dams, ICOLD, Montreal, Canada.

[17] Zhang, C., Pan, J. and Wang, J., (2009), "Influence of Seismic Input Mechanisms and Radiation Damping on Arch Dam Response", Soil Dynamics and Earthquake Engineering, 29, pp.1282-1293.

Permissions

All chapters in this book were first published in AJCE, by Science Publishing Group; hereby published with permission under the Creative Commons Attribution License or equivalent. Every chapter published in this book has been scrutinized by our experts. Their significance has been extensively debated. The topics covered herein carry significant findings which will fuel the growth of the discipline. They may even be implemented as practical applications or may be referred to as a beginning point for another development.

The contributors of this book come from diverse backgrounds, making this book a truly international effort. This book will bring forth new frontiers with its revolutionizing research information and detailed analysis of the nascent developments around the world.

We would like to thank all the contributing authors for lending their expertise to make the book truly unique. They have played a crucial role in the development of this book. Without their invaluable contributions this book wouldn't have been possible. They have made vital efforts to compile up to date information on the varied aspects of this subject to make this book a valuable addition to the collection of many professionals and students.

This book was conceptualized with the vision of imparting up-to-date information and advanced data in this field. To ensure the same, a matchless editorial board was set up. Every individual on the board went through rigorous rounds of assessment to prove their worth. After which they invested a large part of their time researching and compiling the most relevant data for our readers.

The editorial board has been involved in producing this book since its inception. They have spent rigorous hours researching and exploring the diverse topics which have resulted in the successful publishing of this book. They have passed on their knowledge of decades through this book. To expedite this challenging task, the publisher supported the team at every step. A small team of assistant editors was also appointed to further simplify the editing procedure and attain best results for the readers.

Apart from the editorial board, the designing team has also invested a significant amount of their time in understanding the subject and creating the most relevant covers. They scrutinized every image to scout for the most suitable representation of the subject and create an appropriate cover for the book.

The publishing team has been an ardent support to the editorial, designing and production team. Their endless efforts to recruit the best for this project, has resulted in the accomplishment of this book. They are a veteran in the field of academics and their pool of knowledge is as vast as their experience in printing. Their expertise and guidance has proved useful at every step. Their uncompromising quality standards have made this book an exceptional effort. Their encouragement from time to time has been an inspiration for everyone.

The publisher and the editorial board hope that this book will prove to be a valuable piece of knowledge for researchers, students, practitioners and scholars across the globe.

List of Contributors

Md. Akhtar Hossain, Md. Shakhawat Hossain and Md. Kamrul Hasan
Department of Civil Engineering, Rajshahi University of Engineering & Technology, Rajshahi, Bangladesh

Hongyan Li and Jia Yu
Department of Architecture, Faculty of Human Settlement and Civil Engineering, Xi'an Jiao-Tong University, Xi'an, Shaanxi, China

Mohammad Ali Rezvani Befrouei
Department of Civil Engineering, Non-profit Institution of Higher Education, Aba - Abyek, Qazvin, Iran

Mohammad Taghipour
Research Assistant of Non-profit Institution of Higher Education, Aba - Abyek, Qazvin, Iran

Ren Li-li
Department of Railway Engineering, Liaoning Railway Vacational and Technical College, Jinzhou, China

Guo Rui-jin
College of Civil Engineering, Liaoning University of Technology, Jinzhou, China

Welvis Furtado da Silva
Department of Environmental Engineering, Metropolitan Institute of Education and Culture, Anápolis, Brazil

Ernane Rosa Martins
Coordination of Information Technology Area, Federal Institute of Education Science and Technology Goiás, Luziânia, Brazil

Ashwin Amarshi Maru
L.E.College, Civil engineering department, Saurashtra University, Rajkot, State: Gujarat, India

Guo Li, Dan Wang and Jian-Min Du
School of Mechanics & Civil Engineering, China University of Mining and Technology, Xuzhou, China

Chelugo Nicholas Chepchieng and Abiero Gariy
Department of Civil, Construction and Environmental Engineering, Jomo Kenyatta University of Agriculture and Technology, Nairobi, Kenya

Mwatela Josphat K. Z.
Department of Geomatic Engineering and Geospatial Information Systems, Jomo Kenyatta University of Agriculture and Technology, Nairobi, Kenya

Hu Kai, Qu Ge and Jiang Li
China Shipbuilding NRDI Engineering Co., Ltd, Shanghai, China

Qi Sanlin and Yang Shengsheng
CCCC Fourth Harbor Engineering Co., Ltd., Huangzhou, China

Lu Zheng
College of Civil Engineering, Tongji University, Shanghai, China

George Abednego, Onungwe Ishmael, Oba Achemie, Akpan Paul and Sarogoro Samuel
Department of Civil Engineering, Rivers State Polytechnic, Bori-Ogoni, Rivers State, Nigeria, West Africa

Joseph Ignatius Teye Buertey and Felix Atsrim
Department of Built Environment, Pentecost University College, Kaneshie, Accra, Ghana

Daniel Amofa
Department of Project Management, Accra Institute of Technology, Private Mail Bag, Accra, Ghana

Onungwe Ishmael, Sarogoro Samuel, George Abednego and Akpan Paul
Department of Civil Engineering, Rivers State Polytechnic, Bori-Ogoni, Rivers State, Nigeria, West Africa

Mehdi Babaei
Department of Civil Engineering, Faculty of Engineering, University of Zanjan, Zanjan, Iran

Tao Lianjin, Guo Fei, An Junhai, Liu Chunxiao and Bian Jin
Key Laboratory of Urban Security and Disaster Engineering of Ministry of Education, Beijing University of Technology, Beijing, China

Wang Guofeng
Beijing Railway Construction and Management Co.Ltd., Beijing, China

Huang Jun
CCTEB Civil Defense and Underground Space Institute Wuhan, China

Namdeo Hedaoo
Department of Civil Engineering, Govt. College of Engineering, Pune, Maharashtra, India

Namdeo RautDepartment of Applied Mechanics, Govt. Polytechnics, Yavatmal, Maharashtra, India
Laxmikant Gupta
Department of Applied Mechanics, Visvesvaraya National Institute of Technology, Nagpur, Maharashtra, India

Joseph Ignatius Teye Buertey and Felix Atsrim
Department of Built Environment, Pentecost University College, Accra, Ghana

Emmanuel Dadadzogbor
School of Engineering, Accra Institute of Technology, Accra, Ghana

Vahid Rezvani and Hassan Saghi
Department of Civil Engineering, Hakim Sabzevari University, Sabzevar, Iran

S. Mohammad S. Kolbadi
Department of Civil Engineering, Shahid Beheshti University of Technology, Tehran, Iran

Nemat Hassani and M. R. Shiravand
Department of Civil Engineering, Shahid Beheshti University, Tehran, Iran

S. Mahdi S. Kolbadi
Department of Civil Engineering, K. N. Toosi University of Technology, Tehran, Iran

Fatemeh Monjezi
Department of Geography, Payame Nor University of Shushtar, Shushtar, Iran

Hasan Beik Mohammadi
Research Centre of Shakhes Pajoh, University of Isfahan , Isfahan, Iran

Hassan Saghi and Abbas Ansari Aval
Department of Civil Engineering, Hakim Sabzevari University, Sabzevar, Iran

Luo Qiyang, Liu Yongjian, Zhou Yaoyao, Wu Jiansheng and Xie Zhikun
School of Civil and Transportation Engineering, Guangdong University of Technology, Guangzhou, China

Saeid Azmoodeh and Nasser Arafati
Department of Civil Engineering, Tafresh University, Tafresh, Iran

Hasan Dehghanzadeh Najm Abad and Amir Ahmad Hedayat
Department of Civil Engineering, Azad University, Kerman, Iran

Eisa Salajegheh
Department of Civil Engineering, Bahonar University, Kerman, Iran

Saeid Azmoodeh and Nasser Arafati
Department of Civil Engineering, University of Tafresh, Tafresh, Iran

Hamid Shahrabadi
Department of Civil Engineering, Chabahar Maritime University, Chabahar, Iran

Yao George C.
Department of Architecture, National Cheng Kung University, Tainan, Taiwan

Kuo Keng-Chang
Department of Construction Engineering, National Kaohsiung First University of Science and Technology, Kaohsiung, Taiwan

Hsueh Cheng-Luen
Department of Architecture, National Cheng Kung University, Tainan, Taiwan

Tang Yao
Institute of Education, National Cheng Kung University, Tainan, Taiwan

S. O. Odeyemi, M. A. Anifowose, M. O. Oyeleke, A. O. Adeyemi and S. B. Bakare
Department of Civil Engineering, Federal Polytechnic Offa, Offa, Nigeria

Abolfazl Ezzati and Kimiya Ajayebi
Department of Geology, Islamic Azad University, Science and Research Branch, Tehran, Iran

Reza Mehrnia
Department of Geology, Payam Noor University, Tehran, Iran

Sayem Hossain Md
State Key Laboratory of Geomechanics and Geotechnical Engineering, Institute of Rock and Soil Mechanics, Chinese Academy of Sciences, Wuhan, Hubei, China

Department of Geological Sciences, Jahangirnagar University, Dhaka, Bangladesh

Kong Ling-wei and Yin Song
State Key Laboratory of Geomechanics and Geotechnical Engineering, Institute of Rock and Soil Mechanics, Chinese Academy of Sciences, Wuhan, Hubei, China

Hassan Saghi and Nima Arefizadeh
Department of Civil Engineering, Hakim Sabzevari University, Sabzevar, Iran

Masood Heshmati and Seyed Mahdi Seyed Kolbadi
Department of Civil Engineering, K. N. Toosi University of Technology, Tehran, Iran

Seyed Mohammad Seyed Kolbadi
Departmet of Civil Engineering, Lame'i Gorgani University, Tehran, Iran

Index